Sustainable Civil Infrastructures

Sustainable Civil Infrastructures (SUCI) is a series of peer-reviewed books and proceedings based on the best studies on emerging research from all fields related to sustainable infrastructures and aiming at improving our well-being and day-to-day lives. The infrastructures we are building today will shape our lives tomorrow. The complex and diverse nature of the impacts due to weather extremes on transportation and civil infrastructures can be seen in our roadways, bridges, and buildings. Extreme summer temperatures, droughts, flash floods, and rising numbers of freeze-thaw cycles pose challenges for civil infrastructure and can endanger public safety. We constantly hear how civil infrastructures need constant attention, preservation, and upgrading. Such improvements and developments would obviously benefit from our desired book series that provide sustainable engineering materials and designs. The economic impact is huge and much research has been conducted worldwide. The future holds many opportunities, not only for researchers in a given country, but also for the worldwide field engineers who apply and implement these technologies. We believe that no approach can succeed if it does not unite the efforts of various engineering disciplines from all over the world under one umbrella to offer a beacon of modern solutions to the global infrastructure. Experts from the various engineering disciplines around the globe will participate in this series, including: Geotechnical, Geological, Geoscience, Petroleum, Structural, Transportation, Bridge, Infrastructure, Energy, Architectural, Chemical and Materials, and other related Engineering disciplines.

SUCI series is now indexed in SCOPUS and EI Compendex.

Ayman S. Mosallam · Brahim El Bhiri ·
Safae Merzouk
Editors

Civil Engineering, Material and Smart Buildings: New Technologies in Cities' Infrastructures

Proceeding of the 5th International Conference on Advanced Technologies and Humanity (ICATH'2023)

Springer

Editors
Ayman S. Mosallam
Department of Civil and Environmental Engineering
University of California, Irvine (UCI)
Irvine, CA, USA

Brahim El Bhiri
Harmony Technology
Rabat, Morocco

Safae Merzouk
EMSI
Casablanca, Morocco

ISSN 2366-3405 ISSN 2366-3413 (electronic)
Sustainable Civil Infrastructures
ISBN 978-3-031-76556-8 ISBN 978-3-031-76557-5 (eBook)
https://doi.org/10.1007/978-3-031-76557-5

This Springer imprint is published by the registered company Springer Nature Switzerland AG
The registered company address is: Gewerbestrasse 11, 6330 Cham, Switzerland

Contents

New Ideas and Innovations in Cities' Structures

Upgrading of Tailings from the Mbembele Manganese Mine in Gabon: Manganese Enrichment Trial Using Gravimetric Separation or Flotation

Rudy Bekoung Assoumou(✉), Souad. M'Rabet, and Mohamed Tayebi

Geosciences Laboratory, Department of Geology, Faculty of Science, Ibn Tofail University, Kenitra, Morocco

{rudy.bekoungassoumou,tayebi.mo}@uit.ac.ma

Summary. Two beneficiation techniques were tested in the tailings dam at the Mbembele mine in response to the geochemical instability of the tailings ponds in the mining zone, favored by the hot, humid climate. The work was carried out mainly on a composite from the Manganese tailings treatment dam, with a titration of:

Ech	Al_2O_3	CaO	Fe_2O_3	K_2O	MgO	MnO_2	Na_2O	P_2O_5	SiO_2	TiO_2	P-Fire	Total
G1 (%)	10,19	0,06	18,39	0,39	0,03	32,6 7	0,00	0,65	0,15	0,45	19,84	83,4 1
G2 (%)	4,65	0,26	21,10	1,66	0,12	31,9 9	0,06	0,25	0,12	0,17	20,45	81,6 9

*This is the analysis of the raw tailings taken from each zone of the dike prior to any treatment.

Gravity separation and flotation separation were chosen for tailings recovery. Tailings samples processed on a shaking table (gravity separations) show that enriched ore (55.97%) can be recovered from the concentrated fraction with a feed of 15.99% in the first wash. In contrast, flotation treatment of tailings, which even treats reject sludge, recovers Mn contents of 54.59%.

The significance of these results underlines the need to continue developing responsible mining practices at the Mbembele mine, focusing on methods that ensure the sustainable exploitation of mineral resources while preserving the integrity of the environment. It is imperative to promote approaches that do not compromise the health of local ecosystems and that meet sustainability standards, thereby contributing to the positive evolution of the mining industry towards more planet-friendly practices.

Keywords: environment · flotation · Gravity separation · Mine tailings · Manganese · pollution · soil

A. S. Mosallam et al. (Eds.): ICATH 2023, SCI, pp. 3–14, 2024.
https://doi.org/10.1007/978-3-031-76557-5_1

1 Introduction

Manganese pollution in the environment comes from both natural and anthropogenic sources. Volcanic eruptions, mineral weathering and erosion are the main natural sources [1–3], while mining is an anthropogenic source [4, 5]. The use of manganese in the steel industry and in the manufacture of batteries means that it is increasingly sought after and exploited worldwide. Many manganese-producing countries in Africa have a duty to protect the environment from manganese pollution, and to reduce as far as possible the quantities discharged from mine tailings into the dykes of mining areas.

Gabon, the world's second-largest producer of high-grade manganese (Mn) [6], after South Africa and ahead of Brazil [7], has several active manganese quarries. Environmental impact assessment reports in the mining sector indicate that abandoned mine tailings release potentially toxic elements into the environment [4, 8]. In addition, anthropogenic modifications to the relief of exploited areas create unbalanced landscapes that increase the environment's exposure to the risk of erosion [9, 10].

The Mbembele tailings storage area in Gabon presents a high risk of contamination of the surrounding waterways. Under the influence of climatic conditions and the vulnerability of the natural environment, these areas can become a source of pollution. As far as dykes are concerned, a large number have collapsed worldwide due to geotechnical instability problems.

The reasons for dike failure have always been linked either to the high-water content of the concentrator's discharges, which generates high pore pressures in the retention dikes, or to leakage problems, liquefaction or poor design of the warehouse site [11]. Dyke discharges always retain a quantity of metals which, in the event of dyke failure, would contaminate the water and the natural environment.

The valorization trial of manganese content in mine tailings and its treatment is an explorable means of enrichment in Mn [12]. Two methods of treating residues from the manganese washing plant were selected: gravimetric separation and flotation.

The aim of this study is to evaluate the ability of these two methods to recover the manganese present in the tailings storage areas of the Mbembele mine, to obtain, from these tailings, concentrations of additional manganese that meet the requirements of the extractive metallurgy industry and, consequently, to reduce the quantities of residual manganese as much as possible.

2 Materials and Methods

2.1 Study Area

The Mbembelé mine quarry is located in central Gabon, in the Moyen-Ogooué province, 22 km from the town of Ndjolé (latitude 0° 0′45.03″ North and longitude 10°46′29.91″ East) at an altitude of between 190 and 750 m (Fig. 1).

The study area has a transitional equatorial climate, with an average annual rainfall of 1,500 mm (Lambaréné meteorological station, 2006), one of the lowest in the country. The area's rainfall pattern comprises two maxima (March-April and October) and two minima, with a short dry season from June to August. There is also a short dry season in December and January, which is in fact only an attenuation of rainfall (in excess of

100 mm). The average annual temperature is 26 °C, with maximum variations of 30 to 31 °C (April) and minimum of 20 to 21 °C (July). The humidity values that characterize Gabon as a whole also apply to this sector, notably 90% (reaching 98%) at night and 75% during the day [13].

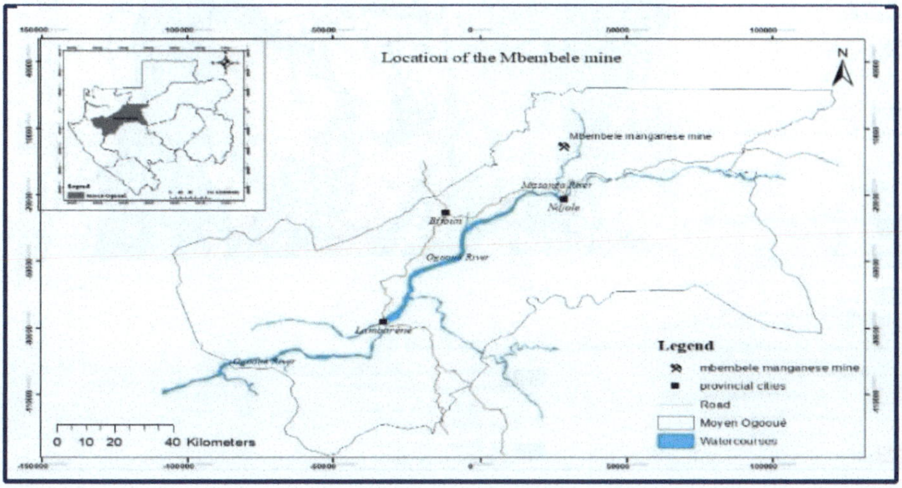

Fig. 1. Geographical location of the Mbembelé manganese quarry in Moyen-Ogooué province.

The geological formations in the study area are composed of Palaeoproterozoic rocks generally characterised by metavolcanites, quartzites, ampelites, schists and metaquartzites (Fig. 2).

2.2 Characterisation of Mn Samples

Three methods are used to characterise the ore: observation of thin sections of mining samples under a polarizing microscopy, x-ray fluorescence analysis and x-ray powder diffraction analysis. Elemental chemical analyses are performed using inductively coupled plasma-atomic emission spectrometry (ICP-AES).

To dissolve the Mn, 250 mg of powder from each soil sample was acid digested with a mixture of 39.5% hydrofluoric acid, 37% hydrochloric acid and 50% nitric acid in Teflon beakers heated to 250 °C on a hot plate. The cooled contents were transferred to 1000 ml flasks and diluted with distilled water to the limit for ICP-AES analysis.

2.3 Gravity Separation

2 kg of manganese tailings taken from the plant's tailings dam at point G1 (N0°03′73.6188′ E10°46′28.17012′) and 2 kg of mine tailings taken from the saturated zone of the same dam at point G2 (N0°04776012 E10°4625734) are processed on a shaking table. Once the shaker table has been set up and secured for a sample, only particles sieved to 1 mm are processed. The gravity separation protocol is summarized in Fig. 3.

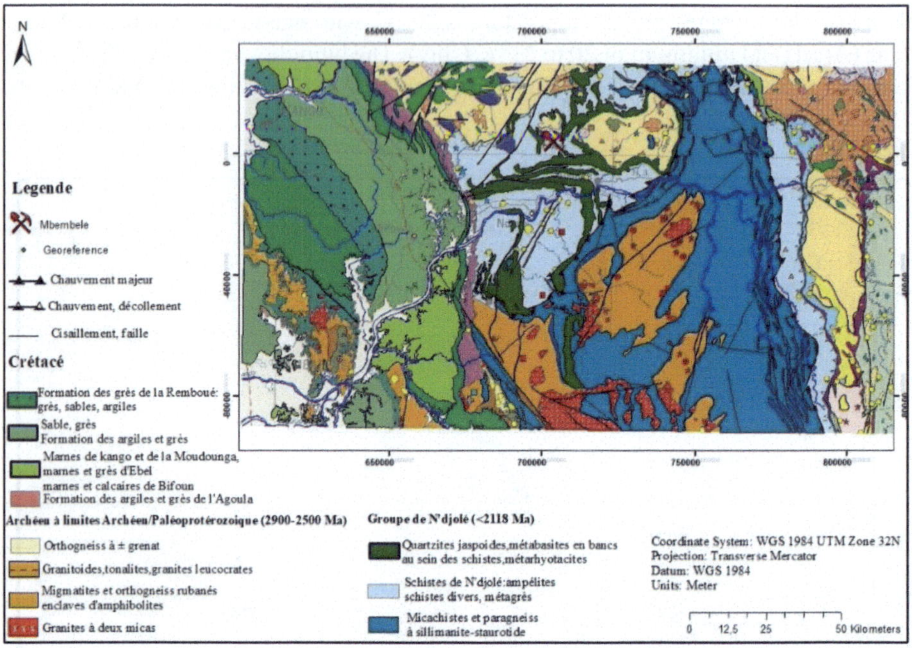

Fig. 2. Geological formation of the Mbembelé manganese mine in Moyen-Ogooué province

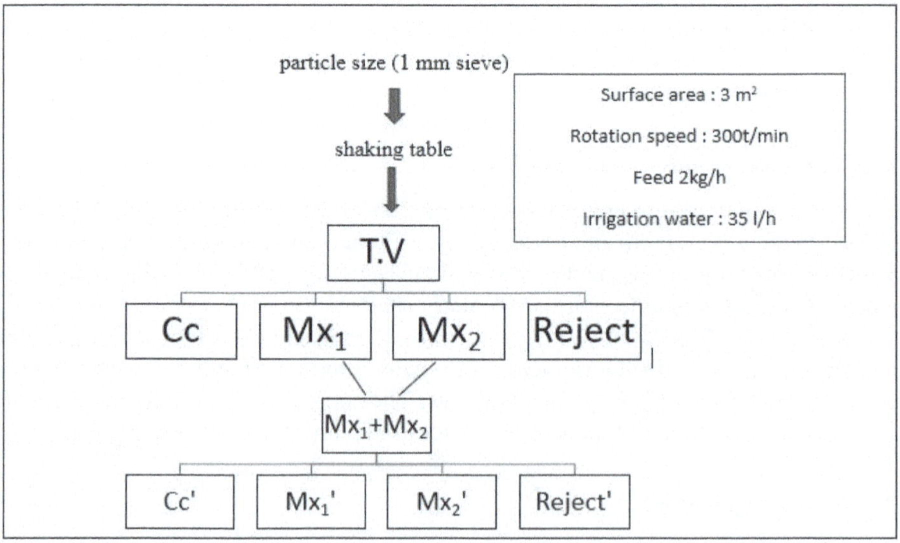

Fig. 3. Gravity separation diagram on a shaking table: Depletion data 1: TV (all comers), Cc (Concentrate), Mx_1 ($Mixed_1$), Mx_2 ($Mixed_2$), Reject. Depletion data 2 ($Mixed_1 + Mixed_2$): Cc' (Concentrate'), Mx_1 '($Mixed_1$'), Mx_2 '($Mixed_2$') and Reject'.

2.4 Floating

For the flotation tests, 500g of prepared ore are treated. The products used for the flotation treatment: Dextrin, Oleic acid, Sodium carbonate. The pH of the treatment was maintained at 8 (Fig. 4).

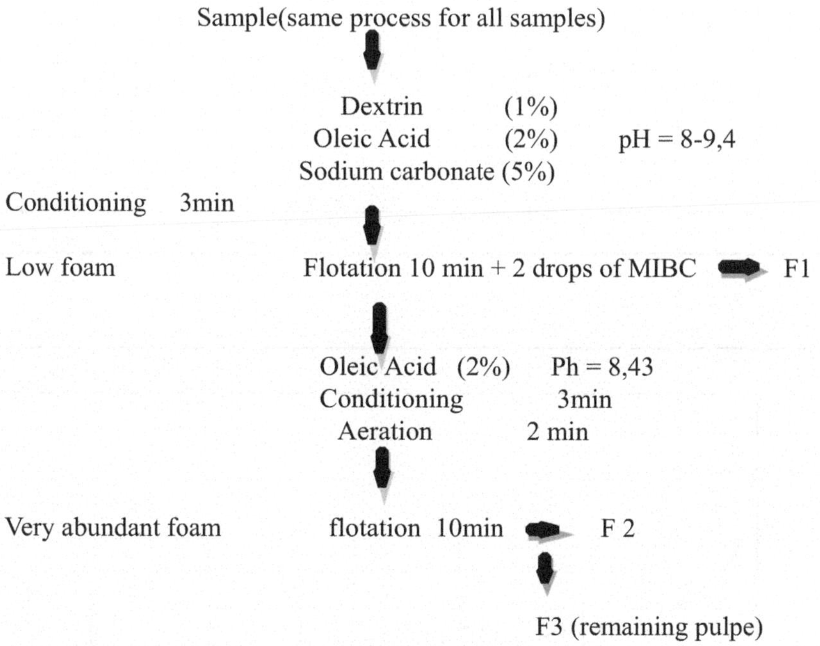

Fig. 4. Manganese flotation process sample taken at points G1 and G2.

3 Results

3.1 Mineralogical and Chemical Compositions

The x-ray diffraction results (Fig. 5) show the minerals present in each sample taken from the Mbembelé mine tailings. These results are confirmed by polarising microscopy (Fig. 6) and x-ray fluorescence analysis (Fig. 7).

Vernadite and hematite are the main minerals in these samples. More than 80% of the minerals are in oxide form (Fig. 5, Fig. 7).

Figure 6 illustrates the detailed mineralogical composition observed in the samples taken at points G1 and G2. Polarising microscopy showed the presence of plagioclase microlites. The average extinction angle of these feldspars is 25° corresponding to andesine (Fig.B.b1). The second mineral observed is nepheline. It has hexagonal sections and xenomorphic patches invaded at their edges by ferruginous margins, which in places show clear micaceous alteration.

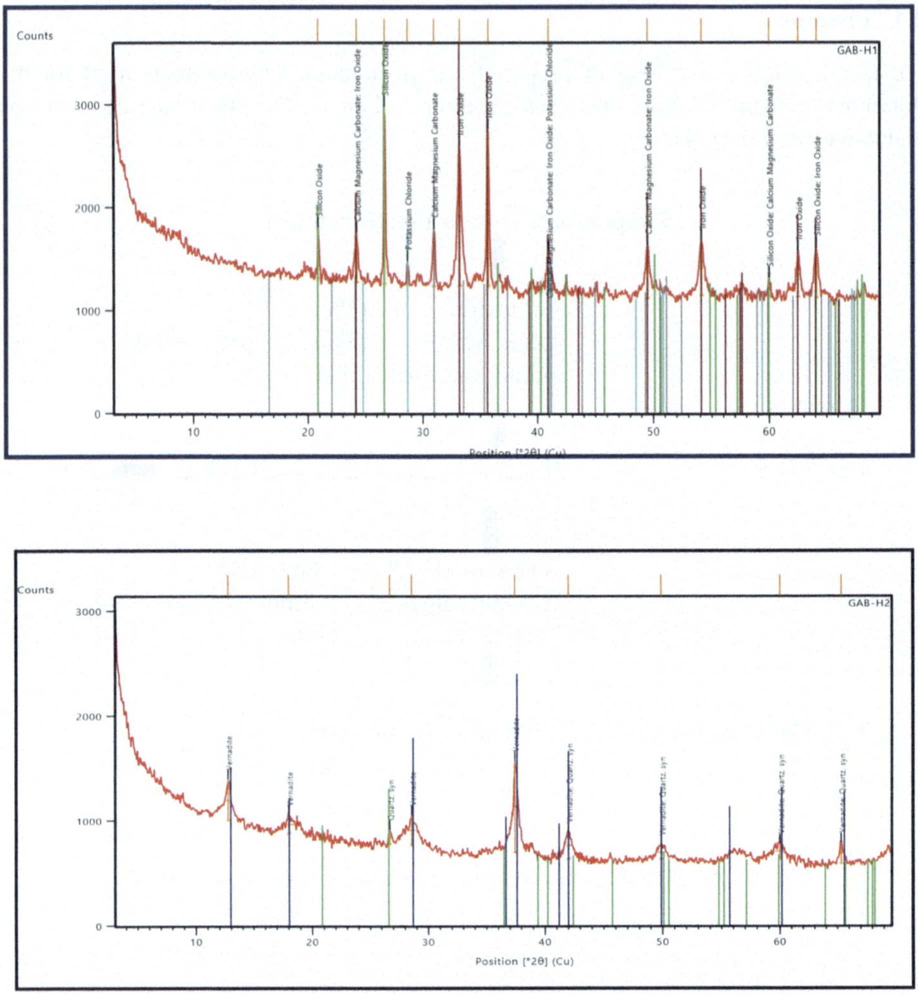

Fig. 5. Mineralogical phases of tailings from the Mbembelé manganese mine processed on a shaking table.

It is originally a volcanic rock and the presence of nepheline, plagioclases corresponding to andesine and the iron oxides observed would make the rock correspond to tephrite or phonotephrite. The vernadite, determined more precisely by x-ray diffraction (Fig. 5), shows the propagation of the haematite observed in epitaxially on the vernadite at certain points under a polarising microscope. A heamatite rim is also visible around the felspathoids.

Four sample beads were analysed by X-ray fluorescence, and the average chemical composition of the samples is shown in Fig. 7.

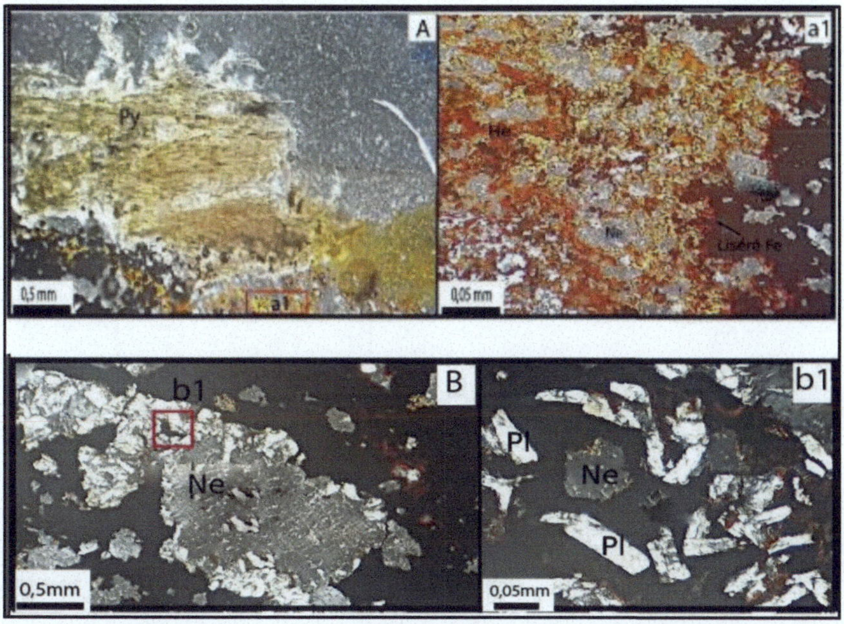

Fig. 6. Photograph of minerals observed with a G x 10 polarising microscope in polarised light (L.P). A: Pyrolusite (Py), a1: Iron streak (He) around nepheline (Ne). B: Nepheline (Ne), b1: plagioclase (Pl) microlites corresponding to andesine.

3.2 Gravity Separation

Gravimetric separation tests on a shaking table (Table 2) show the results of the treatment of samples taken at points (G1 and G2). For the treatment of sample at point G1, it was possible to obtain Mn contents of 17.5% in the reject fraction, 15.2% and 10.67% in the two mixed fractions (1 and 2) and a concentrate of 16.22% for a feed of 15.99% in the first wash. The washing of the mixed fractions (1 and 2) shows that it is possible to improve the Mn content of the concentrate fraction. There was a gradual decrease in Fe_20_3 and other major elements in the sample in the concentrate fraction and an increase in these elements in the reject fraction. This observation suggests that the enrichment of the reject fraction is probably linked to the size of the fine particle.

Processing of samples taken at point G2 shows Mn enrichment in the following fractions: reject (11.05%), $Mixed_1$ (8.55%), $Mixed_2$ (13.19%) and concentrate (32.19%) with a feed of 13.71% in the first wash. The low iron content (2.28%) of these samples is noticeable in all treatment fractions.

The different sampling points of the treated samples reveal a variation in the main constituents depending on the sampling point. These variations affect the content and distribution of the constituent chemical elements in each treatment fraction. Samples taken at point G2 (washey residue reception dam) show a good distribution of recovered manganese in the concentrate fraction (i.e. a distribution of 31.8%). As with the samples taken at point G1, a second washing of the fractions would increase the concentration of recovered Mn in the concentrate.

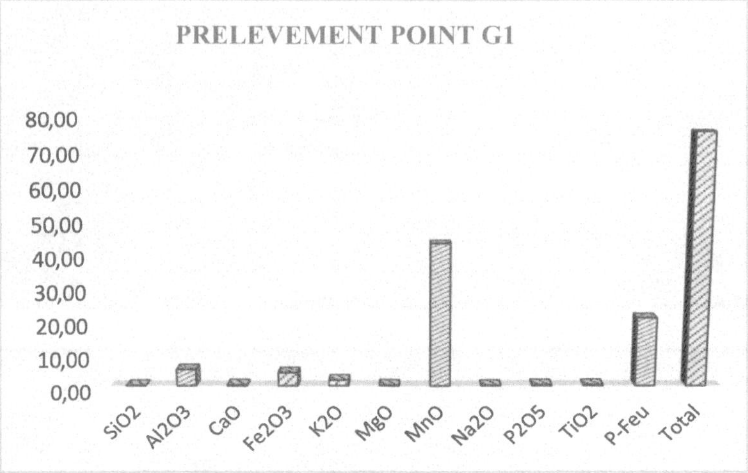

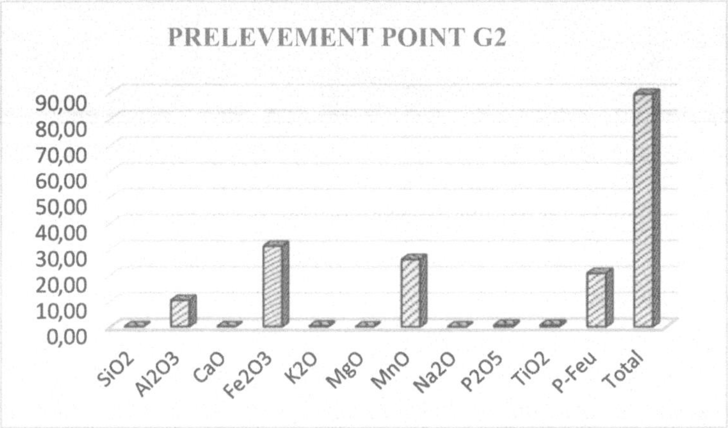

Fig. 7. Chemical analysis of tailings samples taken at points G1 and G2 of the Mbembelé manganese mine.

With regard to Mn and Fe content, processing of the results by one-way ANOVA (Fig. 8) for the fractions (Reject, Mx1, Mx2, Cc) shows a significant difference between the fractions studied. A Turkey test is used to assess the significant differences in Fe and Mn concentration in each fraction after treatment with a 95% confidence interval. The concentrated fraction significantly accumulates recoverable Mn. In the concentrated fraction, iron depletion is progressive with washing, while in the reject and mixed fractions, iron enrichment is noticeable. However, although the concentrated fraction removes more than 40%, iron oxides simultaneously lead to the loss of Mn: 17.5% (reject 1er G1 wash); 9.62% (reject 2ème G1 wash), 11.05% (reject 1er G2 wash); this at relatively high levels compared to the feed of 15.99% at 1st G1 wash; 9.96% at 2nd G1 wash and 13.71% at 1st G2 wash (Table 1).

Table 1. Summary of the sieving of samples taken at points G1 and G2, 1 mm fraction of sieve treated for gravimetric separation.

Fraction	Poids	Poi	P	p_2O	Na	Na_2	Si	SiO	K	Kcl	**M**	Mn	Fe	**Fe_2**	Ca%	Ca	Al	Al_2
Rejet	985,4	51,	0,3	**72,7**	0,0	**99,9**	0,0	**61,0**	1,3	**81,**	**17,**	**55,**	28,	**59,**	0.10	**75,**	5.1	**71,**
Mx_2	552,8	28,	0,1	**17,1**	0	**0,02**	0,0	**22,8**	0,3	**11,**	**15,**	**27,**	23,	**27,**	0.03	**12,**	2.5	**20,**
Mx_1	206,2	10,	0,1	**5,4**	0	**0**	0,0	**8,52**	0,2	**2,9**	**10,**	**7,1**	17,	**7,4**	0.04	**6,3**	1.6	**4,6**
Cc	182,6	9,4	0,1	**4,78**	0	**0**	0,0	**7,55**	0,3	**3,6**	**16,**	**9,6**	16,	**6,3**	0.04	**5,6**	1.3	**3,5**
Sommes des fractions	1927, 15	10 0	0,2 2	**100**	0,0 1	**100**	0,0 3	**100**	0,8 6	**10 0**	**15, 99**	10 0	**24, 63**	**10 0**	**0,07**	10 0	**3,6 5**	10 0
Sample G1 particle size fraction 1,25mm (washing $Mx_1 + Mx_2$)																		
Fraction	Poids	Poi	p%	p_2O	Na	Na_2	Si%	SiO	K	Kcl	Mn	Mn	Fe	Fe_2	Ca%	Ca	Al	Al_2
Rejet'	280,5	38,	0	**0**	0,0	**55,2**	0,0	**68,7**	0,5	**53,**	9,6	**36,**	32,	**44,**	0,03	**33,**	1,3	**63,**
Mx_2'	337,3	45,	0	**0**	0,0	**33,2**	0	**0**	0,3	**42,**	8,2	**37,**	30,	**50,**	0,04	**54,**	0,5	**27,**
Mx_1'	63,83	8,6	0	**0**	0,0	**6,29**	0,0	**31,2**	0,0	**2,1**	13,	**11,**	9,3	**2,9**	0,03	**7,7**	0,5	**5,3**
Cc'	52,49	7,1	0,0	**100**	0,0	**5,17**	0	**0**	0,1	**2,1**	19,	**13,**	9,0	**2,3**	0,02	**4,2**	0,4	**3,6**
Sommes des fractions	734,1	10 0	0	100	0,0 1	100	**0,0 1**	100		10 0	9,9 6	10 0	**27, 65**	10 0	**0,03**	10 0	**0,8 3**	10 0
Sample G2, particle size fraction 1,25mm (first																		
Fraction	Poids	Poi	P%	P_2O	Na	Na_2	Si%	SiO	K	Kcl	**M**	Mn	Fe	**Fe_2**	Ca%	Ca	Al	Al_2
Rejet	896	45,	0,0	**63,4**	0,0	**100**	0,0	**72,4**	2,6	**79,**	**11,**	**36,**	3,7	**74,**	0,15	**72,**	1,9	**73,**
Mx_2	488,8	24,	0,0	**25,9**	0	**0**	0,0	**19,7**	0,8	**14,**	**8,5**	**15,**	1,5	**17,**	0,09	**23,**	0,7	**15,**
Mx_1	310,1	15,	0,0	**8,24**	0	**0**	0,0	**4,18**	0,3	**3,3**	13,	**15,**	0,8	**5,7**	0,01	**1,6**	0,5	**6,7**
Cc	265,4	13,	0,0	**2,35**	0	**0**	0,0	**3,58**	0,3	**2,9**	**32,**	**31,**	0,4	**2,4**	0,01	**1,4**	0,3	**4,1**
Sommes des fractions	1960, 4	10 0	**0,0 6**	100	**0,0 1**	100	**0,0 4**	100	**1,5 4**	10 0	**13, 71**	10 0	**2,2 8**	10 0	**0,09**	10 0	**1,2 2**	10 0

3.3 Floating

The results of the flotation tests on the samples taken at points (G1 and G2) show that (Table 2) it is possible to achieve Mn contents of 8.44% in the F1 phase, a Mn content of 54.59% in the F2 phase and a concentration of 16.02% for the sample remainder in the G2 pulp. Processing of the samples taken at point G1 revealed Mn contents of 8.44% in phase F1, 40.23% in phase F2 and 11.02% in phase F3. Phase two results were positive, as was the effect of adding methyl isobutyl carbinol (MIBC).

4 Discussion

Once extracted from the mine, manganese is generally divided into three units according to grade: metallurgical grade, chemical grade and battery grade. These three units represent the main uses of manganese worldwide.

Vernadite, a hydrated manganese oxide mineral in our case, is in fact a variety of birnessite [14–16]. The structural formula can be represented by MnO_2. The positive grades obtained by gravimetric separation in samples taken at points G1 (16.22% concentrated fraction) and G2 (32.19% concentrated fraction) testify to the high manganese

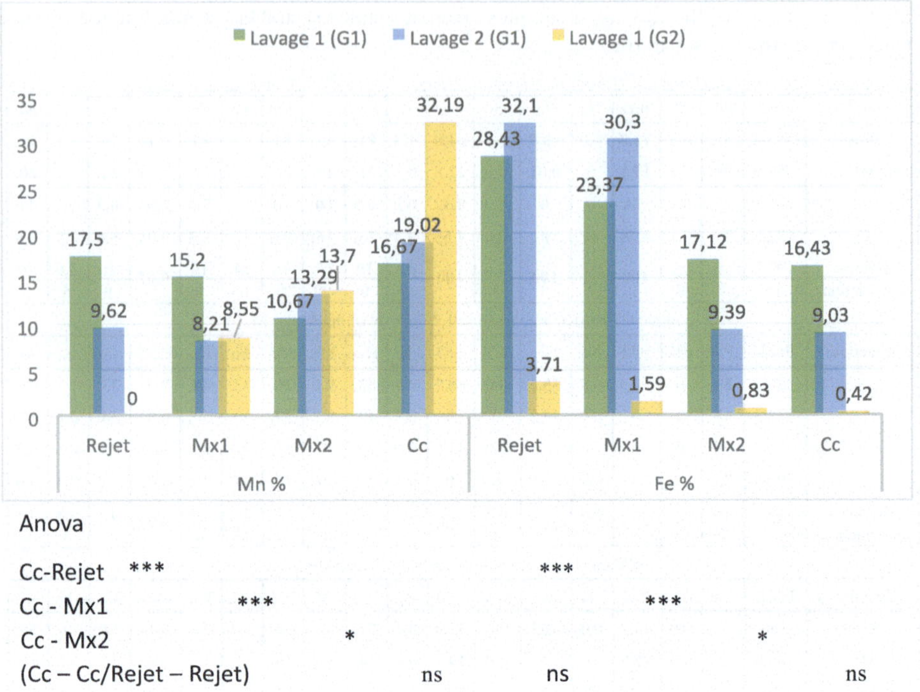

Fig. 8. ANOVA test at 95% confidence interval for fractions (Reject, Mx1, Mx2, Cc) and Turkey test. Significant differences *: $p < 0.05$, **: $p < 0.01$, ***: $p < 0.001$ and ns: not significant. (Lavage = Wash)

Table 2. Manganese mineral flotation test on samples taken from points G1 and G2 of the tailings storage areas 1 mm

G2	P %	Mn% content	Repart Mn%
F1	11,58	8,44	2,55
F2	60,3	**54,59**	85,72
NF3	28,12	16,02	11,73
Power supply	100	38,4	100

G1	P %	Mn% content	Repart Mn%
F1	15,64	8,44	5,23
F2	50,06	**40,23**	79,79
NF3	34,3	11,02	14,98
Power supply	100	25,24	100

contents in the Mbembelé tailings. However, the predominance of Mn in the concentrated fraction suggest that Mn is preferentially associated with this fraction. In addition, washing of the Mixed fractions shows Mn enrichment in the concentrated fraction.

Further processing of the Mbembelé tailings by flotation test shows that the manganese present in the tailings can be enriched to levels of:

- 54.59% Mn in 250 mg of samples taken at point G2 by flotation (Table 2)
- 40% Mn in 250 mg of samples taken at point G1 by flotation (Table 2)

The manganese concentrated in the flotation test is linked to phase F2 (flotation 2 of the treatment).

Manganese's designation as an industrial mineral is due to its non-metallurgical use in products as diverse as fertilisers, glass, perfumes, bricks and water purification systems. These areas make manganese a mineral that can be used at both low and high levels. On the international market, high-grade manganese ores (> 35% Mn) are mainly used in alloys, while lower-grade ores (35% Mn) are added during the production of cast iron [17].

The positive result obtained by flotation and gravimetric separation would confirm the possibility of recovering exploitable grades of Mn from the Mbembelé manganese tailings. Manganese deposits are often associated with iron deposits, because their modes of formation are fairly similar. The abundance of iron in the reject fraction suggests that manganese does not predominate in this fraction and could be linked to the oxide fraction. This finding is consistent with the work of Morio et al. in 1981 [18]. The abundance of iron in the reject fraction (28.43% sample taken at point G1, 32.1% washing of the mixed fraction 1 and mixed fraction 2 of the enchantment sample taken at point G1) at the origin of the red mud in the reject fraction should not be overlooked. Boyarintsev's work on red mud in 2021 [19] has made it possible to evaluate a way of recovering it [20]. A complex approach to red mud treatment involves recovering aluminium as a useful product [21] and other elements. This shows that the enrichment in other elements that could be achieved by recovering manganese from the Mbembelé mine tailings can be reprocessed to reduce pollution.

Although flotation allows 7% more Mn to be recovered than separation by gravimetry, it is still very harmful to the environment. In addition, the Mbembelé mine has the advantage of having a large watercourse for washing the Mixed and Reject fractions, for better recovery of Mn in the tailings from the tailings dam at the washing plant.

5 Conclusion

Solving the problem of mine tailings in dykes is an important application, the solution to which will considerably improve the ecological state of the environment. The treatment of tailings from the Mbembele/Gabon mine by gravimetric separation on a shaking table for samples from the mine's washing plant resulted in:

- High levels of manganese are concentrated in the concentrated fraction,
- Improving the content and distribution of the concentrated fraction by a second washing of the mixed fractions,
- The increase in Fe content 0_{23} in the reject fraction.

The results of the flotation tests on the Mbembelé tailings also revealed fairly high levels of manganese.

The presence of watercourses near the mine favours manganese recovery by gravimetric separation. The trial of manganese recovery by flotation produced a concentrate richer in Mn, but at the expense of environmental pollution.

However, the mine needs to evolve and develop its in-situ extraction logistics in order to make the most of the Mn content left in the tailings.

References

1. Ali, H., Khan, E., Ilahi, I.: J. Chem. Art. ID **6730305** (2019)
2. Walker, M.J.C., et al.: J. Quaternary Sc. **27**(7), 649659 (2012)
3. Masindi, V., Muedi, K.L.: In Heavy MetalsIntechOpen 115–133 (2018)
4. Cerón, J.C., Grande, J.A., de la Torre, M.L., Borrego, J., Santisteban, M., Valente, T.: Hydro Sc J. **lOpen 59**(6), 12131224 (2014)
5. El Adnani, M., et al.: Env. Technol. **28**(1), 71–82 (2007)
6. Comilog, R.: Annu. **68**(2018) (2017)
7. Boupassia, C.: Chem. Corpus ID: 126633394 (2004)
8. Boukhalfa, C., Blake, G., Mennour, A.: Env. Technol. **27**(8), 885890 (2006)
9. Noma, A.S., Abdelali, G., Touré, A.A., Daoudi, L.G.: Relief. Processus. Env. **28**(2), 7792 (2022)
10. Lehmkuhl, F., Stauch, G.: Geomorphology.**421**, 108522 (2023)
11. Aubertin, M., Bussière, B., Bernier, L.: Presses Internationales Polytechnique (2002)
12. Morizot, G., Lesueur, H., Zeegers, H.: Industry and Env. **23**, 7882 (2000)
13. Lambaréné weather station data. **645510 (FOGR)** (2006)
14. Post, J.E.: Proceedings of the National Academy of Sciences of the United States of America **96**, 34473454 (1999)
15. Giovanoli, R.: Mineralium Dep. **5**(2), 251253 (1980)
16. Atkins, A.L., Shaw, S.. Peacock, C.L.: Geoch et Cosmochimica Acta **189**, 158183 (2016)
17. Chalmin, P., Jégourel, Y.: CYCLOPE **49** (2019)
18. Morio, M., Brecl, Y., Gougis, S.: SOGEREM **16** (1981)
19. Boyarintsev, A., Ye Aung, H., Stepanov, S., Shoustikov, A.: E3S Web Conferences **258** 08026 (2021)
20. Stepanov, S.I., Aung, M.M., Boyarintsev, A.V., Goziyan, A.V., Chekmarev, A.M.: Actual Issues of Obtaining and Applying REM and RM, Moscow, Russia (2017)
21. Pyagay, I.N., Kozhevnokov, V.L., Pasechnik, L.A., Skachkov, V.M.: J. Mining Inst. **218** (2016)

Field Evaluation of HMA Pavements with Recycled Base Layers

Hani H. Titi(✉), Rawan Aqel, Mohammad Sooman, and Habib Tabatabai

Department of Civil and Environmental Engineering, University of Wisconsin-Milwaukee, Milwaukee, USA
hanititi@uwm.edu

Abstract. This paper presented field evaluation of existing HMA pavements constructed on RAP base layers by performing visual distress surveys and determining the corresponding PCI values and measuring the pavement surface profile and determining the corresponding IRI values. Observed distresses included cracks (fatigue, longitudinal, transverse, block, and edge), rutting, and flushing. The study showed that fatigue/alligator cracks are commonly observed in the investigated HMA pavement with RAP bases. The calculated PCI values were variable with the majority fall in the range of "fair" classification category. The evaluation of the investigated pavements' profiles shows what can be considered as a "smooth" ride quality with an average IRI of 96 inch/mile. The performance of the HMA pavements with RAP base layers is considered satisfactory as compared with HMA pavements constructed with crushed aggregates base course layers as reported in the literature.

Keywords: RAP base layer · HMA pavement · PCI · IRI · distress

1 Introduction

The use of recycled pavement materials can help reduce energy use in multiple ways. For instance, using recycled pavement materials instead of virgin materials saves energy on extraction/quarrying, processing/sizing, and transportation of the virgin materials. This will result in substantial energy savings. For example, a large amount of energy is needed to produce asphalt cement as well as crushed aggregates.

Additionally, the energy used for construction can be decreased by employing recycled pavement material. One significant source of energy use, for instance, can be the transportation of materials to and from the construction site. Utilizing locally available recycled materials can eliminate the distance over which the materials must be transported, thereby reducing energy consumption. Moreover, the production of virgin materials frequently results in the release of significant volumes of carbon dioxide and other greenhouse gases; therefore, using recycled materials can also assist in lowering greenhouse gas emissions. Reducing a project's overall carbon footprint is feasible using recycled materials.

A. S. Mosallam et al. (Eds.): ICATH 2023, SCI, pp. 15–21, 2024.
https://doi.org/10.1007/978-3-031-76557-5_2

The use of pavement recycled materials must not be at the expense of pavement performance, service life, and cost. This paper presents the results of a comprehensive field-testing program to investigate the performance of HMA pavement with RAP base course layers. Pavement sections were tested using pavement surface profile measurements and visual pavement distress surveys.

2 Research Methodology

Existing HMA pavements with RAP base course layers that were eight years old were the subject of this investigation.

The study consisted of performing visual/manual distress surveys of pavement surfaces. The objective is to identify/quantify the various types of pavement surface and to obtain data necessary to evaluate pavement performance using the pavement condition index (PCI). At each pavement location, a distress survey was carried out for a single 528-foot length. The chosen part is thought to be a good representation of the general state of the pavement. At the examined pavement sites, surface distresses were visually recognized, measured, and documented. In accordance with the FHWA manual on distress identification in which the types, extents, and severity levels of pavement distress were determined.

In addition, the surface profile of the investigated pavements was measured with a CS8800 profiler (Surface Systems and Instruments, Inc.). At each investigated pavement test segment, a 600-foot profile measurement was taken on the inside wheel path, outside wheel path, and center of the lane. The system has a GPS unit and software installed, which enables the presentation of measured profiles in real time. Additionally, the system enables data files to be saved in formats compatible with ProVAL and MS Excel software. Measurement of pavement surface profile is shown in Fig. 1.

Fig. 1. Measurement of pavement surface profile.

3 Results and Analysis

3.1 Visual Distress Surveys on Pavement Surfaces and Profile Measurements

For all examined pavement test sections with RAP base layer, the research team performed a visual distress survey analysis. In order to represent each pavement under investigation, the 528-foot pavement test section's 25-foot-long subsections were to have their PCI values determined. Fatigue (alligator), longitudinal, transverse, block, and edge cracks as well as flushing (bleeding), rutting, and heave of pavement edges were the most frequently seen pavement surface distress in the examined pavement test sections. Figure 2 shows the several types of pavement surface distress that were seen in the field.

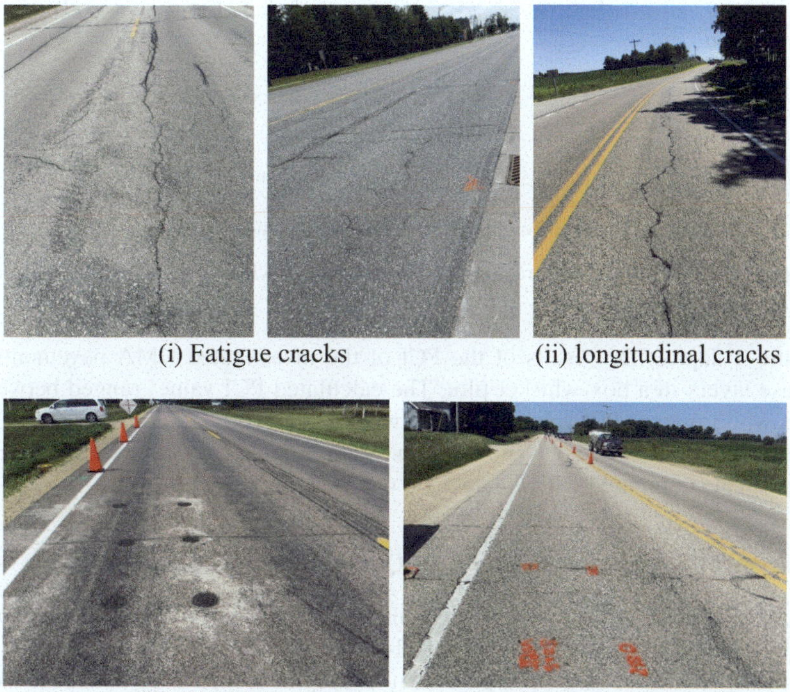

(i) Fatigue cracks (ii) longitudinal cracks

(iii) Flushing (iv) transverse cracks

Fig. 2. Various distresses seen on the investigated HMA pavements with RAP bases.

The computer application MicroPAVER was used to evaluate the data from the visual distress survey and determine the related PCI values. The findings and the accompanying pavement condition categorization for the pavement test subsections at STH 22 with HMA pavement and RAP base layer are shown in Fig. 3. The figure also includes a description of the pavement surface based on the PCI evaluation scale, which is good (PCI between 85 and 100), satisfactory (PCI between less than 85 and 70), etc. For instance, STH 22's PCI varied from 30 to 69, with a classification ranging from very poor to fair. The 25-foot pavement subsections were the bases for the PCI values determined here.

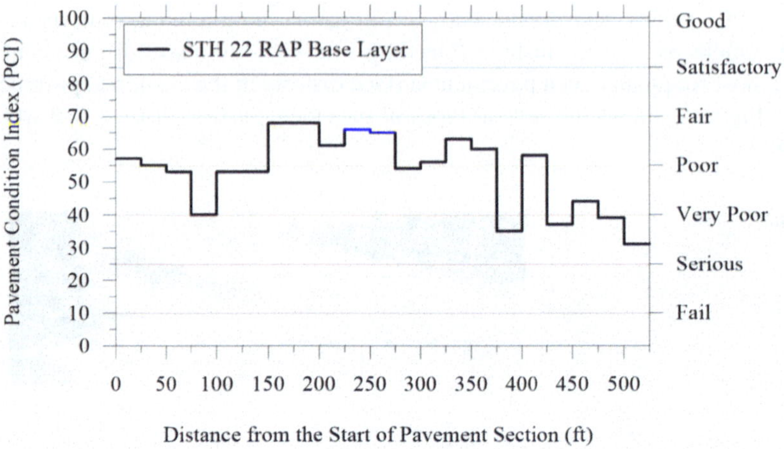

Fig. 3. Results of visual distress surveys conducted on STH 22 with RAP base layer.

Figure 4 depicts the results of the PCI of the investigated HMA pavements with RAP base layers in a box-whisker plot. The calculated PCI values ranged between 25 and 100, while average PCI values varied between 38 and 100. The age of the investigated pavements ranged between one and eighteen years. The PCI scale presented in Fig. 4 is a "rough" scale compared with the detailed one presented in Fig. 3.

3.2 Pavement Surface Profile and Ride Quality Evaluation

For every examined pavement test segment, measurements of the pavement surface profile were made on the outside/right wheel-path (OWP), inside/left wheel-path (IWP), and center-of-the-lane (CL). Each pavement test segment measured 600 feet in length, and the ProVAL program was used to determine the International Roughness Index, IRI, values for 25-foot pavement subsections. The results of the ride quality evaluation of the investigated HMA pavement with RAP bases are shown in Fig. 5.

Upon examining Fig. 5, it is evident that there is a significant degree of variation in the IRI values among the inside/outside wheel paths and also lane center. For the same pavement test segment, the OWP profile measurement generally revealed greater IRI values than the CL and IWP. In contrast to the inside wheel path and lane center, the outside wheel path is closer to the pavement edge or shoulder and typically has insufficient lateral support. Higher rutting and roughness values would result from this, as the research team saw when conducting fieldwork.

Figure 6 shows the average IRI values for the investigated HMA pavement with RAP bases. The results demonstrate variability within the investigated pavements, with the range of IRI average between 42 and 204 in/mile.

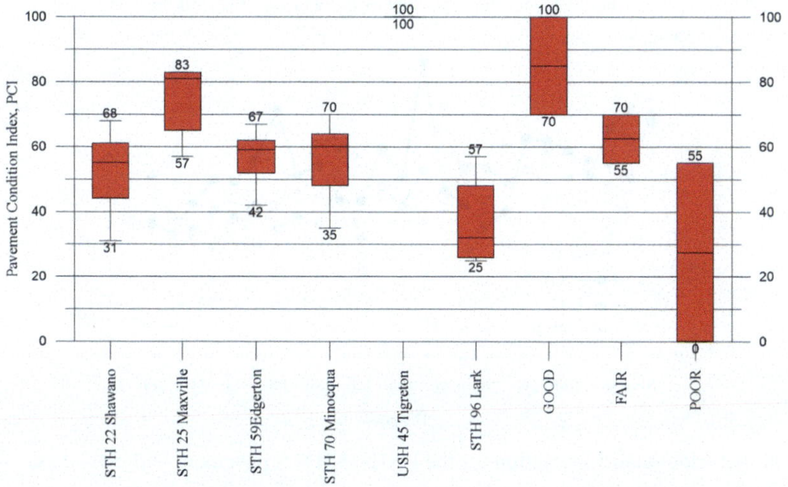

1. PCI variation within the 528-foot sections based on 25-foot subsections

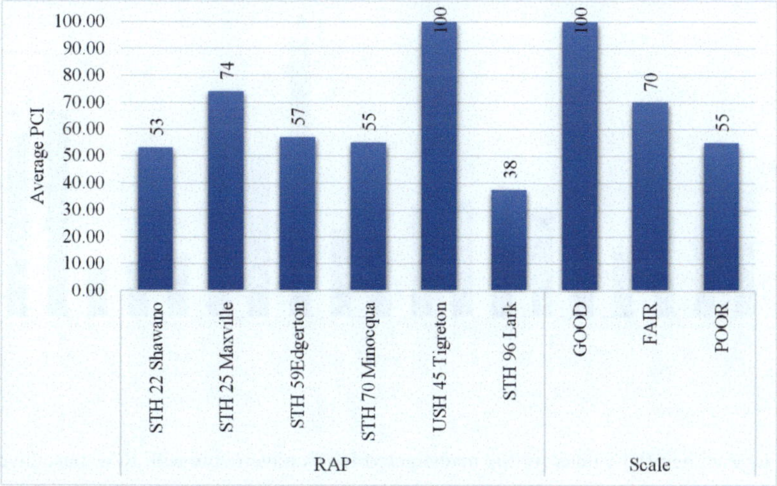

2. PCI average based on the 528-foot sections

Fig. 4. Results of visual distress surveys conducted on STH 22 with RAP base layer.

The FHWA ride quality scale identifies "good ride quality" when the IRI threshold is 95 inch/mile and "acceptable" when the IRI is greater than 95 and less than or equal to170 inch/mile (FHWA 2008). A typical IRI ranges between 52 and 66 inch/mile for a newly constructed HMA pavements (Merritt et al., 2015).

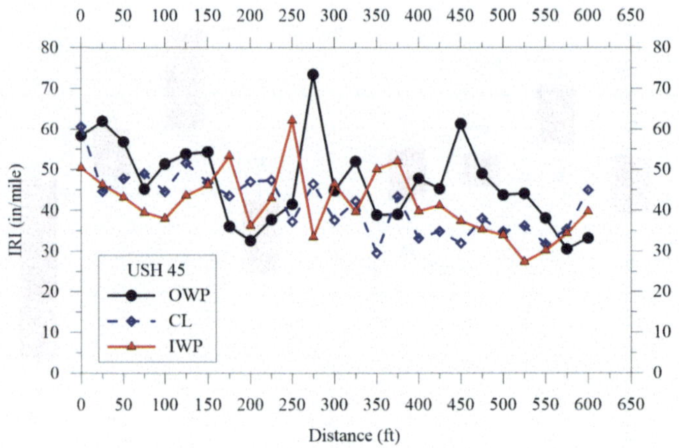

Fig. 5. IRI (ride quality) variation for the USH-45 HMA pavement with RAP base layer.

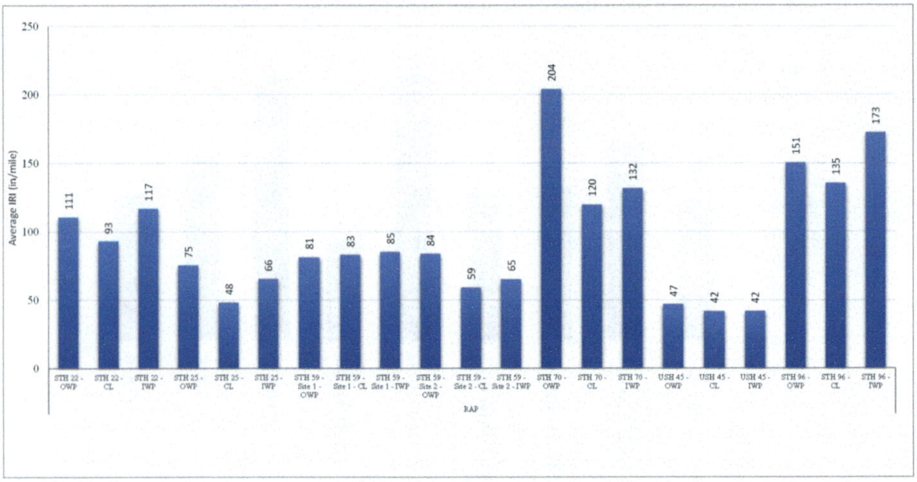

Fig. 6. Average IRI values for the investigated HMA pavements with RAP base layers.

4 Conclusions

This paper presented field evaluation of existing HMA pavements constructed on RAP base layers with various ages. The study consisted of visual distress surveys of 528-foot sections to identify the most common pavement surface distresses and the determination of the corresponding PCI. In addition, it also included ride quality measurements based on 600-foot test sections and the corresponding IRI determination using the walking behind pavement surface profiler.

Observed distresses included cracks (fatigue, longitudinal, transverse, block, and edge), rutting, and flushing. The study showed that fatigue/alligator cracks are commonly observed in the investigated HMA pavement with RAP bases. The calculated PCI values, based on the 528-foot pavement test sections, were variable with the majority fall in the range of "fair" classification category.

The evaluation of the investigated pavements' profiles shows what can be considered as a "smooth" ride quality with an average IRI of 96 inch/mile. Only two of the investigated pavement sections have an "unacceptable" ride quality with IRI values greater than 170 inch/mile.

The research team concludes that the performance of the HMA pavements with RAP base layers is considered satisfactory as compared with HMA pavements constructed with crushed aggregates base course layers as reported in the literature.

Acknowledgments. This study was supported by Wisconsin Highway Research Program and Wisconsin Department of Transportation through project ID 0092–17-01.

References

1. Federal Highway Administration (FHWA). User Guideline for Byproducts and Secondary Use Materials in Pavement Construction. FHWA Report FHWA-RD-97–148, FHWA, VA (2008)
2. Federal Highway Administration (FHWA). The Long-Term Pavement Performance (LTPP) Program (2018)
3. Merritt, D.K., Chang, G.K., Rutledge, J.L.: Best practices for achieving and measuring pavement smoothness, a synthesis of state-of practice. FHWA/LA.14/550. Louisiana Transportation Research Center, Baton Rouge, LA (2015)

Quasi-Periodic Tiling in Moroccan and Andalusian Decorative Art: A Harmonious Symphony of Mathematics and Aesthetics

Zouhair Ouazene(✉)

Sidi Mohamed Ben Abdellah University, Fez, Morocco
zouhair.ouazene@usmba.ac.ma

Abstract. In the vast tapestry of design, the enigma of quasi-crystals has weaved intricate debates regarding their exceptional structures. Emanating like a symphony from the annals of history, echoes of these designs resonate from Islamic art of yesteryears. This modern-day revelation has magnetized scholars, drawing them into a dance of discovery, as they navigate the labyrinth of quasi-periodic patterns scattered across the Islamic tapestry, with a pronounced focus on Moroccan terrains. The non-repetitive cadence of these quasi-periodic tiles, endowed with an expansive order, unfolds as an opulent canvas, dripping with aesthetic allure and rhythmic grace. Traversing both time's corridors and today's contemporary vistas, this scholarly quest seeks to unravel the mathematical enigmas and the cultural heartbeats of these designs, painting a vivid tale of Moroccan and Andalusian artistic legacies. The alchemical process conjures the magic of Islamic quasi-periodic motifs, grounded on a foundational tiling, accentuated with innate nuances. This artistry elegantly fragments its rhombs into harmonious right-angled triangles, bestowing upon them their archetypal symbols. Through the ballet of reflections, the soul of the tile emerges. The grand finale sees these fragmented wonders coalescing into a holistic canvas. It's a poetic interplay of polygons, orchestrated by ancient patterns. Adorned by rosettes, each with their distinct symmetrical whispers, they come alive under the expert hands of Moroccan craftsmen, a craft baptized as Tastir. And as a testament to this artistic renaissance, the canvas will soon unveil fresh quasi-periodic masterpieces, born from this union of art and science.

Keywords: Architecture design · Tastir · Quasi-periodic · Zellig mosaics · Islamic art

1 Introduction

Moroccan and Andalusian decorative arts are celebrated for their intricate geometrical patterns, transcending the mere aesthetic to capture the profound harmony and order underlying the universe. As one delves into the mathematics and geometry behind these designs, the concept of quasi-periodic tiling emerges as an unexpectedly modern nuance amidst these ancient motifs see Fig. 1. This introduction aims to elucidate the enthralling nature of quasi-periodic tiling within the context of Moroccan and Andalusian decorative traditions, offering a portal into the mathematical richness of these age-old designs [1–3].

A. S. Mosallam et al. (Eds.): ICATH 2023, SCI, pp. 22–37, 2024.
https://doi.org/10.1007/978-3-031-76557-5_3

Tracing back to the early Islamic civilizations, it is evident that the decorative arts played a pivotal role in architectural and craft traditions. Especially within the regions of Morocco and Andalusia (modern-day southern Spain), which were hubs of cultural, scholarly, and artistic exchanges during the medieval period, the evolution of patterns and motifs reached its zenith. Whether one gazes upon the intricate ceilings of the Alhambra or the expansive courtyard mosaics of Moroccan riads, the prevalence of complex geometry is undeniable.

In conventional periodic tiling, a single unit or set of units is repeated in a regular manner across a plane. This repetition is both translational and rotational, resulting in a pattern that can be predicted at any point on the plane. However, quasi-periodic tiling introduces a layer of complexity. Rather than exhibiting strict periodicity, these tilings display a form of order where patterns are nearly repeated but not in a strictly regular manner. The result is a design that maintains coherence and structure without the repetitiveness found in conventional tiling [4–7].

The seminal discovery of quasi-crystals in the 1980s by the physicist Dan Shechtman fundamentally altered our understanding of solid-state physics. Here, for the first time, materials exhibited diffraction patterns indicative of quasi-periodicity. This discovery, while not directly linked to the decorative arts, draws striking parallels to the tiling patterns seen in Moroccan and Andalusian motifs. One could argue that these artisans, though perhaps not consciously aware of the mathematical intricacies, were intuitively applying principles akin to quasi-periodicity in their designs.

A profound example can be seen in the Alhambra, a palace and fortress complex located in Granada, Spain. The ornate patterns adorning its walls and ceilings, especially in spaces such as the Hall of the Abencerrajes, exhibit tiling that refuses to settle into simple, predictable repetitions. Instead, the designs flow and morph, presenting sequences that suggest periodicity but evade exact repetition. The zellige tilework of Morocco offers another enchanting canvas of quasi-periodic design. These intricate mosaics, hand-cut from colorful tiles, grace many historical and contemporary Moroccan structures. On close inspection, while there are discernible motifs and patterns, the layout of these tiles often defies strict periodic repetition, aligning more closely with the quasi-periodic description.

What's captivating about quasi-periodic tiling in Moroccan and Andalusian decorative art is not merely its visual appeal but the underlying mathematical sophistication. These patterns, intentionally or not, push boundaries, challenging the viewer's perception of order and repetition. It stands as a testament to the profound intellectual endeavors of these civilizations, where artisans were not just craftsmen but profound thinkers, synthesizing mathematics, art, and spirituality. In contemporary times, the study of quasi-periodic tiling in this context offers insights not just for historians and mathematicians but for architects, designers, and artists seeking to bridge the ancient and the modern. The fusion of art and science in these patterns reminds us that the quest for understanding and expression is timeless, resonating across cultures and epochs.

The quasi-periodic tiling observed in Moroccan and Andalusian decorative arts stands as a testament to the intellectual and artistic prowess of the civilizations that birthed them. As we continue to uncover the depths of these designs, we are reminded of the intricate tapestry of knowledge, creativity, and spirituality that these patterns represent. The subsequent sections of the article are organized in the following manner: Sect. 2 provides an overview of prominent related works. In Sect. 3, we delve into an analysis of Islamic geometric patterns and offer theoretical background essential for comprehending the forthcoming technique. Our proposed methodology is articulated in Sect. 4. Finally, the article reaches its conclusion in Sect. 5.

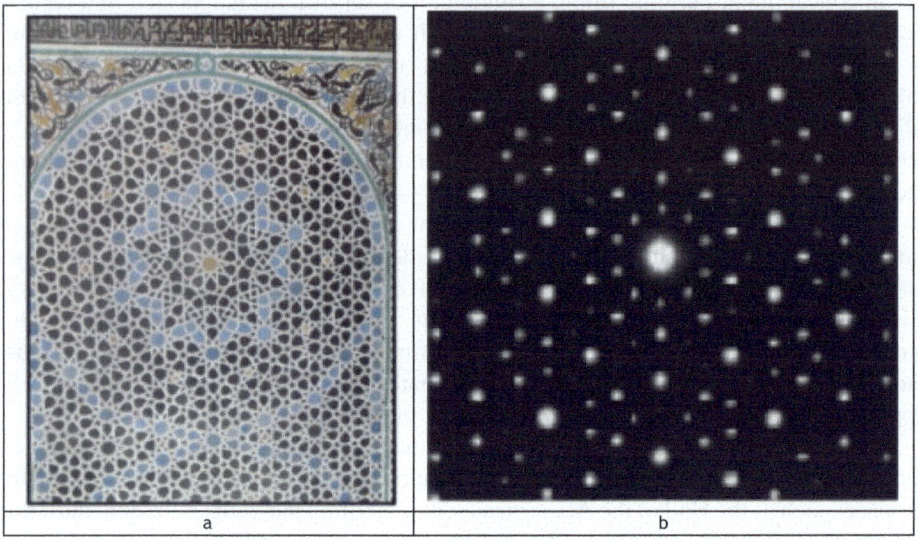

Fig. 1. (a) The board of the Madrasa Attarine (Fez, Morocco) (b) Diffraction pattern showed by D. Shechtman in an Aluminum- Manganese crystal

2 Related Work

Islamic art, characterized by its intricate geometric patterns, has long captivated both artists and mathematicians. Particularly, the symmetries witnessed in some ancient Islamic art patterns bear a resemblance to quasi-periodic structures. This intriguing connection has prompted numerous studies dedicated to deciphering the mathematical constructs underpinning these ornate designs. A significant breakthrough in this domain was presented by Lu and Steinhardt. They identified that tessellations employing Girih tiles, when adorned with lines, facilitated the formation of complex periodic Girih patterns [8]. These patterns can be observed across various historical Islamic sites and artifacts. Notably, at the Darb-I Imam shrine, located in Isfahan, Iran, built in 1453, Lu and Steinhardt pinpointed that combining these tessellations with self-similar transformations could birth nearly impeccable quasi-periodic Penrose patterns. Validating the presence of quasi-periodic patterns in Islamic designs, Saltzman focused on the Gunbad-i Kabud's design and a significant part of the Darb-i Imam design. His findings confirmed that both these designs were rooted in quasi-periodicity [9]. Insights from these analytical explorations are instrumental in understanding these patterns' organizational dynamics and unravelling the underlying principles that drive their creation. Makovicky's research further augmented the existing body of knowledge. Investigating the design of the Blue Tower of Maragha, he unearthed a correlation between the Penrose and Maragha tiling. This discovery led him to conceptualize a novel variation of the Penrose tiling [10–12]. Alongside Fenoll Hach-Alí, Makovicky also turned his attention to Islamic octagonal patterns, predominantly found in Andalusia. Their collective study discerned that these patterns were inherently anchored on a quasi-lattice composed of nested octagons [13]. Further enriching the study sphere, multiple research [14, 15] pieces drew parallels between decagonal quasi-periodic patterns in Andalusia and the Penrose tiling systems. Noteworthy among these was the 2011 analysis centered on the dodecagonal cartwheel in the mausoleum Moulay Idriss II, situated in Fez. This study postulated a compatibility between the design and the Amman quasi-lattice [16]. The Gunbad-e-Qabud in Maragha, Northwest Iran, was another focal point in Makovicky's work, where he sought to comprehend the genesis of the quasi-periodic tiling characterizing the site. Supplementing the body of research [14] Al Ajlouni proposed a model, termed the Hierarchical Framework Exemplary (HFM), explicitly crafted to elucidate the long-range global order observed in decagonal quasi-periodic Islamic art formations. Anchored on a foundational grid of nested decagrams, this framework demarcates the centers of star units. The regions interspersed between the stars are draped using a combination of overlapping star units, positioned in accordance with specific intersection points of the grid. This method, while pioneering, is primarily focused on decagonal and pentagonal patterns, characterized by their distinct five and ten-fold rotational symmetry. Each of Al Ajlouni's model iterations demanded an extensive search to unearth various possible formations that bridge stars.Expanding upon these insights, Aboufadil and his team introduced a technique to craft Moroccan quasi-periodic patterns. Their method, rooted in the multigrid and the Hasba technique traditionally employed by artisans, however, was not amenable for computational applications and demanded a substantial degree of hands-on expertise [17]. Lastly, Kaplan's proposition centered on the creation of non-periodic patterns [18, 19]. Employing quasi-periodic tiling formed by rhombs as a foundational guide, he envisaged a method where regular polygons were strategically placed to yield a non-periodic tiling. This system was further refined with rosettes inscribed within the polygons, while the remaining tiles were methodically filled leveraging an inference algorithm. In conclusion, the exquisite geometry of Islamic art, intertwined with the complexity of quasi-periodic tiling, offers an unparalleled canvas for exploration, blending artistry with mathematical precision. Through concerted research efforts, the mysteries of these designs continue to unfold, enriching our appreciation and understanding of this age-old artistic tradition.

3 Background

3.1 Study of a Historic Pattern

Islamic decorative art, heralded for its symmetrical designs and elaborate patterns, exhibits unparalleled mathematical elegance. One such paragon of this artistic brilliance graces the board of the Madrassa Attarine in Fez, Morocco. Recognized among the most intricate geometric configurations in Islamic art, this design has garnered significant scholarly interest [20–22]. A comprehensive analysis of this pattern unveils its origin and underlying structures, amplifying our understanding of its aesthetic and mathematical constructs. Central to the Madrassa Attarine's pattern is its intricate formation, which, upon examination, can be attributed to two specific Penrose tiles - fat and thin rhombs. These tiles, emblematic in their geometric shape, serve as the foundational elements of the design, as exemplified in Fig. 2.

Delving into the specifics of the tiles, a prominent motif commonly encountered in Islamic decorative designs emerges. Termed as the 'rosette', this motif takes a pronounced stance within the fat rhomb, occupying its entirety (as seen in Fig. 2(e)). A more dissected representation of the rosette is visible within both the fat and thin rhombs. In this distribution, each angle of the rhomb hosts a fraction of the rosette, contributing to the motif's segmented presence (as depicted in Fig. 2(f)).

The study of the Madrassa Attarine's board unravels several pivotal conclusions:

- Motif Integration: The quasi-periodic tiles, specifically the rhombs, can seamlessly integrate the rosette and star motifs, offering a harmonious blend of geometry and design.
- Motif-Rhomb Alignment: A discernible relationship exists between the positioning/order of the rosette or star motif and the specific angles of the rhombs in which they are situated.
- Centrality of the Motif: The vertices of the rhombs can serve as the central points for positioning the rosette or star motifs.
- Motif Completeness: While each rhomb's angle houses a segment of the rosette, a holistic view reveals that the amalgamation of these segments culminates in a full rosette.
- Linearity Assurance: A critical aspect of the design involves ensuring that the pattern lines maintain their continuity across the boundaries of adjacent rhombs, preserving the coherence of the design.
- Pattern Derivation: Islamic quasi-periodic designs, as intricate as they are, can be effectively conceived by leveraging existing quasi-periodic tiles.

Given these insights, it becomes evident that the Islamic quasi-periodic patterns, in their grandeur, can be meticulously derived from quasi-periodic tiles dominated by rhombs. This essentially means that the intricate art of crafting an Islamic quasi-periodic design can be distilled down to determining the content within the tiles that constitute its corresponding quasi-periodic tiling.The geometric patterns adorning the Madrassa Attarine in Fez stand as a testament to the mathematical prowess and artistic brilliance inherent in Islamic decorative art. By deconstructing the design's elements and understanding their interconnectedness, we appreciate not just its visual appeal but also the structured logic that underpins its creation. Such explorations serve as bridges, connecting art's beauty with the rigor of mathematics, reaffirming the timeless allure of Islamic geometric patterns.

3.2 Quasi-Periodic Tiling

The intricate world of geometric patterns and tiling has been a subject of keen interest for mathematicians and artists alike. One such work that provides a comprehensive view of this domain is Martin Gardner's scientific article, which delves deeply into Roger Penrose's tiling innovations. Published in 1977, Gardner's article emphasized Penrose's second tile pattern, popularly recognized today as "kites and darts". While the spotlight was on this second pattern, Gardner made a passing reference to Penrose's initial tiling design. This original design comprised a set of six distinctive aperiodic tiles. However, for simplicity and clarity, it has often depicted using only four primary tiles: the pentagon, star, boat, and rhombus. For ease of reference, this set is referred to as P1, as illustrated in Fig. 3. Penrose's journey in the realm of tiling was driven by an aspiration: to curate a tiling pattern using as few aperiodic tiles as possible while still maintaining the ability to tile a plane non-periodically. From this quest emerged his iconic second pattern. Using just two aperiodic tiles, the kite and dart, he crafted what came to be known as the P2 pattern, as depicted in Fig. 4a.

Though Gardner meticulously detailed the P2 pattern's properties and intricacies, he also hinted at its inherent adaptability. Specifically, he mentioned the potential to extract other tile patterns from P2. This could be achieved by subdividing tiles and then strategically recombining them, thereby forging new tile configurations. Following this principle, various tile pairs, bearing properties reminiscent of the original P2 tiling, could be synthesized. An exemplar of this adaptability is provided in the form of Penrose's own innovation. Gardner elucidated how Penrose adapted the P2 to derive a two-rhombus tiling design. This consists of two aperiodic tiles, specifically a "fat" and a "skinny" rhombus, which is subsequently labeled as the P3 pattern, as showcased in Fig. 4c.

Broadening the horizon beyond Penrose's designs, references [23] and [24] provide a comprehensive description of quasi-periodic tiling anchored in the foundations of pentagrids. Over time, various methodologies have emerged to craft these quasi-periodic tiling's. Prominent among these is the 'cut-and-projection' technique, which derives its methodology from higher-dimensional spaces, as elucidated in references [25] and [26]. Another noteworthy approach is the generalized dual method. Central to these designs is the quasi-periodic tiling by rhombs (as captured in Fig. 4). This foundational grid is typically accentuated by adorning its tiles with ornate rosette or star motifs, adding an additional layer of aesthetic depth. The world of tiling, as presented through the lens of Martin Gardner's analysis of Roger Penrose's work, reveals a tapestry of mathematical elegance interwoven with artistic flair. The evolution from P1 to P3 showcases not just the brilliance of Penrose's mathematical mind but also the infinite adaptability and potential of tiling patterns in the realm of geometric design.

In the expansive field of geometric designs and tiling, De Bruijn's 1981 work introduced a revolutionary approach: the multigrid process for crafting quasiperiodic tiling. The essence of this method, as elucidated by De Bruijn, is the derivation of dodecagonal (12-sided) tiling from what is termed a 6-grid, as visualized in Fig. 5. What sets this quasiperiodic tiling apart is its unique composition. Two distinct rhombic tiles characterize the design, each differing in their vertex angles. One tile has vertex angles of 30 degrees and 60 degrees, while the second is a perfect square tile, as illustrated in Fig. 6. To delve deeper into the geometric nuances of these tiles, let us consider the square tile.

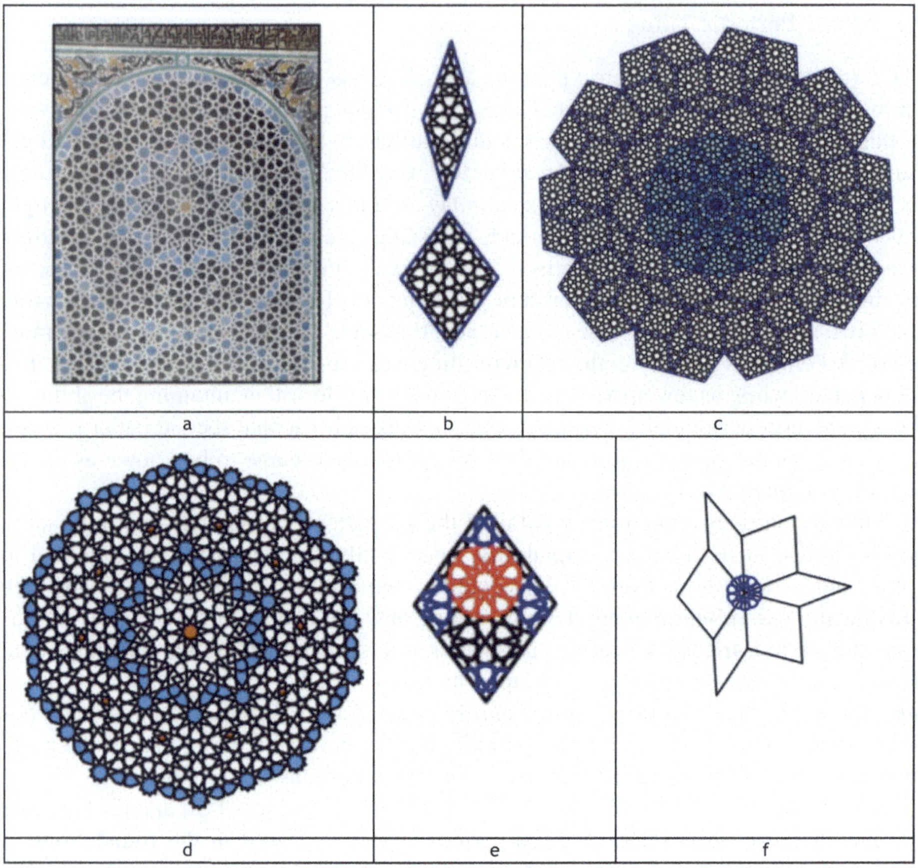

Fig. 2. (a) The board of the Madrasa Attarine (Fez, Morocco), (b) the matching decorated tiles, (c) organization of the decorated tiles, (d) the similar part to the panel, (e) arrangement of the rosettes in the fat rhomb, and (f) a 10-pointed rosette centered on a shared vertex between six rhombs.

Fig. 3. Penrose's first tile pattern in 3 colors

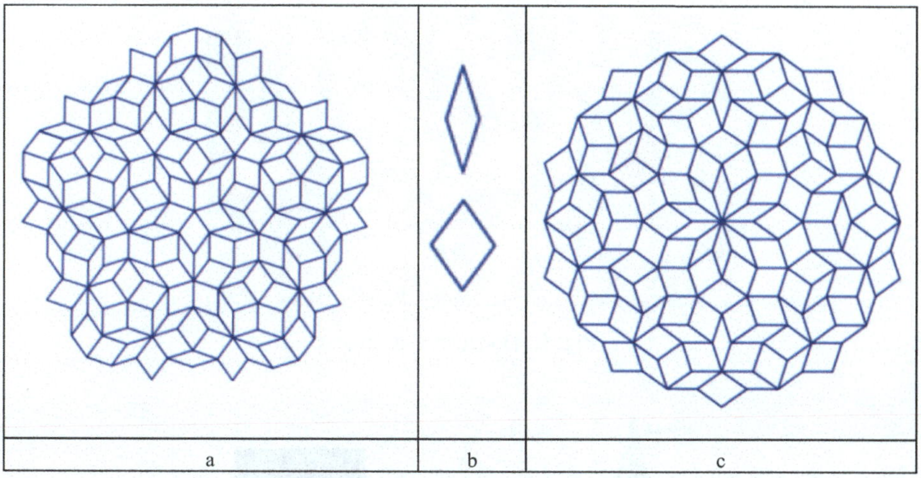

Fig. 4. (a) P2 tile pattern, kites and darts (c) P3 tile pattern, two-rhombuses (b) their associated tiles.

Let us denote its side length as L0. With this definition, the diagonal of this square can be computed as twice the value of L1, as showcased in Fig. 7-c.

The intricacies of the rhombic tiles can also be broken down in a similar fashion. The larger, or "thick", rhombus is essentially a pairing of two equilateral triangles. Each side of these triangles matches the length L0, which can be visualized in Fig. 7-b. The slimmer, or "thin", rhombus, on the other hand, is the combination of two isosceles triangles. These triangles have sides of length L0 and a base designated as L2. This geometric configuration is depicted in Fig. 7-a.

However, De Bruijn's multigrid process was not solely about the creation of tiles; it extended into the realm of ornamental design. In this context, the thin rhombus is adorned with a pattern referred to as T2. The more expansive, thick rhombus is embellished with the T1 pattern. Meanwhile, the square, central to this tiling system, holds its own unique design, as highlighted in Fig. 7-c. In summation, De Bruijn's 1981 exploration into the world of quasiperiodic tiling using the multigrid process presents a harmonious blend of mathematics and art. Through systematic breakdowns and precise geometric calculations, he introduces a method that not only creates a unique tiling pattern but also highlights the potential for ornamental design within each tile. This confluence of structure and aesthetics underscores the timeless appeal and depth of geometric tiling, reminding us of the endless possibilities it holds.

4 Methods

The realm of geometric patterns in design, particularly in the context of quasi-periodic tiles, is both intricate and fascinating. Central to this is the method of creating these tiles through the progressive stacking of polygons, which moves outward from a chosen center towards the periphery. This intricate process offers an illuminating glimpse into the harmonious blend of art and mathematics, and its deep roots in the rich tapestry of

Fig. 5. Dodecagonal tiling

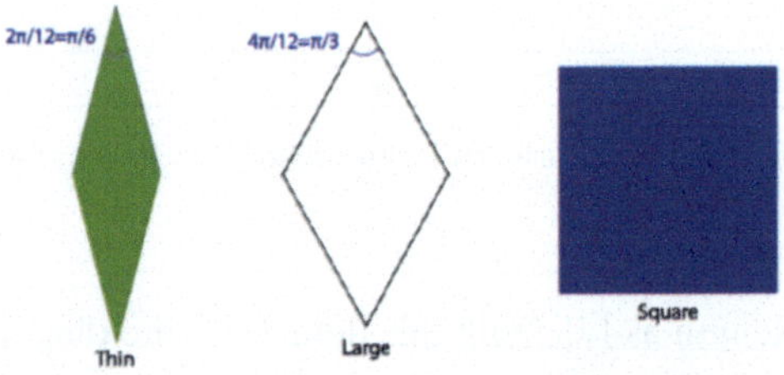

Fig. 6. Three types of tiles that build up the quasiperiodic lattice

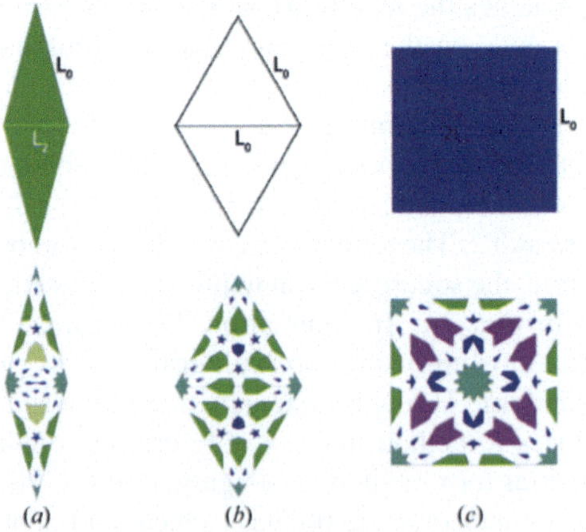

Fig. 7. Decoration of the three types of shapes by using the overlapping tiles.

artistic traditions. The foundational step in this process involves the careful extraction of rhombs from the established quasi-periodic tiling. Following this extraction, each individual tile undergoes a process of division, resulting in the formation of symmetrical right triangles, as visualized in Fig. 9. Providing a more detailed illustration, as shown in Fig. 8, the rhombs are meticulously divided into four congruent right triangles. In contrast, the squares within this geometric configuration are dissected into eight uniform right triangles. This nuanced approach to dividing the tiles ensures a sense of coherence and symmetry across the entire design.

Zouhair and his colleagues, in their works [22, 27, 29], delve deeper into this methodology, offering a comprehensive exposition on constructing the content of these tiles. This method, rooted in mathematical precision, finds echoes in the time-honored techniques employed by artisans, underscoring its significance in traditional craftwork. A particularly intriguing facet of this domain is the presence of octagonal quasi-periodic tilings. These tilings stand out due to their distinct non-repetitive geometric configurations, which, while lacking in conventional periodic repetition, exude a long-range order marked by octagonal symmetry, as elaborated upon in [7, 28]. Distinguishing characteristics of these tilings include the absence of traditional translational symmetry, a hallmark of regular periodic tiles. Instead, in octagonal quasi-periodic tilings, tiles are positioned following a set of mathematical regulations and constraints. This ensures the creation of a design that is not only visually enticing but also maintains a sense of harmony and balance. An integral part of these tilings encompasses a melange of geometric shapes—octagons, squares, and an assortment of other polygons. These are artfully arranged to ensure that they cover the entire plane without any unintended overlaps or gaps. Furthermore, while the tiles within this configuration might vary in size or proportion, their relationship and positioning with respect to each other remain consistent, thereby preserving the pattern's integrity. It's imperative to understand that octagonal quasi-periodic tilings aren't monolithic. Specific designs can exhibit unique attributes and variations. Factors influencing these variations can range from the intricacies of the design itself to the constraints imposed upon it and even the cultural context within which it is realized.

The creation and study of quasi-periodic tiles, especially through progressive polygons stacking, offer a window into a world where art and mathematics converge. Whether it's the meticulous division of tiles into symmetrical triangles or the exploration of octagonal quasi-periodic designs, each step in this journey resonates with both aesthetic beauty and mathematical rigor. As we continue to study and appreciate these designs, we are reminded of the timeless allure of geometry and the endless possibilities it presents in the realm of art and design.

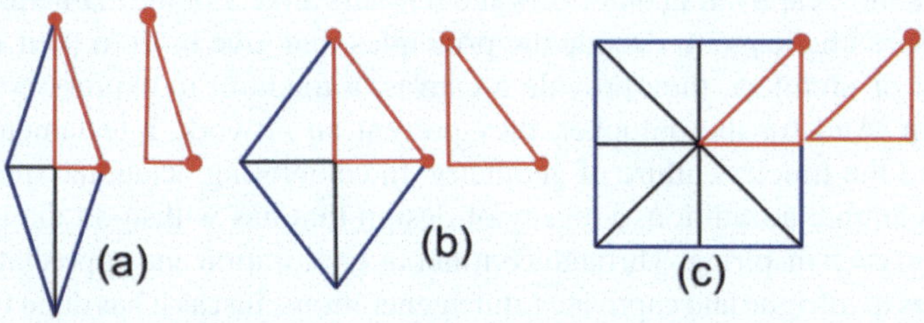

Fig. 8. (a, b) Division of the rhombs into right triangles and (c) division of the square into right triangles. The rosettes can be centered on the highlighted red corners.

In the intricate tapestry of design and mathematical patterns, octagonal quasi-periodic tilings stand out as a captivating subject. These tilings, as illustrated in Fig. 10, manifest across various domains, encapsulating the realms of architectural design, decorative arts, and profound mathematical explorations. Their presence is not just a modern-day phenomenon; they echo through history, adorning our built environment and serving as a testament to human craftsmanship and ingenuity. Historically, these tilings can be observed in architectural marvels, particularly in Islamic structures where they bring forth the rich legacy of traditional geometric patterns. Not just limited to grand architectural endeavors, these patterns can also be seen intricately woven into decorative tilework, adding splendor to interior spaces and elevating their aesthetic appeal. This ancient art form has managed to transcend time, with its principles remaining just as relevant and awe-inspiring today. The evergreen allure of octagonal quasi-periodic tilings has piqued the curiosity of mathematicians and researchers who are keen to delve deep into the nuances of their formation, eager to decipher the mathematical rationale behind these elaborate designs.

At its core, the study of octagonal quasi-periodic tilings is a convergence of several mathematical ideas. One of the foundational concepts is that of aperiodic tiling. Unlike periodic tiling, which is characterized by a regular and repeating pattern, aperiodic tilings lack this repetitive symmetry. Instead, they fill space without periodic repetition, showcasing a unique blend of order and complexity. Further, the study delves into the realm of symmetry groups. These groups help categorize patterns based on their inherent symmetries, providing a systematic framework to understand the repetitive elements within a design and how they interact with one another. Substitution systems further embellish the understanding of these tilings. As the name suggests, this concept involves the substitution of tiles with predefined sets of smaller tiles. Through iterative substitution processes, one can generate increasingly complex and larger patterns, which are yet governed by the original tiling rules. By intertwining these concepts, one can begin to grasp the mathematical scaffolding that underpins octagonal quasi-periodic tilings. This intricate structure is, in essence, a harmonious assembly of a rhomb (specifically angled at $\pi/4$) and a square. This simple yet profound combination crafts the basis for a myriad of designs, each more captivating than the next.

Octagonal quasi-periodic tilings, in all their geometric splendor, are more than just patterns; they represent the culmination of artistry, history, and mathematical brilliance. From grand architectural facades to the nuanced strokes of tilework, they manifest as visual symphonies, captivating onlookers and scholars alike. For the mathematician, they offer a puzzle, a challenge to decode the principles that give birth to their complexity. For the artist or architect, they provide a canvas, a medium to express creativity and craftsmanship. And for the onlooker, they present an artwork, a testament to human capability and the timeless allure of geometry. In celebrating octagonal quasi-periodic tilings, we celebrate a tradition, a legacy of design that has withstood the test of time, ever-evolving, ever-inspiring. Through continuous exploration and appreciation, this art form promises to intrigue and captivate future generations, just as it has done for centuries past.

At the confluence of mathematical rigor and artistic splendor lies the realm of decagonal quasi-periodic tilings. These intricate patterns, rather than repeating in a predictable manner as with periodic structures, sprawl across surfaces in a non-repetitive yet ordered fashion, centered on a decagonal motif. Such tiling's, while complex, strike a harmonious balance between chaos and order, leading to designs that are as mathematically intriguing

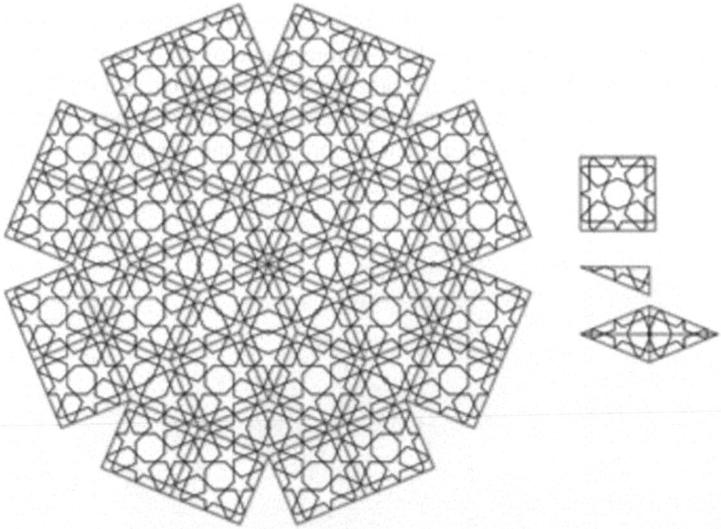

Fig. 9. Octagonal patterns suitable for the method.

as they are aesthetically pleasing. At the heart of decagonal quasi-periodic tiling's lies a unique combination of shapes. Dominating this tiling landscape is the dodecagonal composition, which harmoniously brings together a square and two distinct rhombs. These rhombs differ in their angular dimensions, with one being characterized by an angle of $\pi/6$, and its counterpart boasting an angle of $\pi/3$. However, the world of decagonal tilings isn't just restricted to these foundational shapes. Woven into these patterns are a myriad of other geometrical forms, ranging from regular to irregular decagons and even encompassing triangles and other multifaceted polygons. The magic of these designs stems from the meticulous arrangement of these shapes. Following a stringent set of rules and constraints, they come together to craft a visually compelling tapestry that resonates with harmony and balance. One of the most distinguishing features of decagonal quasi-periodic tilings is their ten-fold rotational symmetry. This symmetry, while adding layers of complexity, also infuses the pattern with a unique visual rhythm. Every rotation, every twist, and every turn is a dance of geometry, making these designs dynamically fluid and yet consistently symmetrical.

Decagonal quasi-periodic tiling's, as represented in Fig. 11, have since then become a focal point of study, drawing interest from both researchers and artists. On one hand, there is the mathematical allure. Delving deep into the realm of these tilings reveals a world rich in properties, symmetries, and intricate geometric relationships. The algorithms and rules that underpin these designs perennially intrigue scholars and researchers. From analyzing the inherent symmetries to understanding the self-similarities that ripple through these patterns, the mathematical exploration is as endless as it is fascinating. Concurrently, on the other hand, is the aesthetic and cultural exploration. Artists, drawn to the visual cadence of these tilings, often find inspiration in their intricate designs. Beyond mere aesthetics, these patterns often hold deep cultural significance, resonating with historical narratives and traditional craftsmanship.

Decagonal quasi-periodic tilings stand as a testament to the human quest for understanding and beauty. They represent a harmonious blend of mathematical precision and artistic expression, each feeding into the other. As we continue to study, appreciate, and draw inspiration from these designs, they serve as a reminder of the limitless potential that lies at the intersection of art and science, geometry and design, chaos and order.

Fig. 10. Octagonal pattern with a main rosette of order 16, an 8-pointed rosette placed on the sides of the two right-angled triangles and an 8-pointed star placed on the free vertex of the right-angled triangle of the square.

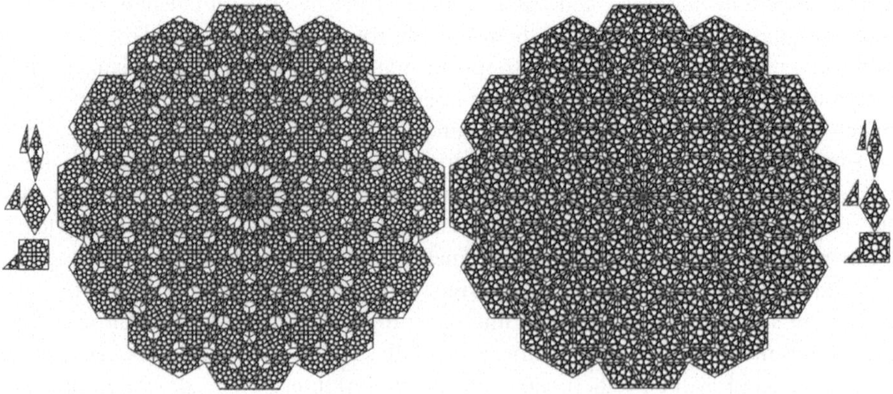

Fig. 11. Decagonal Islamic quasi-periodic patterns obtained by the same process

5 Conclusion

The enchanting world of Islamic geometric designs has captured the imagination of many, reflecting the rich tapestry of history, culture, and mathematics intertwined. In this detailed exposition, we endeavor to shed light on a novel methodology we have introduced to craft and rejuvenate Islamic quasi-periodic patterns, drawing inspiration from existing architectural masterpieces.

At the very heart of our technique lies a fundamental principle: to draw upon established quasi-periodic tilings from contemporary times as a foundational framework. Rather than reinventing the wheel, we aimed to enhance and reinterpret these patterns, imbuing them with renewed aesthetic appeal. This endeavor to revive ancient art forms by meshing them with modern designs stands as a testament to the timeless allure of Islamic patterns.

A significant aspect of our methodology was a deliberate shift in focus to the most basic geometric unit - the triangle. We chose to concentrate our construction efforts on the smallest triangles extracted from the tiles of the chosen quasi-periodic tiling. By considering these triangles as elemental units within an overarching quasi-periodic scaffold, we were able to lay the groundwork for intricate designs with a strong foundation.

The construction of these patterns wasn't a singular, linear process. Instead, it involved a series of meticulously planned steps, each contributing to the final design:

- Design Components: The core of our patterns comprises two elemental designs – stars and rosettes. By adorning the right triangles with these motifs, we introduced a layer of complexity and visual interest to the design.
- Isometric Transformations: Once the foundational motifs were in place, we turned to the mathematical tool of isometric transformations. Applying these transformations to our template motifs, we birthed the contents of the rhombs, or what we refer to as the 'unit motifs'.
- Populating the Quasi-Periodic Tiling: With our unit motifs ready, the next step was akin to assembling a jigsaw puzzle. These motifs were meticulously replicated and positioned within the larger quasi-periodic tiling, ensuring coherence and harmony in the overall design.

The culmination of our efforts was a fresh array of quasi-periodic patterns. But these weren't just any patterns. They resonated with the echoes of ancient architectural wonders while simultaneously reflecting contemporary design principles. This synthesis of old and new offers a fresh perspective on Islamic geometric designs, ensuring their relevance and appeal in modern times.

In presenting this technique, our aim was to bridge the past and the present, to offer a pathway through which the timeless beauty of Islamic geometric designs could be reimagined for contemporary aesthetics. By focusing on minimal triangles, leveraging the power of isometric transformations, and meticulously crafting patterns, we have introduced a method that respects tradition while embracing innovation. As we look at the new variety of quasi-periodic patterns birthed from this technique, we are reminded of the endless possibilities that lie at the intersection of art, history, and mathematics.

References

1. Abas, S.J., Salman, A.S., Moustafa, A., Atiyah, M.: Symmetries of Islamic Geometrical Patterns. World Scientific (1994). https://doi.org/10.1142/2301
2. Castera, J.M., Jolis, H. : Géométrie Douce. éditions 61/2. Paris (1992)
3. El-Said, I., El-Bouri, T.: Islamic Art and Architecture: the System of Geometric Design. Ithaca Press (1993)
4. Khamjane, O., et al.: The Polygonal Technique for Constructing Islamic Geometric Patterns, pp. 521–30 (2023). https://doi.org/10.1007/978-3-031-29860-8_53
5. Djibril, M.O., Thami, R.O.H.: Islamic geometrical patterns indexing and classification using discrete symmetry groups. J. Comput. Cult. Herit. **1**(2), 95–112 (2008). https://doi.org/10.1145/1434763.1434767
6. Wichmann, B.: Symmetry in Islamic geometric art. Symmetry Cult. Sci. **19**(2–3), 95–112 (2008)
7. Al Ajlouni, R., Al Ajlouni, R.: Octagon-based quasicrystalline formations in islamic architecture. Aperiodic Cryst. 1197, 49–57 (2013). https://doi.org/10.1007/978-94-007-6431-6_7
8. Lu, P.J., Steinhardt, P.J.: Decagonal and quasi-crystalline tilings in medieval Islamic architecture. Science (80) **315**(5815), 1106–1110 (2007). https://doi.org/10.1126/science.1135491
9. Khamjane, A., Benslimane, R.: Generating islamicquasi-periodic patterns: a new method. J. Comput. Cult. Herit. **11**(3), 13:113:18 (2018). https://doi.org/10.1145/3127090
10. Al Ajlouni, R.A., Al Ajlouni, R.A., Al Ajlouni, R.A.: A long-range hierarchical clustering model for constructing perfect quasicrystalline formations. Philos. Mag. **91**(19–21), 2728–2738 (2011). https://doi.org/10.1080/14786435.2010.532515
11. Makovicky, E., Pérez, F.R., Hach-Alí, P.F.: Decagonal patterns in the Islamic ornamental art of Spain and Morocco. Bol. la Soc. Esp. Mineral. **21**, 107–127 (1998)
12. Thalal, A., Aboufadil, Y., Elidrissi Raghni, M.A., Jali, A., Oueriagli, A., Ait Rai, K.: Symmetry in art and architecture of the Western Islamic world. Crystallogr. Rev. **24**(2), 102–130 (2018). https://doi.org/10.1080/0889311X.2017.1343306
13. Makovicky, E.: Another look at the Blue Tomb of Maragha, a site of the first quasicrystalline Islamic pattern. Symmetry Cult. Sci. **19**, 127–151 (2009)
14. Ajlouni, R.: An ancient rule for constructing dodecagonal quasiperiodic patterns. J. Phys. Conf. Ser. **809**, 012028 (2017). https://doi.org/10.1088/1742-6596/809/1/012028
15. Makovicky, E., Makovicky, N.M.: The first find of dodecagonal quasiperiodic tiling in historical Islamic architecture. J. Appl. Crystallogr. **44**(3), 569–573 (2011). https://doi.org/10.1107/S0021889811013744
16. Hargittai, I., Makovicky, E.: 800-year-old pentagonal tiling from Marāgha, Iran, and the new varieties of aperiodic tiling it inspired. In: Fivefold Symmetry, Singaport-New Jersey etc.: World Scientific Publishing Co Pte Ltd, pp. 67–86 (1992). https://doi.org/10.1142/9789814439497_0004
17. Aboufadil, Y., Benatia, J., Jali, A., Raghni, M.A.E., Thalal, A.: Practical realization of Moroccan geometric pattern. Acta Crystallogr. Sect. A Found. Crystallogr. **65**(a1), s191–s192 (2009). https://doi.org/10.1107/S0108767309096044
18. Kaplan, C.S.: Islamic star patterns from polygons in contact. In: Proceedings - Graphics Interface, pp. 177–185 (2005). http://doi.acm.org/10.1145/1089508.1089538
19. Kaplan, C.S.: Computer Graphics and Geometric Ornamental Design. University of Washington (2002)
20. Ouazene, Z., Khamjane, A., Benslimane, R.: Relationship Between Eight-Fold Star and Other Tiles in Traditional Method 'Tastir,' **110**. Springer International Publishing (2022). https://doi.org/10.1007/978-3-030-94188-8_33

21. Khamjane, A., Benslimane, R.: A computerized method for generating Islamic star patterns. Comput. Des. **97**, 15–26 (2018). https://doi.org/10.1016/j.cad.2017.11.002
22. Khamjane, A., Benslimane, R., Ouazene, Z.: Method of construction of decagonal self-similar patterns. Nexus Netw. J. **22**(2), 507–520 (2020). https://doi.org/10.1007/s00004-019-00461-4
23. Bonner, J.: The historical antecedents, initial development, maturity, and dissemination of islamic geometric patterns. In: Islamic Geometric Patterns, Springer, pp. 1–152 (2017)
24. Critchlow, K.: Islamic Patterns. Thames and Hudson London (1976)
25. Lalvani, H.: Coding and generating complex periodic patterns. Vis. Comput. **5**(4), 180–202 (1989). https://doi.org/10.1007/BF02153749
26. d'Avennes, P. : L'art arabe d'après les monuments du Kaire depuis le VIIe siècle jusqu'à la fin du XVIIIe: Atlas, **4**. A. Morel et Cie, (1877)
27. Ouazene, Z., Benslimane, R., Khamjane, A.: CAD geometry for the traditional Tastir style Moroccan and Andalusian 16-fold rosettes. Symmetry Cult. Sci. **33**(1), 55–68 (2022). https://doi.org/10.26830/symmetry_2022_1_055
28. Makovicky, E., Makovicky, N.M.: Nonperiodic octagonal patterns from a jali screen in the mausoleum of muhammad ghaus in gwalior and their periodic relatives. Nexus Netw. J. **19**(1), 101–120 (2017). https://doi.org/10.1007/s00004-016-0316-6
29. Khamjane, A., Abderazzak, T., Zouhair, O., Rachid, B.: Computer graphics for generating islamic geometric periodic and quasi-periodic patterns. In: Proceedings - 2019 International Conference on Intelligent Systems and Advanced Computing Sciences, ISACS 2019, pp. 391–400 (2019). IEEE. https://ieeexplore.ieee.org/document/9068879/

Thermal Characterization of Novel Multi-Layer Panel for Building Thermal Insulation

Hafida Er-Rradi[1(✉)], Zineb Samouh[2], Imane Bouchefra[3], Chakib Bojji[4], Abdelilah Jilbab[5], and Rachida Idchabani[6]

[1] L3GIE, Mohammadia School of Engineer, Mohamed V University, Rabat, Morocco
h.er-rradi@enim.ac.ma

[2] IRESEN - Solar Energy and New Energies Research Institute, Rabat, Morocco

[3] Laboratory of molecular chemistry, materials and catalysis, Sultan Moulay Slimane University, Beni Mellal, Morocco

[4] School of Engineering Sciences and Technologies, Rabat, Morocco

[5] National School of Arts and Crafts, Rabat, Morocco

[6] National School of Mine, Rabat, Morocco

Abstract. This paper explores the thermal properties of multi-layer panels crafted from sustainable and local materials, with a focus on agglomerated cork cores and bio-composite external layers. It delves into both symmetric and asymmetric designs, with varying cork thickness. The study assesses the panels by evaluating density, thermal conductivity, specific heat through the use of an HFM apparatus, and calculates thermal diffusivity and effusivity from these measurements. Findings reveal the panels are lightweight with superior thermal insulation capabilities, highlighting thermal conductivity rates between 0.071 and 0.102 W/mK. The research also compares measured values of thermal conductivity to theoretical predictions, highlighting the accuracy of the multi-layer thermal analysis method and the strong cohesion among the panel layers. The newly developed panels offer thermal efficiency and contribute to reduced energy consumption in buildings. Suitable for use as partitions, interior cladding, and false ceilings, these panels can also be painted or covered with adhesive paper to satisfy aesthetic demands.

Keywords: Eco-sandwich · Thermal properties · Wood waste · Plaster

1 Introduction

Morocco's primary energy consumption is still almost 90% fossil fuels and the demand of energy is increasing especially with the increase of total population. Subsequently, research on saving and conservation energy has become indispensable more than ever before. In Morocco building is the second consumer sector after transportation of which it represents 33% of the country's total energy consumption [1] and this energy consumption is expected to increase rapidly due to the evolution of construction project and as a consequence of the increase in energy demand the negative environmental impact are becoming more important due to the increase of CO_2 in the atmosphere. The use of thermal insulation in building can enhance the thermal performance and reduce the energy consumption in building [2].

A. S. Mosallam et al. (Eds.): ICATH 2023, SCI, pp. 38–49, 2024.
https://doi.org/10.1007/978-3-031-76557-5_4

Eco-friendly insulation materials prioritize sustainability. They are derived from natural resources like agricultural and forestry byproducts, as well as recycled materials, contributing to their environmentally conscious nature [3]. Enhancing thermal building insulation with bio-based materials represents a significant focus in the ongoing research endeavors of numerous scientists and researchers. Recently, many scientific researches have delved into the utilization of eco-friendly materials for thermal building insulation. Included among these substances are cork, cotton, fibers derived from wood, wood wool, hemp, jute, coconuts, and wool from sheep.

Extensive research has focused on sustainable materials, with hemp being a primary subject of interest in several studies [4–6]. Enrico Sassoni et al. [7] combined hemp waste with an innovative hybrid adhesive to produce a material exhibiting efficient thermal characteristics. L.Freivalde et al. [8] focused their research on the fire resistance properties. J. Lazko et al. [9] developed insulating biocomposites from fibers and linseed oil, investigating their mechanical and thermal characteristics. Xiao-yan Zhou et al. [10] further explored and engineered cotton stalk fibers as insulation, examining their capabilities in terms of thermal conductivity and mechanical strength. In their study, Bouchefra et al. [11] examined the effects of adding doum fiber to clay-based compressed earth bricks (CEB) on both thermal conductivity and compressive strength. The outcomes confirmed that doum fiber inclusion notably enhances the thermal insulation capability of CEB. The production of thermal insulation materials typically takes the form of solid boards/panels, raw biomass materials, and multi-layered constructions, including sandwich [12]. These sandwich configurations, known for their light structural framework, are increasingly popular in diverse fields like automotive, aerospace, marine, and energy [13, 14]. Kawasaki et al. [15] created and examined sandwich panels crafted from veneer-overlaid fiberboards bonded with industrial resin. These panels, notable for their light weight at a thickness of 12 mm, offer superior thermal insulation and mechanical strength. In a subsequent research, Kawasaki et al. [16] innovated sandwich panels with a plywood facade and a low-density fiberboard core (PSW), which showed better insulation properties, lighter weight, and improved thermal retention than standard boards. Touil et al. [17] demonstrated that the sandwich structure's thermal efficiency is higher than models where the fibers are dispersed randomly within the main matrix. An experimental research on different sandwich samples was carried out by Castro et al. [18]. They discovered that sandwich panels made of cork have excellent mechanical and thermal conductivity. A. Limam et al. [19] investigated the use of cork as insulation within a sandwich structure. They found a thermal conductivity between 0.063 W/mK and 0.068 W/mK. Reis et al. [20] examined the mechanical strength of a multi-layers incorporating inner layer cork and carbon/epoxy outer layer. Their findings indicated that smaller grain sizes and higher densities yield the most favorable results. An eco-sandwich made of natural materials including cork, flax fibers, and bio-based epoxy resin was produced by La Rosa et al. [21]. The eco-sandwich has a thermal conductivity between 0.074 and 0.081 W/m K. They discovered that applying the eco-sandwich concept improves environmental performance.

In the current study, a novel sandwich panel made with ecological components has been assessed experimentally. The samples were made from eco-composite materials for the outer layers and a cork core for the inner layer, utilizing plaster mixed with either granular cork or wood waste (chips and sawdust) for the composites. Two configurations were tested: a symmetric configuration with identical outer layers, and an asymmetric configuration with two different outer materials. Cork core thicknesses of 2 cm and 4 cm were examined. The experimental characterization included measurements of density, thermal conductivity, and specific heat, with thermal effusivity and diffusivity calculated from these findings.

2 Materials and Methods

2.1 Wood Residue

Wood residue refers to the byproducts generated during the sawmill process. These byproducts typically include wood chips and sawdust, which are collected and separated through a sieving process. Wood chips are larger pieces of wood typically produced by chipping or shredding larger wood pieces, while sawdust consists of fine particles created during the cutting or milling of wood.

2.2 Cork

Granular cork waste is a byproduct that arises from the manufacturing process of agglomerated cork panels. This granulates are used to produce eco-composite materials based on plaster matrix.

Cork panel used as inner layer are obtained from an industrial companies specializing in cork processing and manufacturing. Used samples of cork panel care in square shape measuring 300 mm*300 mm with a thicknesses of 2 and 4 cm, exhibiting a density range spanning from 207 to 310 kg/m^3.

2.3 Gypsum Plaster

Gypsum, when heated, produces plaster, a natural substance utilized in creating decorative elements like false ceilings for the construction sector. The plaster applied in this study was supplied by a Moroccan plaster manufacturing company, with a water-to-plaster ratio of 0.7.

2.4 Production of Samples

The samples produced are aimed at assessing their thermal properties, utilizing a square mold measuring 300 mm by 300 mm. The exterior layers of these sandwich structures incorporate a trio of composite materials comprising plaster, cork granulate and wood residues in the form of sawdust and wood chips. The inner layer of sandwich is agglomerated cork, in thicknesses of 2 cm and 4 cm, leading to the creation of two types of configurations for analysis:

- Symmetric Configuration: This setup involves cores of both 2 cm and 4 cm thicknesses, intended to investigate how the core's thickness influences the thermal characteristics of the studied samples.

- Asymmetric Configuration: This design, featuring a 4 cm thick cork core, aims to explore how an asymmetrical arrangement of the outer layers affects the material's thermal properties.

Table 1 details the composition and thickness dimensions of the sandwiches produced. The constituents and specific attributes of the composite materials forming the skins are detailed in Table 2.

To assemble the final sandwich structure, the procedure starts with pouring a 1 cm thick layer of the plaster-based composite into the mold, followed by the placement of the agglomerated cork core. Subsequently, the second composite layer is added, and the surface is evened out with a trowel. Once removed from the mold, the samples are dried in an oven set at 40 °C until their weight stabilizes, which prepares them for dry analysis.

Table 1. Prepared specimens Composition

Symmetric samples	Skin 1	Core	Skin 2	Total thickness
2WC-4-2WC	2WC (1 cm)	Cork (4 cm)	2WC (1 cm)	6 cm
3S-4-3S	3S (1 cm)	Cork (4 cm)	3S (1 cm)	6 cm
2C-4-2C	2C (1 cm)	Cork (4 cm)	2C (1 cm)	6 cm
2WC-2-2WC	2WC (1 cm)	Cork (2 cm)	2WC (1 cm)	4 cm
3S-2-3S	3S (1 cm)	Cork (2 cm)	3S (1 cm)	4 cm
2C-2-2C	2C (1 cm)	Cork (2 cm)	2C (1 cm)	4 cm
Asymmetric samples				
3S-4-2WC	3S (1 cm)	Cork (4 cm)	2WC (1 cm)	6 cm
2C-4-2WC	2C (1 cm)	Cork (4 cm)	2WC (1 cm)	6 cm
3S-4-2C	3S (1 cm)	Cork (4 cm)	2C (1 cm)	6 cm

Table 2. Materials Designation and thermal conductivity for outer layers [22]

	Designation	λ (W/mK)
2C	Plaster + 20% cork	0.197
3S	Plaster + 30% sawdust	0.279
2WC	Plaster + 20% wood chips	0.219

2.5 Method for Thermal Characterization

The HFM 446 instrument is utilized for determining the thermal conductivity and specific heat of materials. This device, designed for the empirical assessment of these thermal properties, features a configuration of two plates, each fitted with a heat flux sensor.

The setup arranges the heated plate above and the cooled plate below, as shown in Fig. 1. Positioning the sample between these plates (Fig. 2) enables the achievement of a stable state, crucial for precise thermal property analysis. The specimens prepared for this process measure 30 cm by 30 cm, with variations in thickness at 4 cm and 6 cm, respectively.

The operational principle of the HFM, illustrated in Fig. 1, revolves around reaching a thermal equilibrium state. At this point, the thermal conductivity (λ) of the specimen is derived through the application of the Fourier heat flow equation, q (W/m22), which is expressed as:

$$q = \frac{Q}{A} = \lambda.\frac{dT}{dx} \tag{1}$$

The heat flux transducers (Fig. 1) sense heat flow only in the central portion of the sample in order to minimize the measurement error caused by lateral heat losses, and produce a voltage V proportional to the heat flux in the measurement area and the constant of proportionality N can be used to convert the voltage signal to heat flux:

$$q = N.V \tag{2}$$

Combining Eqs. (1) and (2) and solving for λ produces the expression for the specimen thermal conductivity:

$$\lambda = N.V.\frac{dx}{dT} \tag{3}$$

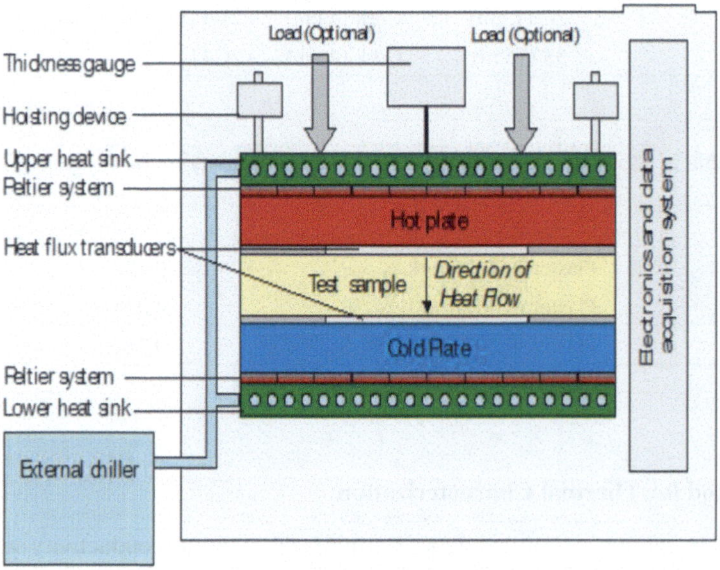

Fig. 1. Measuring principle for HFM

The following equation is used to get the specific heat value:

$$C_p = \frac{N_\lambda.A.(Q - Q_E)}{m.dT} \tag{4}$$

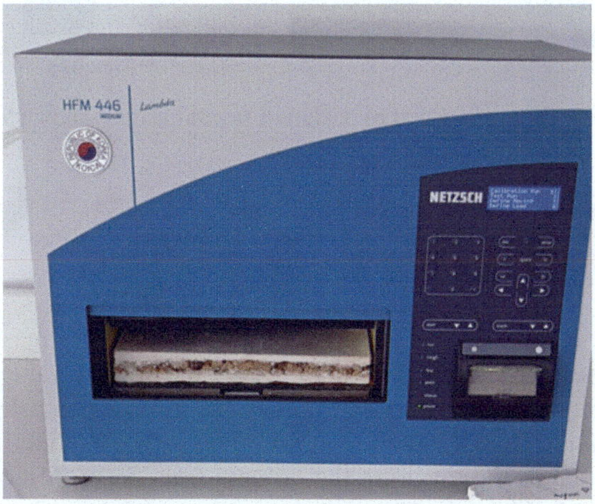

Fig. 2. Sample placed within the HFM

Thermal effusivity (E) can be calculated using Eq. 5

$$E = \sqrt{\rho.\lambda.C_P} \tag{5}$$

Thermal diffusivity (α) is calculated using Eq. (6):

$$\alpha = \frac{\lambda}{\rho.C_P} \tag{6}$$

After obtaining the experimental measurements of λ, the theoretical equivalent thermal conductivity λ_{eq}, is estimated using the series model depicted in Fig. 3. This theory states that the combined thermal resistance of all the layers results in the resulting multi-layer's thermal resistance:

$$\text{Rsandiwch} = \text{R}_1 + \text{R}_2 + \text{R}_3 \tag{7}$$

$$\text{Rsandwich} = \frac{e_t}{\lambda_{eq}} \tag{8}$$

where layer 1, core, and layer 2 thermal resistances are denoted by R_1, R_2, and R_3, respectively. e_t is the multi-layer panel's entire thickness.

Employing Eqs. (7) and (8), the equivalent thermal conductivity is calculated using the following equation:

$$\lambda_{eq} = e_t/(\frac{\text{e}_1}{\lambda_1} + \frac{\text{e}_2}{\lambda_2} + \frac{\text{e}_3}{\lambda_3}) \tag{9}$$

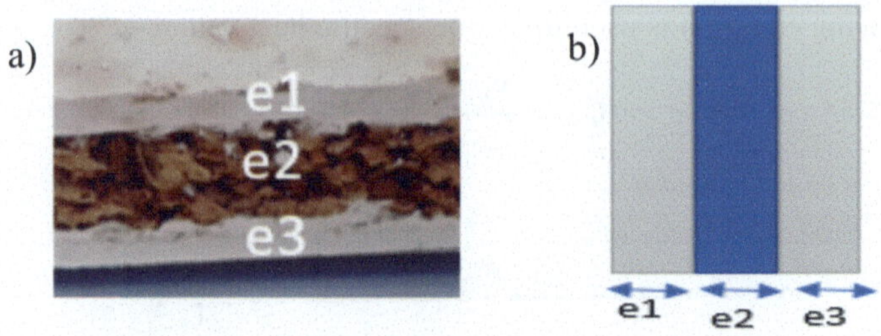

Fig. 3. a) View and b) design of several layers sandwich

3 Findings and Analysis:

3.1 Density

The density of the samples was assessed by calculating their dry mass and dimensions. The findings are presented in Fig. 4, which details the density values. The integration of cork into eco-composites made from plaster has resulted in a lightweight panels, with densities spanning from 423 to 572 kg/m^3. This indicates that the panels' reduced weight is directly tied to the lowering the volume fraction of different layers and the augmentation of cork's thickness. This outcome aligns with the research by Lakreb et al. [23], which demonstrated an increase in the density of sandwich panels with an increase in the layer count. Additionally, both configurations at 4 cm thick cork core display comparable densities, attributable to the uniform volume percentage of the core and the outer layers.

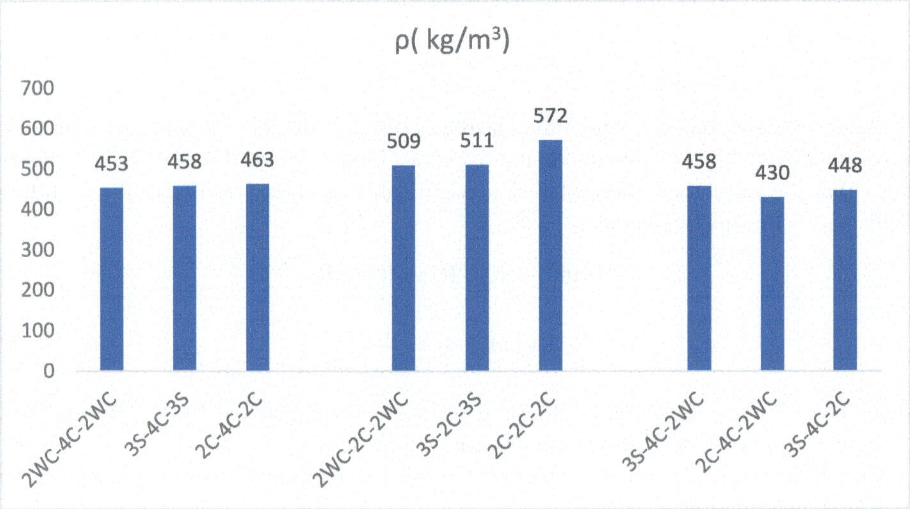

Fig. 4. Density of prepared samples

3.2 Thermal Conductivity

Figure 5 presents the thermal conductivity outcomes. The thermal conductivity of cork, as determined by the HFM446 device, stood at 0.056 W/mK. For both symmetric and asymmetric configurations with a cork thickness of 4 cm, thermal conductivity values were closely grouped between 0.073 and 0.082 W/mK. Given that the cork core constitutes 67% of the overall thickness of the sandwich, it significantly influences the thermal conductivity values. In the case of symmetric configurations, employing a 2 cm cork core led to marginally elevated thermal conductivity values compared to those with a 4 cm core. The findings indicate that the thermal conductivity of the multi-layered constructs escalates with an increase in the outer layer percentage and a reduction in cork thickness. The poorest thermal insulation performance was noted in the symmetric setup with a 2 cm cork core and a 3S skin, attributed to the 3S composite with higher thermal conductivity and the cork's reduced thickness, accounting for 50% of the sandwich's total thickness. An increase in the skin percentage correspondingly raises the λ values [23, 24]. Thermal conductivity is thus influenced by the thickness of both the skin and the cork core. In comparison, Limam et al. [19] observed that thermal conductivity for studied samples was 0.06 W/mK with a 4 cm cork core, and 0.05 W/mK with an 8 cm cork core.

The series model was employed to calculate the equivalent thermal conductivity (Eq. 9), with values displayed in Fig. 5. The difference between experimental and theoretical values remained below 9%, except for the 2C-2C-2C configuration, where it reached 10%. This suggests that the experimental outcomes are reliable and that the bond between layers and cork is excellent. Moreover, the thermal conductivity for any configuration using these materials can be predicted based on the known thermal conductivities and thicknesses of each layer.

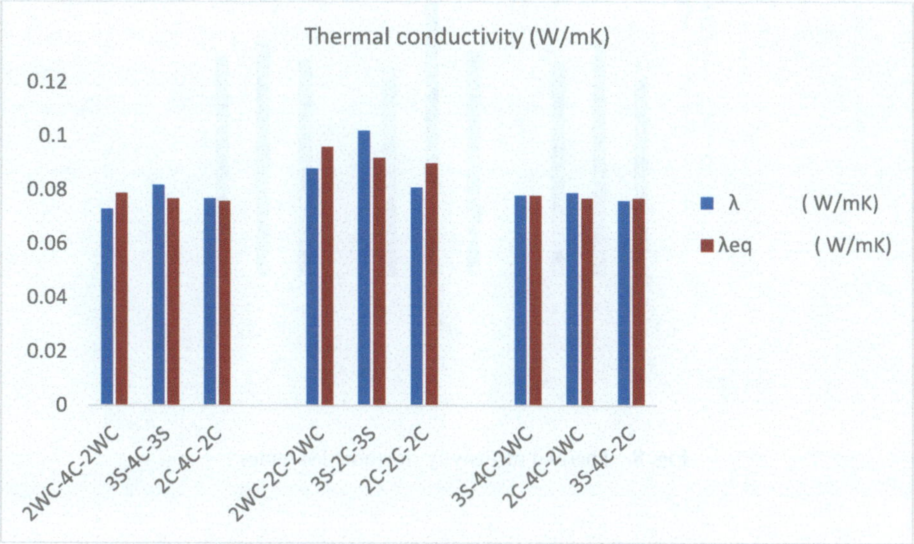

Fig. 5. Thermal conductivity of studied samples

3.3 Specific Heat

Specific heat represents the ability of a material to absorb thermal energy relative to its mass, and it was determined using the HFM device. The values illustrated in Fig. 6 represent the average of three measurements taken at temperatures of 10, 30, and 50 °C. These values exhibit a marginal increase with temperature elevation. The obtained values for the symmetric configuration fall within the cork's maximum to the minimum of the composite skins values. Obtained values are in the range of 1274 and 1513 J K^{-1} kg^{-1}. Essentially, there exists an inverse correlation between the volume percentage of the skin and specific heat; diminishing the thickness of the externals skin leads to a rise in the specific heat of the sandwich structure, as found by El Wardi et al. [24].

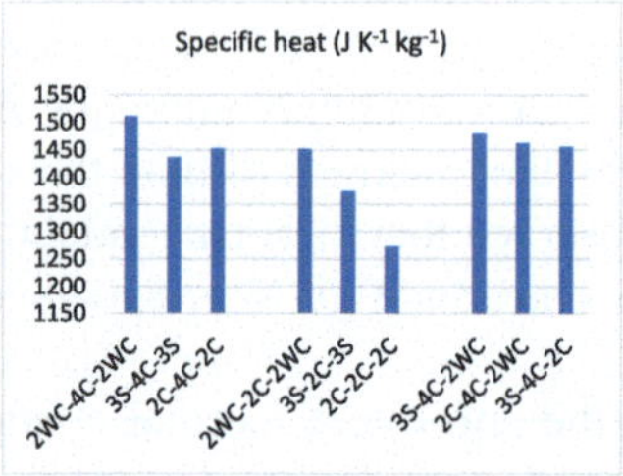

Fig. 6. Specific heat of studied samples

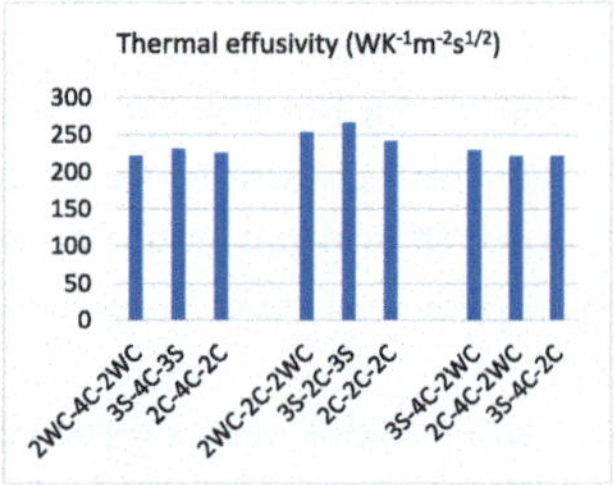

Fig. 7. Thermal effusivity of studied samples

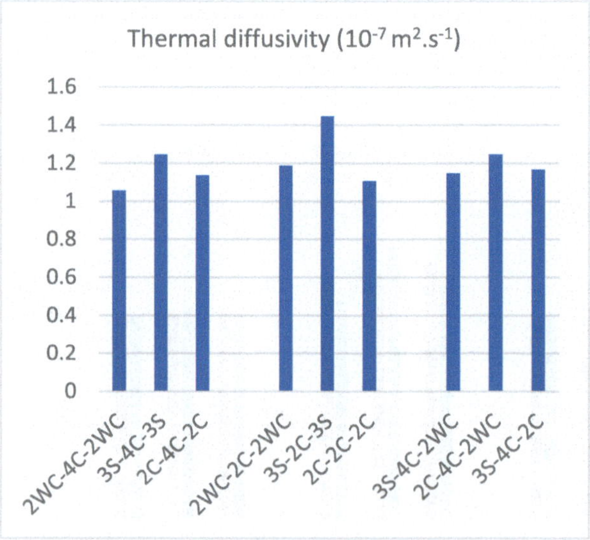

Fig. 8. Thermal diffusivity of studied samples

3.4 Thermal Effusivity

Obtained values of thermal effusivity are illustrated in Fig. 7. In symmetric configurations, there is a minor uptick in thermal effusivity values with a reduction in cork thickness and an augmentation in the percentage of layers, a finding that aligns with El Wardi et al. [24]. According to their study, the thermal effusivity of the examined sandwich panels diminishes as the thickness of the cement and plaster external layers decreases. It's important to note that the type of skin used and the thickness of the cork core do not significantly impact thermal effusivity, with measured values spanning from 203 to 268 $WK^{-1}m^{-2}s^{1/2}$.

3.5 Thermal Diffusivity

Thermal diffusivity measures the speed at which heat disperses through a material by conduction, directly correlating with thermal conductivity λ and inversely with specific heat capacity C_p. Obtained values of thermal diffusivity are illustrated in Fig. 8. In symmetric configurations, thermal diffusivity diminishes as the thickness of the cork increases, with observed values ranging from 1.06 to 1.45 $10^{-7}(m^2.s^{-1})$. Similar findings are reported by [24]. This indicates that the panels produced are effective in delaying heat transmission, an advantageous property that, coupled with low thermal conductivity, plays a crucial role in reducing heat exchange between a building's exterior and interior environments. The attenuation of heat diffusion is attributed to the synergistic effects of the cork's honeycomb microstructure and the reduced thermal diffusivity of composites made from plaster and cork or wood residue, compared to plaster alone.

4 Conclusion

Aiming to enhance building energy efficiency through the use of ecological and locally sourced materials, this study introduces innovative sandwich panels, employing agglomerated cork for the core and bio-composite materials for the outer layers. The thermal characteristics of these panels were evaluated with an HFM device, focusing on measurements of thermal conductivity and specific heat across various configurations. The findings revealed thermal conductivity values ranging between 0.071 and 0.088 W/mK, and specific heat values from 1274 to 1513 J/kg.K. Notably, panels with a 4 cm cork core exhibited superior thermal performance compared to those with a 2 cm core. A comparative analysis between experimental outcomes and theoretical thermal conductivity values yielded a maximum measurement discrepancy of 11%. This accuracy suggests that the thermal conductivity of any series configuration composed of these materials can be reliably predicted using the series model.

Acknowledgement. This work is a part of the project CTAI financially supported by grant from the IRESEN (Institut de Recherche en Énergie Solaire et Énergies Nouvelles), Morocco.

Special thanks are sent to ESSTI (Ecole Supérieure des Sciences et Technologies de l'Ingénierie) as a partner of the CTAI project for their technical support.

GEP (Green Energy Park) is gratefully acknowledged for providing the HFM446 destined for thermal characterization.

References

1. AMEE | la Règlementation Thermique dans le bâtiment (RTCM). https://www.amee.ma/fr/reglementation-thermique
2. Idchabani, R., Garoum, M., Lamzah, A.: Thermal performance and environmental impact of residential buildings' heating in Morocco. Int. J. Sustain. Energy. **35**, 295–307 (2016). https://doi.org/10.1080/14786451.2014.884567
3. Kumar, D., Alam, M., Zou, P.X.W., Sanjayan, J.G., Ahmed, R.: Comparative analysis of building insulation material properties and performance. Renew. Sustain. Energy Rev. **131**, 110038 (2020). https://doi.org/10.1016/j.rser.2020.110038
4. Sutton, A., Black, D., Pete, W.: Hemp Lime. BRE Inf. Pap. 1–6 (2011)
5. Collet, F., Achchaq, F., Djellab, K., Marmoret, L., Beji, H.: Water vapor properties of two hemp wools manufactured with different treatments. Constr. Build. Mater. **25**, 1079–1085 (2011). https://doi.org/10.1016/j.conbuildmat.2010.06.069
6. Nguyen, S.T., Tran-Le, A.D., Vu, M.N., To, Q.D., Douzane, O., Langlet, T.: Modeling thermal conductivity of hemp insulation material: a multi-scale homogenization approach. Build. Environ. **107**, 127–134 (2016). https://doi.org/10.1016/j.buildenv.2016.07.026
7. Sassoni, E., Manzi, S., Motori, A., Montecchi, M., Canti, M.: Novel sustainable hemp-based composites for application in the building industry: physical, thermal and mechanical characterization. Energy Build. **77**, 219–226 (2014). https://doi.org/10.1016/j.enbuild.2014.03.033
8. Freivalde, L., Kukle, S., Andžs, M., Bukšans, E., Gravitis, J.: Flammability of raw insulation materials made of hemp. Compos. Part B Eng. **67**, 510–514 (2014). https://doi.org/10.1016/j.compositesb.2014.08.007
9. Lazko, J., Dupré, B., Dheilly, R.M., Quéneudec, M.: Biocomposites based on flax short fibres and linseed oil. Ind. Crops Prod. **33**, 317–324 (2011). https://doi.org/10.1016/j.indcrop.2010.11.015
10. Zhou, X. yan, Zheng, F., Li, H., Lu, C.: An environment-friendly thermal insulation material from cotton stalk fibers. Energy Build. **42**, 1070–1074 (2010). https://doi.org/10.1016/j.enbuild.2010.01.020
11. Bouchefra, I., EL Bichri, F.Z., Chehouani, H., Benhamou, B.: Mechanical and thermophysical properties of compressed earth brick rienforced by raw and treated doum fibers. Constr. Build. Mater. **318**, 126031 (2022). https://doi.org/10.1016/j.conbuildmat.2021.126031
12. Liu, L.F., et al.: The development history and prospects of biomass-based insulation materials for buildings. Renew. Sustain. Energy Rev. **69**, 912–932 (2017). https://doi.org/10.1016/j.rser.2016.11.140
13. Zenkert, D.: The Handbook of Sandwich Construction **442** (1997)
14. Susainathan, J., Eyma, F., De Luycker, E., Cantarel, A., Castanie, B.: Experimental investigation of impact behavior of wood-based sandwich structures. Compos. Part A Appl. Sci. Manuf. **109**, 10–19 (2018). https://doi.org/10.1016/j.compositesa.2018.02.029
15. Kawasaki, T., Zhang, M., Kawai, S.: Sandwich panel of veneer-overlaid low-density fiberboard. J. Wood Sci. **45**, 291–298 (1999). https://doi.org/10.1007/BF00833493
16. Kawasaki, T., Kawai, S.: Thermal insulation properties of wood-based sandwich panel for use as structural insulated walls and floors. J. Wood Sci. **52**, 75–83 (2006). https://doi.org/10.1007/s10086-005-0720-0
17. Touil, M., Lachheb, A., Saadani, R., Kabiri, M.R., Rahmoune, M.: A new experimental strategy assessing the optimal thermo-mechanical properties of plaster composites containing Alfa fibers. Energy Build. **262**, 111984 (2022). https://doi.org/10.1016/j.enbuild.2022.111984
18. Castro, O., Silva, J.M., Devezas, T., Silva, A., Gil, L.: Cork agglomerates as an ideal core material in lightweight structures. Mater. Des. **31**, 425–432 (2010). https://doi.org/10.1016/j.matdes.2009.05.039

19. Limam, A., Zerizer, A., Quenard, D., Sallee, H., Chenak, A.: Experimental thermal characterization of bio-based materials (Aleppo Pine wood, cork and their composites) for building insulation. Energy Build. **116**, 89–95 (2016). https://doi.org/10.1016/j.enbuild.2016.01.007
20. Reis, L., Silva, A.: Mechanical behavior of sandwich structures using natural cork agglomerates as core materials. J. Sandw. Struct. Mater. **11**, 487–500 (2009). https://doi.org/10.1177/1099636209104523
21. La Rosa, A.D., et al.: Environmental impacts and thermal insulation performance of innovative composite solutions for building applications. Constr. Build. Mater. **55**, 406–414 (2014). https://doi.org/10.1016/j.conbuildmat.2014.01.054
22. Er-rradi, H., et al.: Experimental study of composite materials based on ecological wastes. J. Compos. Mater. **56**, 3295–3306 (2022). https://doi.org/10.1177/00219983221113029
23. Lakreb, N., Şen, U., Toussaint, E., Amziane, S., Djakab, E., Pereira, H.: Physical properties and thermal conductivity of cork-based sandwich panels for building insulation. Constr. Build. Mater. **368** (2023). https://doi.org/10.1016/j.conbuildmat.2023.130420
24. El Wardi, F.Z., Khabbazi, A., Cherki, A.B., Khaldoun, A.: Thermomechanical study of a sandwich material with ecological additives. Constr. Build. Mater. **252**, 119093 (2020). https://doi.org/10.1016/j.conbuildmat.2020.119093

Advancing Concrete Segregation Detection: Leveraging Explainable AI (XAI) for Improved Accuracy and Transparency

Hafsa Matich(✉) and Hajar Mousannif

Computer Science Department, Faculty of Sciences Semlalia, Cadi Ayyad University, Marrakech, Marocco
h.matich.ced@uca.ac.ma

Abstract. In the aftermath of the devastating seismic event that struck Morocco in September 2023, resulting in the collapse of numerous fragile structures and a significant loss of human life, the urgency to strengthen disaster preparedness and response capabilities has never been more critical. This juncture provides a unique opportunity to leverage the transformative potential of cutting-edge technologies, particularly Artificial Intelligence (AI) with a specific emphasis on Explainable AI (XAI) methods.

Concrete, being an essential construction material valued for its notable strength, durability, and versatility, sees extensive use in various infrastructure projects. However, the presence of concrete segregation presents a substantial challenge in both the construction and upkeep of concrete structures. Concrete segregation entails the uneven distribution of aggregates within the concrete matrix, which can lead to adverse effects. It jeopardizes structural integrity, diminishes performance, and potentially poses safety hazards.

Given these significant concerns, it is imperative to identify and quantify instances of concrete segregation. In response to this requirement, the current investigation introduces a novel method for assessing concrete segregation using an image processing-based technique, incorporating Deep Learning and Explainable Artificial Intelligence. This research not only improves the precision and reliability of detecting concrete segregation but also ensures transparency in the decision-making process. This innovative approach has the potential to transform concrete construction and maintenance practices, ensuring the structural integrity and performance of infrastructure projects are upheld while minimizing safety risks linked to concrete segregation.

Keywords: Explainable Artificial Intelligence · Layer-wise relevance Propagation · Affinity propagation · Image classification · Concrete segregation

1 Introduction

Taking a historical perspective, the seismic catastrophe that unfolded in Morocco on September 8, 2023, stands out as a watershed moment, laying bare the inherent vulnerabilities of numerous homes and structures across the country [1]. This devastating

A. S. Mosallam et al. (Eds.): ICATH 2023, SCI, pp. 50–62, 2024.
https://doi.org/10.1007/978-3-031-76557-5_5

earthquake served as a stark reminder of the critical importance of implementing robust construction practices in regions prone to such natural disasters [2].

One sobering observation arising from this event is that natural disasters tend to exert a more profound impact on fragile buildings [3]. Homes and structures constructed with substandard or fragile materials often bear the brunt of the devastation. This fact underscores the compelling need to address issues such as concrete segregation, a phenomenon that can exacerbate structural vulnerabilities [4].

Leveraging AI for concrete segregation detection has shown remarkable progress, particularly the application of deep learning methods, including Convolutional Neural Networks (CNNs). Introduced a deep reinforcement learning framework [5], achieving high accuracy in identifying concrete cracks and defects. Additionally, GAN-based approaches, as demonstrated by [6], integrated GANs with deep belief networks for effective segregation detection and visualization.

To enhance model interpretability, we have focused on integrating explainable Artificial Intelligence (XAI) techniques with deep learning models. Saliency mapping, attention mechanisms, and Layer-wise Relevance Propagation (LRP) have been employed to highlight crucial regions in concrete images contributing to segregation detection. Incorporated XAI techniques into their CNN-based model, making the detection process more transparent and interpretable [7].

The evaluation of XAI models involves metrics such as accuracy, precision, recall, F1-score, and area under the ROC curve. Utilized precision-recall curves and ROC analysis to assess their deep graph convolutional networks, demonstrating the robustness of their model [8]. In presenting these advancements, this article aims to showcase XAI techniques as a powerful solution to address the challenges in concrete segregation detection.

2 Overview of Literature

In recent years, researchers have intensified efforts to develop models for estimating construction segregation Table 1, a critical concern impacting the structural integrity of concrete structures. The traditional reliance on manual inspection and visual assessments for segregation estimation lacked accuracy, prompting a shift towards advanced and objective techniques.

The advent of computer vision and image analysis, fueled by technological advancements, has transformed segregation estimation. Digital methods, [9] particularly image processing and machine learning algorithms, have garnered attention. Image processing techniques enhance concrete image quality, while machine learning algorithms, including CNNs, SVMs, and decision trees, classify concrete images into segregation categories [10]. AI techniques have been integrated to provide insights into model decisions, enhancing transparency [11].

3 Devices and Methodology

In the concrete defect detection domain, the combined use of XAI (Explainable Artificial Intelligence), LRP (Layer-wise Relevance Propagation), and AP (Affinity Propagation) enhances transparency and interpretability. XAI techniques, including saliency mapping

Table 1. Examples of research in which models were developed to make segregation detection.

Source	Model	Output
"Segregation Detection of Fresh Concrete Based on Feature Learning with Convolutional Neural Networks" (2021)	(CNNs)	The model utilizes feature learning with CNNs to automatically detect and classify segregation patterns in freshly mixed concrete samples. The output is the classification of concrete samples into different segregation categories, indicating the severity of segregation in the fresh concrete
"A Segregation Detection System for Fresh Concrete Based on Deep Convolutional Neural Networks" (2020)	(CNNs)	The model based on deep CNNs is designed to automatically detect and assess segregation in freshly mixed concrete. The output is a prediction of the segregation level, providing insights into the uniformity of aggregate distribution within the fresh concrete samples
"Concrete Segregation Detection Using Graph Convolutional Networks" (2022)	(GCNs)	The model utilizes graph convolutional networks to process the graph-structured data of concrete samples, capturing the spatial relationships between aggregates. The output is the classification of concrete samples into different segregation categories, indicating the level of segregation present
"Concrete Segregation Detection Using Ensemble Learning with XGBoost" (2019)	Ensemble Learning with XGBoost	The model employs ensemble learning techniques, particularly XGBoost, to combine multiple weak learners and make predictions on concrete samples. The output is the classification of concrete samples into different segregation categories, providing insights into the presence and severity of segregation

and attention mechanisms, are employed to pinpoint anomalies in concrete images. The model, trained on a dataset comprising over 6,000 unlabeled images sourced from a private concrete research laboratory, covers various types of segregation, ensuring the model's generalization. This dataset, fostering the development of a segregation detection model using XAI, aims to advance concrete research and improve detection reliability. The model assigns anomaly scores to new images based on the probability of defects, with XAI revealing the specific locations of the detected anomalies.

4 Models' Development

Artificial neural networks (ANN) are gaining popularity, particularly in tasks related to image classification. Demonstrating remarkable performance in large- scale competitions [12]. A pivotal factor contributing to their success is their capability to efficiently train with millions of samples contemporaneously on GPU tackle. This enables the application of expansive training datasets, aligning with the substantial number of parameters in deep neural networks. Despite these achievements, unraveling the corridor of an image that impact a bracket decision remains an unanswered question for neural networks and numerous othernon-linear models. To attack this challenge, the work of [13] introduced LRP as a result for expounding the significance of pixels in an image concerning a categorization decision. Originally applied to neural networks, this approach was also extended to encompass bag- of- word models [14, 15] and latterly further extended to Fisher vectors in posterior work [16]. This paper proposes an extension of LRP to neural networks with nonlinearities that surpass generally used phrasings. A case of similar nonlinearities includes original renormalization layers, which standard LRP [13] doesn't effectively handle. The presented approach is predicated in the first (or advanced) order Taylor expansion. The study concentrates on a bracket setup with real- valued labors, where a classifier f is a mapping of an input space $f\ X \rightarrow R$, and $f(x) > 0$ signifies the presence of a specific group.

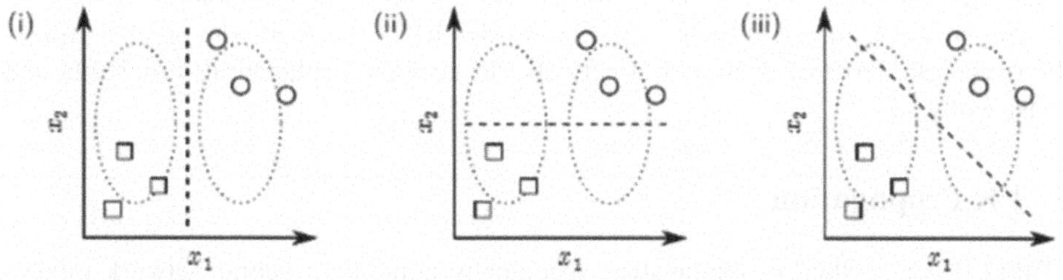

Fig. 1. Demonstration of the challenge posed by spurious correlations in high-dimensional data.

Utilizing input variables effectively is crucial for accurate predictions. A straightforward example is provided in Fig. 1, where the model achieves perfect classification by employing either feature x1, feature x2, or both, it's crucial to note that only the utilization of the first option guarantees accurate generalization to unseen data. Not accurately identifying the input features may result in predictors resembling the 'Clever Hans' phenomenon [17]. Feature selection [18] arises as a possible remedy by providing limiting the learning machine to only a select few 'good' input characteristics. However, implementing this strategy can be complex, particularly in fields such as recognizing images, where the importance of different pixels varies.

Approaches to Explainable ML [19–21] approaches the issue differently. Initially, a model is trained without a strong emphasis on feature selection.

Using explanatory feedback about the input features that the neural network has learned; we can identify and remove undesirable features. After this, the model can undergo retraining with the improved dataset. [15]. A straightforward technique, Taylor Decomposition [13, 23], generates insights through the process of expanding the prediction f(x) using a Taylor expansion at a nearby reference point x:

$$f(x) = f(\widetilde{x}) + \sum\nolimits_{i=1}^{d} (x_i - \widetilde{x}_i) \cdot [\nabla f(\widetilde{x})]_i + \ldots \tag{1}$$

The method of utilizing first-order terms to quantify the relevance of input features for predictions, forming the foundation of explanations, faces instability in the context of deep neural networks. This instability arises from well-known limitations:

- **Disrupted Gradients:** Even though the value of the function f(x) is usually correct, the gradient becomes unreliable or noisy.
- **Adversarial Instances:** changing the input x can lead to alterations in the function value f(x).

These challenges complicate the identification of a significant source point with a substantial grade $\nabla f(x)$, impeding the development of a reliable clarification [21]. In response to the intricacies of deep neural networks, various explanation techniques have emerged. Some improve explanations by integrating numerous local gradient estimates [24, 25], while others substitute a less detailed approximation of the gradient using a broader assessment of the effect [60], such as the model's reaction to perturbations resembling patches [26]. Additional methods involve optimizing a local surrogate model [27] or refining the explanation itself [28]. It's essential to note that these techniques often entail multiple evaluations of the neural network, leading to computational expenses. Moving forward, our attention is directed towards LRP [13], a methodology that utilizes the structural layout of deep neural networks to generate explanations efficiently and consistently.

5 LR Propagation

LRP [13] is a method of explanation specifically crafted for neural network models and can be applied to diverse inputs, including videos, images, or text [13, 29, 30]. LRP functions by backward propagation of the forecast f(x) through the neural network, employing specific local rules. Notably, LRP adheres to a conservation property, reminiscent of Kirchhoff's laws in electrical circuits and observed in other explanation works [31–33].

This involves redistributing information received by a:

$$R_j = \sum_k \frac{z_{jk}}{\sum_j z_{jk}} R_k. \tag{2}$$

The quantity Z_{jk} quantifies how much neuron j contributes to making neuron k relevant. to ensure the significance of neuron k. The denominator is used to maintain a preservation principle, and the propagation concludes when reaching the input features. Applying this rule to all neurons in the network ensures the layer-wise conservation property $\sum_j \mathrm{R}_j \sum_k R_k$, and consequently, [Fig. 2] The overall conservation property $\sum_j \mathrm{R}_j = \mathrm{f(x)}$.

Despite its distinction from the simple Taylor decomposition approach, reveals that every stage of the propagation process can be depicted as a separate Taylor decomposition conducted on original amounts within the graph [34]. LRP has found applications in uncovering prejudices in frequently employed machine learning models and datasets [35, 36]. It has also been employed to gain observations from proficient machine learning models, such as in face expression recognition [37, 38]. In diverse industrial applications, LRP has been used to identify relevant features for tasks like audio source localization [39], pinpoint points of interest inside channel traces [40] and explain decisions in brain-computer interfaces by identifying EEG patterns [41]. In the industry domain, LRP has been applied to...

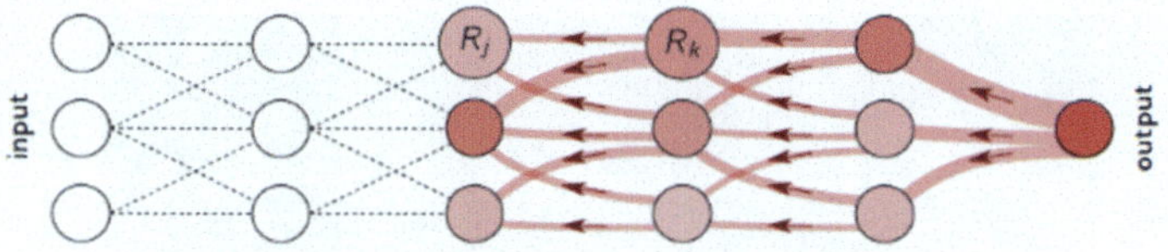

Fig. 2. Demonstration of the LRP procedure

The essence of LRP (LRP) lies in its fundamental principle: every neuron within the neural network redistributes relevance to the lower layer, ensuring that it allocates to the extent it has been impacted by the higher layer. This mechanism aligns with a conservation property, akin to the principles observed in Kirchhoff's laws in electrical circuits. In simpler terms, LRP ensures that the importance or relevance attributed to each neuron is appropriately passed down to the subsequent layer, contributing to the overall understanding of how input features influence the model's predictions. This approach facilitates a comprehensive and interpretable analysis of neural network behavior, enhancing the transparency and interpretability of complex models.

As a broad propagation framework, LRP (LRP) provides considerable flexibility in determining the rules for each level and setting the parameters λ and γ. Optimally opting LRP parameters necessitates a measure of illustration quality, which remains an active research topic [42]. While a comprehensive discussion of this topic exceeds the scope of this chapter, we approach LRP in the context of two widely acknowledged desirable properties for explanations: fidelity and understandability [19]. In essence, an explanation should accurately represent the specific output neuron and should be easily understandable for a person.

Composite LRP mitigates the drawbacks of the approaches. It accurately identifies and fully highlights the features of the pillar, ensuring both fidelity and understandability in the explanation.

6 Image Processing

The proposed algorithm for concrete classification based on aggregate uniformity distribution utilizes image processing techniques to assess segregation levels in concrete samples. Validation tests using known segregation patterns demonstrate the effectiveness of the approach in identifying potential issues in buildings or structures. By implementing these techniques, construction professionals can gain valuable insights into concrete quality, enabling early identification and mitigation of risks associated with aggregate segregation. The integration of image processing and classification algorithms offers a comprehensive approach, empowering professionals to make informed decisions for enhancing the durability and performance of concrete structures.

6.1 Enhancing Clarity in Concrete Defect Localization

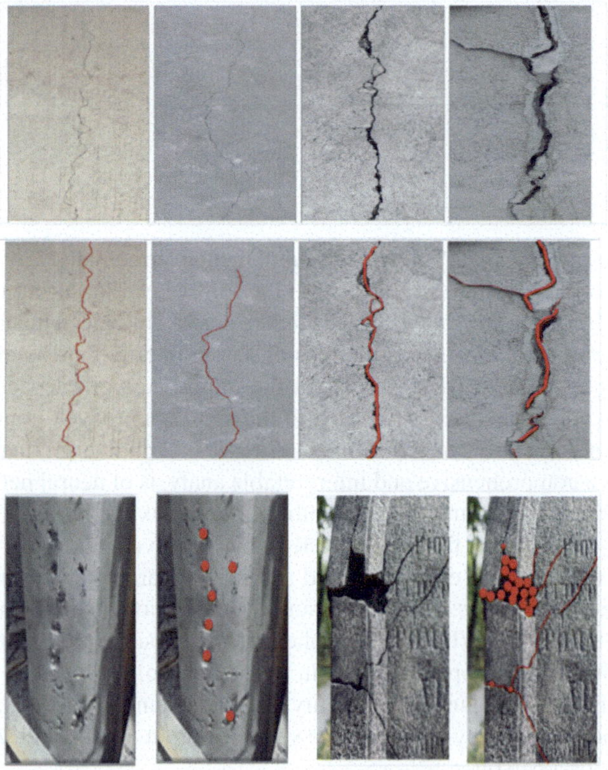

Fig. 3. Results indicate the successful identification of the segregation of different aggregate types during localization.

By combining computer vision and RL techniques, these studies demonstrate the potential of RL in accurately [Fig. 3] detecting and localizing defects in concrete structures. RL enables the agent to learn from its interactions with the environment, adapting its decision-making process to improve defect detection performance over time.

In conclusion, the application of RL for detecting the location of defects in concrete represents an exciting research direction. By merging computer vision and RL techniques, researchers are developing systems that can autonomously identify and locate defects in concrete structures. Continued advancements in RL algorithms and training methodologies are expected to further enhance the accuracy and efficiency of defect detection in concrete, leading to improved maintenance and durability of structures.

6.2 Assessing the Level of Risk

Clustering techniques have been extensively employed for identifying flaws in concrete structures. In the study by Dinh et al. [23], k-means clustering was applied to partition ground-penetrating radar data, evaluating the corrosion level in reinforced concrete bridge decks. Additionally, clustering methods were utilized for segmenting infrared thermography images to assess defects in reinforced concrete bridge decks [24, 25]. Acoustic emission signals were analyzed, and damage severity in concrete elements was assessed using principal component analysis and self-organizing neural network clustering [26]. In laboratory investigations, hierarchical clustering was employed with impact-echo data for similar purposes [41]. Sajid et al. [42] utilized factor scores from principal components of the FRF in impulse-response tests as indicators of defects, proving to be more effective in defect detection than indices from ASTM C1740. However, the pattern recognition-based procedure involves subjectively selecting the appropriate number of components from the scree plot. Factor scores are then computed for the selected components, and in some cases, defects cannot be solely detected based on the scores of the first component. Analysts may need to resort to scores from higher components for detection, thereby increasing the likelihood of false positives. In this paper, we leverage the affinity propagation (AP) algorithm for the following purposes:

- To categorize and differentiate various types of defects commonly encountered in concrete structures.
- To identify defects by assigning risk levels on a scale from 0 to 4. [Fig. 4] These assigned levels serve as indicators of the potential impact or urgency associated with addressing the identified issues. Higher levels signify more critical problems that require immediate attention.

We present the AP algorithm, a method that facilitates cluster formation through iterative message exchanges among data points until convergence. In contrast to clustering techniques like k-means or k-medoids, AP does not require the predefined specification of the cluster count. Key parameters include preference, which determines the number of exemplars used, and the damping factor, moderating message updates to prevent numerical oscillations. Exemplars, serving as representative samples of the dataset, describe it, while messages convey the suitability of one sample being the exemplar for another. Iterative updates persist until convergence, leading to the selection of final exemplars and defining the clusters.

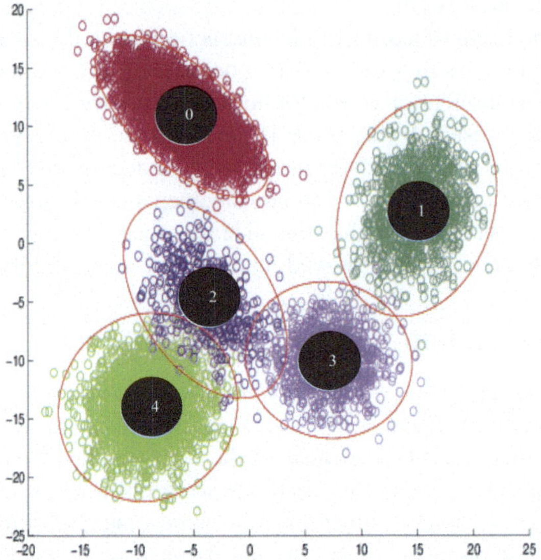

Fig. 4. Assessing AP algorithm performance on the training dataset.

AP relies on pairwise similarity between data points and a set of preference measures. Preference measures indicate the likelihood of being a cluster exemplar. In each iteration, responsibility and availability values are generated through message passing. Responsibility measures how well-suited a data point is to serve as an exemplar for another, reflecting its suitability as a cluster center based on similarity and preferences. Availability quantifies how well data points are allocated to a cluster, considering incoming responsibilities and self-responsibility.

Throughout each iteration, responsibility and availability values are updated, capturing the dynamic nature of the clustering process. This iterative approach allows the algorithm to continually refine its understanding of cluster exemplars and the allocation of data points to clusters across successive steps.

Before diving into the specifics of each risk level Fig. 5, it is essential to understand the overall framework of how concrete risk levels are classified. The classification system operates on a continuum, where each level corresponds to varying degrees of structural integrity and potential defects in concrete. The goal of such classification is to systematically assess the quality of concrete, identify potential risks, and prioritize intervention strategies. By interpreting these levels, engineers and quality control professionals can make informed decisions to ensure the durability, safety, and longevity of concrete structures. This approach allows for both preventative and corrective actions to be tailored to the degree of risk observed.

Level 0 (No Risk - Strong Concrete)

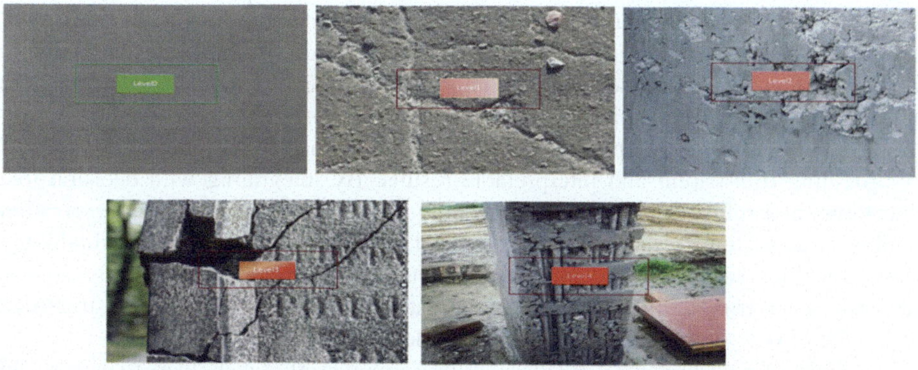

Fig. 5. Results indicate the successful identification level risk of the segregation.

Instances categorized under Level 0 represent concrete samples with no discernible risk. These samples exhibit robust properties, suggesting high strength and durability. Understanding the characteristics of this level can serve as a benchmark for excellence in concrete production.

Levels 1 and 2 (Low to Moderate Risk)

Concrete samples falling into Levels 1 and 2 indicate a low to moderate risk of defects. These levels may encompass minor surface imperfections or early signs of potential issues. Identifying specific characteristics within these levels, such as crack patterns or surface irregularities, will guide proactive measures to mitigate risks and enhance concrete quality.

Level 3 (Moderate to High Risk)

Level 3 suggests a moderate to high risk of concrete defects. Instances within this level may exhibit more pronounced issues that require immediate attention. Analyzing the types and patterns of defects in Level 3 will aid in formulating targeted remediation strategies to prevent further deterioration.

Levels 4 and 5 (High to Critical Risk)

Concrete samples classified under Levels 4 and 5 represent a high to critical risk of defects, signifying significant structural concerns. These levels may include severe cracking, delamination, or other critical issues.

In-depth analysis of these levels is crucial for implementing urgent and effective measures to address the identified defects and ensure structural integrity.

7 Conclusion and Perspectives

The combination of XAI with deep learning models brings numerous benefits, including increased transparency, error identification for model refinement, user-friendly interfaces, compliance with regulations, and collaborative decision-making. Continuous model improvement ensures adaptation to new data and evolving requirements.

For defect localization, deep learning models combined with XAI techniques accurately identify and localize various defects, including cracks, spalling, voids, and anomalies. Trained on labeled datasets with both normal and defective concrete samples, these models learn patterns indicative of defects.

In risk classification, Explainable Artificial Intelligence (XAI) plays a pivotal role by providing transparent and interpretable results. By integrating with decision tree algorithms and reinforcement learning techniques, XAI facilitates the categorization of defects according to their severity or risk levels. The decision tree methodology systematically assesses specific concrete characteristics such as crack visibility, texture intensity, shape irregularity, and texture contrast, allowing for a precise and structured evaluation of risk factors associated with each defect.

XAI ensures interpretable decision-making, empowering stakeholders to understand the reasoning behind risk classifications for informed maintenance and repair actions.

As the initial analysis of Explainable AI (XAI) and image analysis of concrete segregation lays the foundation for a deeper understanding, it also gives rise to several intriguing questions that underscore the ongoing nature of this research. While XAI techniques provide valuable insights into the detection process, there are further inquiries to explore. For instance, how do variations in aggregate size, shape, and grading influence the occurrence and severity of segregation? How does the ratio of fine to coarse aggregates impact the homogeneity of the concrete mixture? Is there a correlation between water content and the propensity for segregation in different concrete compositions?

Moreover, investigating the role of cement types, mineral admixtures, and chemical additives in segregation behavior remains a critical avenue for research. Does the reactivity of certain cementitious materials affect aggregate dispersion and lead to different segregation patterns? How do various admixtures influence the rheological properties of the concrete mix and contribute to or mitigate segregation tendencies?

Understanding the interaction of these material properties with external factors, such as temperature, humidity, and transportation conditions, can further illuminate the underlying mechanisms of segregation. Does temperature variation during the curing process influence aggregate settling? How does the handling of concrete during placement and consolidation affect its uniformity?

These questions point to the multifaceted nature of concrete segregation and emphasize the need for comprehensive investigations into the complex interplay of materials and environmental factors. By embracing the ongoing nature of this research and building upon the first XAI analyses, we can unlock a wealth of knowledge on the real reasons behind concrete segregation, particularly those intricately linked to the first materials and ingredients of concretes. Through collaborative efforts and innovative methodologies, the construction industry can leverage these insights to refine concrete mix designs, optimize construction practices, and ultimately ensure the long-term durability and performance of concrete structures in a sustainable and resilient manner.

References

1. Smith, J.R., et al.: Vulnerability of housing to earthquakes: lessons from the morocco event of 2023. J. Seismic Eng. **10**(2), 237–251 (2023)

2. Hearn, N., Weeks, J.: Artificial intelligence in earthquake engineering: a review. Eng. Struct. **180**, 243–256 (2019)
3. El-Rich, M., El-Hacha, R.: Sustainable concrete with high volume of supplementary cementitious materials for earthquake-resistant applications. J. Clean. Prod. **192**, 595–608 (2018)
4. ACI Committee 301. Specifications for Structural Concrete. American Concrete Institute (2016)
5. Zhang, L., et al.: A deep reinforcement learning framework for efficient concrete crack recognition. Autom. Constr. **100**, 317–327 (2019)
6. Chen, X., et al.: Concrete segregation detection using generative adversarial network and deep belief network. Autom. Constr. **119**, 103368 (2020)
7. Sanchez-Rangel, M.A., et al.: Explainable concrete defect detection using convolutional neural networks and saliency maps. Constr. Build. Mater. **257**, 119598 (2020)
8. Li, et al.: Concrete segregation dataset and segregation detection using deep graph convolutional networks. Autom. Constr. **125**, 103612 (2021)
9. Aksornkij, A., et al.: Automated assessment of concrete segregation in freshly mixed concrete using image analysis. Cement Concr. Compos. **34**(1), 73–78 (2012)
10. Tripathy, D.P., et al.: Segregation of concrete mixes: a review and case study on segregation prediction using decision trees. Constr. Build. Mater. **266**, 121203 (2021)
11. Wang, et al.: Detection and visualization of concrete segregation based on explainable artificial intelligence. Constr. Build. Mater. **278**, 122403 (2022)
12. Krizhevsky, A., Sutskever, I., Hinton, G.E.: Imagenet classification with deep convolutional neural networks. In NIPS, pp.1106–1114 (2012)
13. Bach, S., Binder, A., Montavon, G., Klauschen, F., Müller, K.R., Samek, W.: On pixel-wise explanations for non-linear classifier decisions by layer-wise relevance propagation. PLOS ONE **10**(7), e0130140 (2015)
14. Csurka, G., Dance, C., Fan, L., Willamowski, J., Bray, C.: Visual categorization with bags of keypoints. In: Workshop on Statistical Learning in Computer Vision, ECCV, pp. 1–22 (2004)
15. Van de Sande, K.E.A., Gevers, T., Snoek, C.G.M.: Evaluating color descriptors for object and scene recognition. IEEE Trans. Pattern Anal. Mach. Intell. **32**(9), 1582–1596 (2010)
16. Bach, S., Binder, A., Montavon, G., Klauschen, F., Müller, K.-R., Samek, W.: Analyzing classifiers: Fisher vectors and deep neural networks. In: Proceeding IEEE CVPR (2016)
17. Lapuschkin, S., Wäldchen, S., Binder, A., Montavon, G., Samek, W., Müller, K.R.: Unmasking Clever Hans predictors and assessing what machines really learn. Nature Communications **10**, 1096 (2019)
18. Guyon, I., Elisseeff, A.: An introduction to variable and feature selection. J. Machine Learning Res. **3**(Mar), 1157–1182 (2003)
19. Swartout, W.R., Moore, J.D.: Explanation in second generation expert systems. In: Second Generation Expert Systems, pp. 543–585. Springer-Verlag, New York Inc (1993)
20. Baehrens, D., Schroeter, T., Harmeling, S., Kawanabe, M., Hansen, K., Müller, K.: How to explain individual classification decisions. J. Machine Learning Res. **11**, 1803–1831 (2010)
21. Montavon, G., Samek, W., Müller, K.R.: Methods for interpreting and understanding deep neural networks. Digital Signal Processing **73**, 1–15 (2018)
22. Bazen, S., Joutard, X.: The Taylor Decomposition: A Unified Generalization of the Oaxaca Method to Nonlinear Models. Working papers, HAL (2013)
23. Smilkov, D., Thorat, N., Kim, B., Víegas, F.B., Wattenberg, M.: Smoothgrad: Removing Noise by Adding Noise. CoRR abs/1706.03825 (2017)
24. Sundararajan, M., Taly, A., Yan, Q.: Axiomatic attribution for deep networks. In: Proceedings of the 34th International Conference on Machine Learning, pp. 3319–3328 (2017)

25. Zintgraf, L.M., Cohen, T.S., Adel, T., Welling, M.: Visualizing deep neural network decisions: prediction difference analysis. In: International Conference on Learning Representations (2017)
26. Zeiler, M.D., Fergus, R.: Visualizing and understanding convolutional networks. In: Proc. of European Conference on Computer Vision (ECCV), pp. 818–833 (2014)
27. Ribeiro, M.T., Singh, S., Guestrin, C.: Why should I trust you?: explaining the predictions of any classifier. In: Proceedings of the 22nd ACM SIGKDD International Conference on Knowledge Discovery and Data Mining, pp. 1135–1144 (2016)
28. Fong, R.C., Vedaldi, A.: Interpretable explanations of black boxes by meaningful perturbation. In: IEEE International Conference on Computer Vision, pp. 3449–3457 (2017)
29. Anders, C., Montavon, G., Samek, W., Müller, K.R.: Understanding patch-based learning of video data by explaining predictions. In: Explainable AI: Interpreting, Explaining and Visualizing Deep Learning. Lecture Notes in Computer Science 11700, Springer (2019)
30. Arras, L., Horn, F., Montavon, G., Müller, K.R., Samek, W.: What is relevant in a text document?: an interpretable machine learning approach. PLoS ONE **12**(8), e0181142 (2017)
31. Landecker, W., Thomure, M.D., Bettencourt, L.M.A., Mitchell, M., Kenyon, G.T., Brumby, S.P.: Interpreting individual classifications of hierarchical networks. In: IEEE Symposium on Computational Intelligence and Data Mining, pp. 32–38 (2013)
32. Shrikumar, A., Greenside, P., Kundaje, A.: Learning important features through propagating activation differences. In: Proceedings of the 34th International Conference on Machine Learning, pp. 3145–3153 (2017)
33. Zhang, J., Bargal, S.A., Lin, Z., Brandt, J., Shen, X., Sclaroff, S.: Top-down neural attention by excitation backprop. Int. J. Comput. Vision **126**(10), 1084–1102 (2018)
34. Montavon, G., Lapuschkin, S., Binder, A., Samek, W., Müller, K.R.: Explaining nonlinear classification decisions with deep Taylor decomposition. Pattern Recognition **65**, 211–222 (2017)
35. Lapuschkin, S., Binder, A., Montavon, G., Müller, K.R., Samek, W.: Analyzing classifiers: fisher vectors and deep neural networks. In: Proceedings of the IEEE Conference on Computer Vision and Pattern Recognition, pp. 2912–2920 (2016)
36. Esteva, A., et al.: Dermatologist-level classification of skin cancer with deep neural networks. Nature **542**(7639), 115–118 (2017)
37. Arbabzadah, F., Montavon, G., Müller, K., Samek, W:. Identifying individual facial expressions by deconstructing a neural network. In: 38th German Conference on Pattern Recognition, pp. 344–354 (2016)
38. Lapuschkin, S., Binder, A., Müller, K.R., Samek, W.: Understanding and comparing deep neural networks for age and gender classification. In: IEEE International Conference on Computer Vision Workshops, pp. 1629–1638 (2017)
39. Perotin, L., Serizel, R., Vincent, E., Gúerin, A.: CRNN-based multiple DoA estimation using acoustic intensity features for ambisonics recordings. Journal of Selected Topics in Signal Processing **13**(1), 22–33 (2019)
40. Hettwer, B., Gehrer, S., Güneysu, T.: Deep neural network attribution methods for leakage analysis and symmetric key recovery. IACR Cryptology ePrint Archive 2019, 143 (2019)
41. Sturm, I., Lapuschkin, S., Samek, W., Müller, K.R.: Interpretable deep neural networks for single-trial EEG classification. J. Neuroscience Methods **274**, 141–145 (2016)
42. Samek, W., Binder, A., Montavon, G., Lapuschkin, S., Müller, K.R.: Evaluating the visualization of what a deep neural network has learned. IEEE Transactions on Neural Networks and Learning Systems **28**(11), 2660–2673 (2017)

Construction and Materials

Mechanical Properties of Ferrock Concrete: An Experimental Study

Wael Ibrahim[1(✉)], Ahmed Farouk Deifalla[2], Amr Nagi[1], Amal H. Ibrahim[1], and Waleed Abdallah[1]

[1] Faculty of Engineering, Civil Department, Helwan University, Cairo, Egypt
wael.ibrahim@rwth-aachen.de, amnagi@helwan.edu.eg, {aml_hassanin, waleed_abdallah}@m-eng.helwan.edu.eg

[2] Faculty of Engineering and Technology, Structural Engineering and CM Department, Future University in Egypt, Cairo 11835, Egypt
ahmed.deifalla@fue.edu.eg

Abstract. Nowadays, the whole academic society tends to find ecological solutions to climate change issues. One of the more widely utilised substances all over the world is OPC (ordinary Portland cement), and researchers are trying to discover an alternative or substitution to this material. One of the recently discovered materials was Ferrock; it is just a blend of many by-product materials, such as iron powder, fly ash or recycled glass, limestone, metakaolin, and weak acid. The main goal of this study is to look into the mechanical characteristics of Ferrock concrete. The mechanical properties of regular and Ferrock concrete were assessed using compressive, flexural, and tensile laboratory tests. The impact of adding Ferrock as an admixture percentage on the hardening characteristics of concrete was researched. The experimental results showed that the inclusion of Ferrock increased concrete's flexural and compressive strength at the ages of 7 and 28 days, and the ideal Ferrock percentage was 15 to 20%. Furthermore, results indicated that Ferrock concrete is ductile and has considerable displacement prior to actually collapsing.

Keywords: Ferrock · Green concrete · Sustainability · by-product materials · Iron Powder

1 Introduction

Concrete is among the most widely used building materials all over the world. It is considered probably the second most utilized material after water all over the world, it records for eight percentage points to ten percentage points (8% - 10%) of overall global CO2 emissions [1]. Reactive powder concrete (RPC) was developed by Richard et al. as just a type of ultra-high-performance concrete UHPC [2]. According to fiber reinforced concrete [3]. As mentioned in [4, 5] green development is regarded as one method of achieving ecological sustainability, and green concrete also has emerged as an unavoidable tendency in cement-based composites. RPC typically necessitates a high

A. S. Mosallam et al. (Eds.): ICATH 2023, SCI, pp. 65–76, 2024.
https://doi.org/10.1007/978-3-031-76557-5_6

amount of cement and quartz sand, raising construction costs and releasing large quantities of CO_2 into the atmosphere. RPC can be enhanced in microstructure, strength, water permeability, and resistance to corrosion by replacing a portion of cement with industrial by-products like fly ash, silica fume, and slag powder, as well as many other active additives [6–8]. Thus far, there are numerous researches have been done to obtain physical, chemical, and mechanical characteristics of several environmental and sustainable materials that don't harm the environment try substituting Ordinary Portland Cement binders in the concrete industry. One of these materials is Ferrock material. Our research is focused on replacing cement with Ferrock in different ratios as 5%, 10%, 15%, 20%, and 25% in concrete structures instead of ordinary Portland cement. The effects of different Ferrock ratios on compression stress, splitting tensile stress, and flexural behaviours of cubes, cylinders, and concrete beams respectively are investigated in this study. The behaviour of concrete beams with Ferrock is investigated for the issue of providing some guidelines for the design of such structures. The flexural behaviour is examined for the RC beams with the Ferrock. Their behaviour is evaluated by studying several different Ferrock ratios as mentioned above. The strength is compared with a similar beam reinforced with traditional concrete. The investigation is done experimentally. As stated previously in [9], many researchers around the world are refining approaches for the flexural behaviour of reinforced concrete specimens [10–16].

2 Research Significance

Ferrock is a 95% by-weight recycled substance that has been demonstrated to be less pricey, more strength, and more versatile in construction implementations than original concrete made using ordinary Portland cement. Moreover, this innovative material accelerates the curing process by using pressurized carbon dioxide and does not necessitate additional heat to catalyse its chemical reaction, creating it a carbon-negative preferable option to Ordinary Portland Cement. According to the reviewed studies, the optimal percentage of materials is 60% of iron powder, 20% of fly ash or recycled glass granules, 12% of metakaolin, and 8% of limestone. Hardening products contained between eight and 11% captured CO_2 by weight, according to evaluation (atomic emission spectroscopy). Since ferrock is strong in tension due to the amount of iron powder in it, it improves the compression strength and versatility of sustainable concrete. Ferrock is hence a "carbon negative material," in contrast to ordinary Portland cement, that is one of the most sources of carbon dioxide emissions as well as different atmospheric pollutants during mass production. One of Ferrock's unique properties is that it gets more powerful in aquatic environments. Ferrock does have a particle size of 90 microns and a particular gravity of around 5.29, thus according to IS 383 [17], and that is easily accessible locally (passing through 2.36 mm) with a 4.66 fineness modulus and 2.73 specific gravity. Ferrock concrete was mixed and cured using clean water [18, 19]. As shown in Table 1, All Percentages of components are given by [20], while specifications for each individual material and any comments are taken from [21]. David Stone uses waste steel dust from what is called "shot blasting operations." It is very fine (<50 microns) and is typically dark grey or black in color from the graphite dust in it. In this paper [22], the impact of slag powder is that the values of RPC strength decrease continuously after 3 days and then tend to increase after 7 and 28 days. Whenever the substitute ratio is increased from 0% to 30.0%, the compressive and flexural strengths of the specimens decrease by 6.5% and 9.8%, respectively. Once the slag powder substitute proportion is raised from 25% to 30%, the RPC specimens' compressive and flexural strengths recover by 2.1% and 5.0%, in both. Although some researchers have supposed to add benefits to concrete by using ferronickel slag powder as a cement partial alternative material [23–25].

Table 1. Required Materials for ferrock and their percentages.

Material	Percentage [20]	Specs/Comments [21]
Iron Powder	60%	The size of the particles of By-Product iron powder is 19.03microns
Recycled glass or Fly Ash	20%	Ground glass particles or fly ash with F Class F according to ASTM-C618
Lime Stone	10%	Lime stone powder with a particle size of 0.7μm according to ASTM-C568
Metakaolin	8%	According to ASTM-C618
Oxalic-Acid	2%	In prior studies, oxalic acid was utilized as a catalyst

3 Experimental Program

3.1 Material Testing

3.1.1 Properties of Cement

Ordinary Portland cement (OPC) was used for this project with a grade of 52.5. This's manufactured in accordance with Egyptian standard ES 4756/1–2013 [26] and meets European standard EN 197/1–2011 [27].

3.1.2 Properties of Course Aggregate (CA) and Fine Aggregate (FA)

All of the fine particles (Sand) in this project were siliceous-graded natural sand. It has a specific gravity of 2.62. While, the coarse aggregates used were all crushed dolomite sizes of 1 and 2. The specific gravity of this gravel is 2.6.

3.1.3 Properties of Ferrock

Ferrock is just a cementitious binder which could be used in place of Portland cement in certain applications. Flexural strength, toughness, and fracture resistance are all greater in Ferrock than in Ordinary Portland Cement. It will not degrade at high temperatures (>600 C), unlike Ordinary Portland Cement. It is also more environmentally friendly since it recycles solid waste (steel dust, fly ash, and recycled glass) additionally it absorbs 10% of CO_2 during the curing process.

3.1.4 Properties of Superplasticizer.

ADDICRETE BVF was used as a superplasticizer and high-range water reduction with no retarding impact for all specimens. ADDICRETE BVF is a ready-to-use brown liquid concrete additive as well as, ADDICRETE BVF meets with ASTM-C494 type F [28], and EN934–2 [29]. The dosage used is around 2–3% of cement weight.

3.2 Experimental Plan

The experimental program involves testing Ferrock concrete beam specimens under pure flexural loads to calculate the actual resistance, load-deflection curve, and the mode of failure for all tested beams, as well as testing cubes under compression and cylinders under compression and splitting tensile.

- Thirty-six cubes ($15 \times 15 \times 15$ cm^3) of both Normal concrete (Control) and Ferrock concrete. Eighteen cubes were crushed to determine the 7-day compressive strength, as well as eighteen cubes, were crushed to determine the 28-day compressive strength for the same day during Beams' pouring.
- Twelve cylindrical samples of Normal concrete (Control) and Ferrock concrete (150 mm in diameter and 300 mm in height). To investigate the compression strength of concrete, cylinders were tested under a uniaxial compression test.
- Six cylindrical samples of Normal concrete (Control) and Ferrock concrete (150 mm in diameter and 300 mm in height). To determine the tensile strength of concrete, cylinders were tested under a splitting tensile test.
- Six specimens, five Ferrock concrete with different percentages of Ferrock (5%, 10%, 15%, 20 and 25%) and one specimen of normal concrete (control specimen). All beams had dimensions of 2000 mm in length, 300 mm in depth, and 200 mm in width. Figure 1, shows a typical cross-section, a typical longitudinal section, and longitudinal steel bars. The thickness of the cover was 25 mm. The longitudinal reinforcement was 2Ø16 (high tensile steel) at the beam's bottom and 2Ø10 (high tensile steel) at the beam's top. Using mild steel for stirrups 10Ø8/m' to minimize the probability of shear failure.

3.3 Mix Proportions

For the experimental program there were six patches, each patch (was consisting of one beam, six cubes, and three cylinders) required 66.5 kg cement, 36.6 L water, 129.2 kg of sand, 224.2 kg of gravel, one liter of addicrete BVF super-plasticizer, and different values of Ferrock as per Table 2.

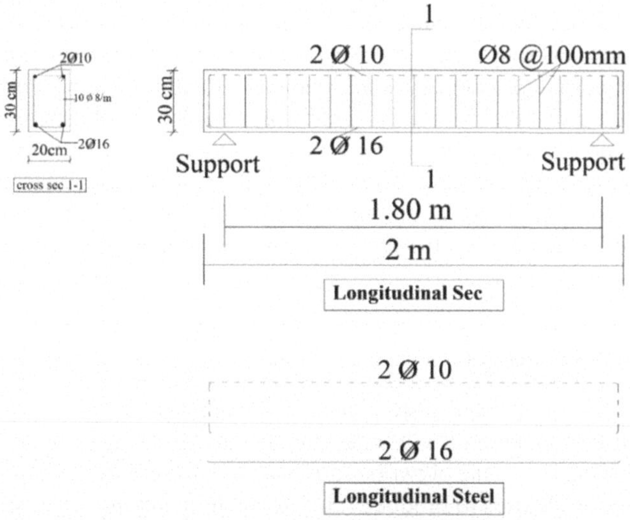

Fig. 1. Cross-section, & Longitudinal section for tested beams.

Table 2. Ferrock % of each design mix.

Item		Cement (Kg)	Water (Litres)	FA (Kg)	CA (Kg)	Superplasticizer (Litres)	Ferrock (Kg)
Beam 1	Control mix	66.5	36.6	129.2	224.4	1	0
Beam 2	5% Ferrock	63.2	36.6	129.2	224.4	1	3.3
Beam 3	10% Ferrock	69.9	36.6	129.2	224.4	1	6.6
Beam 4	15% Ferrock	56.6	36.6	129.2	224.4	1	9.9
Beam 5	20% Ferrock	53.3	36.6	129.2	224.4	1	13.3
Beam 6	25% Ferrock	49.9	36.6	129.2	224.4	1	16.6

3.4 Casting of Specimens

The specimens that were tested were placed in the appropriate shape using a plywood form. The form was clean and with neat dimensions of the specimens. To avoid the absorption of mixing water, the plywood form was washed before casting. The steel bars were attached to the stirrups to produce reinforcement matching what was required for the flexural behaviour of reinforced concrete beams. To facilitate removal after concrete hardening, the forms were covered with a thin coating of oil. The reinforcement bars were then put inside the formworks. Meanwhile, reinforcement cages were lifted by plastic spacers to ensure concrete cover. Dry materials (fine aggregates, coarse aggregates, cement, and ferrock) and water were mechanically combined in a drum mixer for 2–3 mins before being cast in the moulds. Testing the consistency of concrete with a slump test before casting. The fresh concrete was then vibrated using a vibrator. Resulting in a smoother finished concrete surface. After 24 h, the forms were taken from the specimens. And the specimens were bathed continually with water for at least 7 days one time in the morning and another at dusk time before being stored in a laboratory atmosphere for around 4 weeks.

3.5 Experimental Procedure

The specimens that were tested were placed in the appropriate shape using a plywood form. The form was clean and with neat dimensions of the specimens. To avoid the absorption of mixing water, the plywood form was washed before casting. The steel bars were attached to the stirrups to produce reinforcement matching what was required for the flexural behaviour of reinforced concrete beams. To facilitate removal after concrete hardening, the forms were covered with a thin coating of oil. The reinforcement bars were then put inside the formworks. Meanwhile, reinforcement cages were lifted by plastic spacers to ensure concrete cover. Dry materials (fine aggregates, coarse aggregates, cement, and ferrock) and water were mechanically combined in a drum mixer for 2–3 mins before being cast in the moulds. Testing the consistency of concrete with a slump test before casting. The fresh concrete was then vibrated using a vibrator. Resulting in a smoother finished concrete surface. After 24 h, the forms were taken from the specimens. And the specimens were bathed continually with water for at least 7 days one time in the morning and another at dusk time before being stored in a laboratory atmosphere for around 4 weeks.

3.5.1 Compression Strength of Cubes as Shown in Fig. 2

Fig. 2. Test Setup for cubes compressive strength test.

3.5.2 Compression Strength and Splitting Tensile Strength of Cylinders as Shown in Figs. 3-A and 3-B

Fig. 3. (a) Compression test for cylinders; (b) Splitting tensile test for cylinders.

3.5.3 Flexural Strength of Beams.

After 28 days all specimens have been transferred from the casting field to the laboratory. Each specimen was loaded with the crane, and after that laboratory technician put it above the two hinged supports after checking the distance between the two supports which was 180cm, this distance was divided into 3 equal distances of 60cm using a stiff standard I beam to transfer the load from the hydraulic jack to two rods as shown in Figs. 4.a and b. While assuring the Verticality and horizontality of all specimens in addition to plates of loading and supports by using a pocket level. LVDT and the dial gauges' positions were adjusted. Special devices were used to read the hydraulic jack's force and deformation gauges. The loading process was gradually applied until the specimen failed. The apparatus was then turned off, and the beam was moved in preparation for another test.

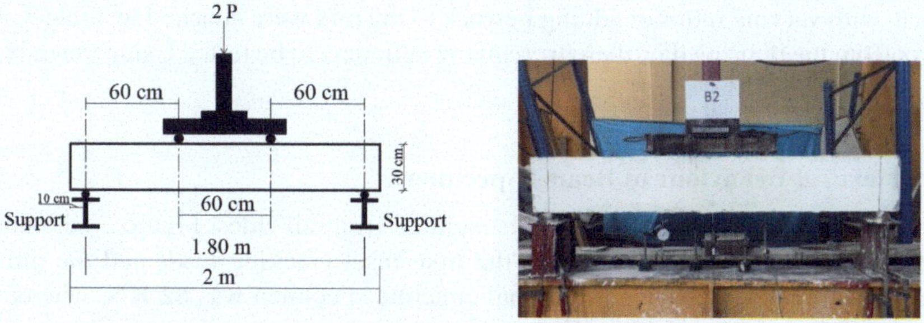

Fig. 4. (a) Test setup with dimensions; (b) A Real photo for test setup.

4 Results and Discussion

Experimental testing gives tangible information and knowledge about a system's behaviour. Furthermore, test findings are critical in calibrating and validating analytical models. An experimental program was implemented to expand knowledge and a better understanding of the physical and mechanical properties of Ferrock concrete due to testing cubes under compression test and cylinders under compression test and splitting tensile test, as well as the flexural behaviour of reinforced Ferrock concrete beams.

4.1 Mechanical and Structural Properties for Ferrock Concrete

4.1.1 Cubes Results

The compressive strength of normal concrete (control mix) and Ferrock concrete with varying ratios of adding Ferrock to the mix were attached in Table 3. Cubes were crushed to obtain the 7-day compressive strength, and then other cubes were crushed to determine the 28-day compressive strength. The optimal Ferrock ratio was 15%, with an incremental value of about 25% more than the control mix's compressive strength value.

4.1.2 Cylinders Results

- Compression Test

The compressive strength of normal concrete (control mix) and Ferrock concrete with varying ratios of adding Ferrock to the mix were attached in Table 3. Cylinders were crushed to obtain the 7-day compressive strength, and then other cylinders were crushed to determine the 28-day compressive strength. The optimal Ferrock ratio was 15%, with an incremental value of about 30% more than the control mix's compressive strength value.

- Splitting Tensile Test

The results of the splitting tensile test of normal concrete (control mix) and Ferrock concrete with varying ratios of adding Ferrock to the mix were attached in Table 3. After 28 days from the pouring day, there were many cylinders to be tested Using the universal testing machine UTM.

4.1.3 Flexural Behaviour of Beams Specimens

The behavior of the specimens should be studied from all sides. Figure 5 summarizes the results of the specimens regarding the first onset cracking loads and the ultimate loads. The ultimate force for a traditional concrete specimen was **82 KN**, whereas the results for all Ferrock concrete specimens were more than the normal concrete specimen, with 92KN, **96KN, 107 KN, 109KN,** and **102KN** for **5%, 10%, 15%, 20%,** and **25%** Ferrock to cement weight specimens. As shown in Fig. 4.1, the cracking force and ultimate force were obtained for each specimen while the cracks appeared instantly, thus the examiner couldn't accurately specify the beginning force value to create the first fracture. Whereas the ultimate forces for each specimen were obtained in those values 82, 92, 96, 107, 109, and 102 for the conventional concrete, Ferrock with 5%, 10%, 15%, 20%, and 25% respectively, observing elevated values of ultimate force to a Ferrock ratio of 20% and then a drop in ultimate force to a Ferrock ratio of 25%.

Deflection was measured by using three dial gauges, which were assembled at the end of the test zone (at the acting load) and at the mid-span. The distance between each dial gauge was 300 mm. These dial gauges have an accuracy of 0.01 mm. Figure 6 (a, b) show the relationship between the load and deflection for each specimen respectively. In Fig. 6.a deflection profile for all specimens at the peak load is figured out with two values the first one just under the concentrated force at 600 mm from the support and the second value at the center of the specimen (at mid-span) 900 mm from the support. As shown, the load-deflection curves are categorized into 3 phases: (1) preliminary cracking, in which the load-deflection curves increase incrementally without any observable cracks; (2) pre-yield, in which cracks begin to appear, crack width increases when the load is increased and Cracks extend all the way to the beam's top, reducing flexural stiffness; and (3) post-yield, that More cracks develop, and the bottom steel yields.

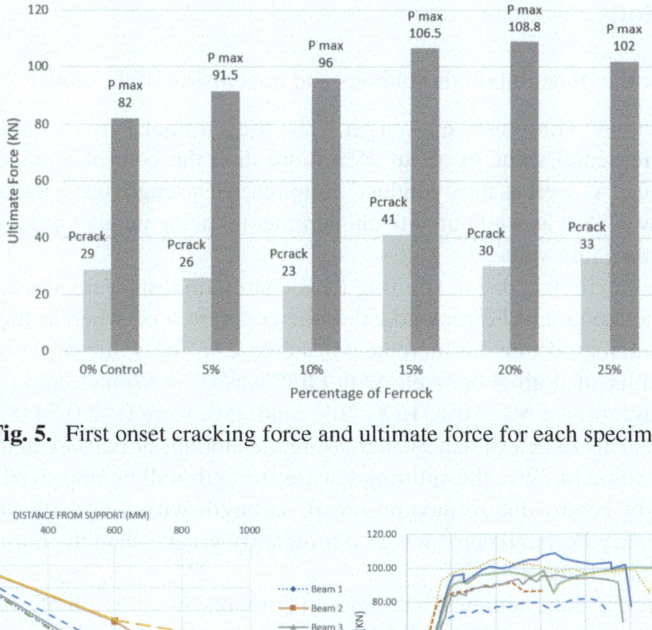

Fig. 5. First onset cracking force and ultimate force for each specimen.

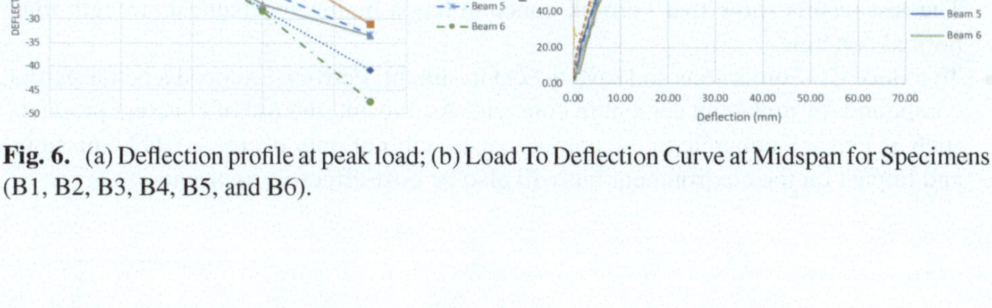

Fig. 6. (a) Deflection profile at peak load; (b) Load To Deflection Curve at Midspan for Specimens (B1, B2, B3, B4, B5, and B6).

Table 3. Cubes Compressive strength results and Cylinders Compressive and splitting strength results.

	Cubes Comp. Strength		Cylinders Comp. Strength		Cylinders Tens. Strength
	After 7 Days (N/mm2)	After 28 Days (N/mm2)	After 7 Days (N/mm2)	After 28 Days (N/mm2)	After 28 Days (N/mm2)
Beam 1	21.4	37.0	12.8	25.8	0.32
Beam 2	24.0	38.9	14.8	28.7	0.32
Beam 3	25.1	39.6	18.2	31.0	0.34
Beam 4	26.6	46.4	18.6	33.6	0.36
Beam 5	22.1	37.9	16.3	28.8	0.37
Beam 6	20.4	35.0	15.4	28.4	0.39

5 Conclusions

Below lines are the most important findings and conclusion of the study:

- While the cubes' compressive strength tests, the optimal Ferrock ratio was 15%, with an incremental value of about 25% more than the control mix's compressive strength value. As well as the cylinders' compressive strength tests, the optimal ratio of Ferrock was 15% as well but with an incremental percentage of around 30% more than the control mix value.
- The results of cylinders due to splitting tensile strength tests were almost effectively related to the amount of Ferrock added to the concrete mix, whereas the percentage of Ferrock increased thus an incremental increase in the value of tensile strength, the control mix of normal concrete with OPC was 0.32 N/mm2, and the results of Ferrock with ratios of 5%, 10%, 15%, 20%, and 25% were 0.32,0.34,0.36,0.37, and 0.39 N/mm2.This indicates that by increasing the amount of Ferrock in the mix to an appropriate value of 22%, the splitting tensile strength will be improved.
- The results of beams due to flexural tensile strength with a Ferrock proportion of 20% of ordinary Portland cement was significantly greater than the normal concrete beam's ultimate force by 33%.
- Ferrock concrete cube achieved a maximum compressive strength of 46.4 N/mm2.
- In Ferrock concrete, 15–20% replacement of cement shows an increase in strength. The test results show that Ferrock concrete has a higher mechanical strength than normal concrete.
- To reduce CO2 emissions and save the environment, Ferrock is indeed a better partial compound for replacing cement in concrete. As a result, the use of concrete products such as Ferrock concrete in the coming years will not only decrease CO2 emissions and impact on the environment but will also be cost-effective to produce.

References

1. Vijayan, D.S., Arvindan, S., Janarthanan, T.S.: Evaluation of Ferrock: a greener substitute to cement. In: Materials Today: Proceedings (2019)
2. Richard, P., Cheyrezy, M.: Composition of reactive powder concretes. Cem. Concr. Res. **25**(7), 1501–1511 (1995)
3. Richard, P., Cheyrezy, M.H.: Reactive powder concretes with high ductility and 200–800 MPa compressive strength. In: Acids, Symposium Paper American Concrete Institute (1994)
4. Boobalan, S.C., Shereef, M., Saravanaboopathi, P., Siranjeevi, K.: Studies on green concrete – a review. In: Materials Today Proceedings, vol. 65, part 2, pp. 1404–1409 (2022)
5. Doo-YeolYoo, Y., Banthia, N.: Mechanical properties of ultra-high-performance fiber-reinforced concrete: a review. Cem. Concr. Compos. **73**, 267–280 (2016)
6. Yazıcı, H., Yardımc, M.Y., Yiğiter, H., Aydın, S., Türkel, S.: Mechanical properties of reactive powder concrete containing high volumes of ground. Cem. Concr. Compos. **32**(8), 639–648 (2010)
7. Tuan, N.V., Ye, K., Breugela, G., Copuroglu, O.: Hydration and microstructure of ultra-high-performance concrete incorporating rice. Cem. Concr. Res. **41**(11), 1104–1111 (2011)
8. Tafraoui, A., Escadeillasa, G., Lebailib, S., Vidal, T.: Metakaolin in the formulation of UHPC. Constr. Build. Mater. **23**(2), 669–674 (2009)
9. Alkhatib, S., Deifalla, A.: Reliability-based assessment and optimization for the two-way shear design of lightweight reinforced concrete slabs using the ACI and EC2. Case Stud. Constr. Mater. **17** (2022)
10. Deifalla, A.F., Awad, A., Seleem, H., Abdelrahman, A.: Investigating the behavior of lightweight foamed concrete T-beams under torsion, shear, and flexure. Eng. Struct. **219**, 110741 (2020)
11. Fattouh, M.S., Tayeh, B.A., Agwa, I.S., Elsayed, E.K.: Improvement in the flexural behaviour of road pavement slab concrete containing steel fibre and silica fume. Case Stud. Constr. Mater. **18** (2023)
12. Alshannag, M.J., Charif, A., Alqarni, A.S., Nasser, S.: Flexural performance and ductility of RC beams made using natural LWA. Case Stud. Constr. Mater. **16** (2022)
13. El-Sayed, T.A., Algashb, Y.A.: Flexural behavior of ultra-high performance geopolymer RC beams reinforced with GFRP bars. Case Stud. Constr. Mater. **15** (2021)
14. Raju, S., Arulpandian, V., Avudaiappan, S., Amran, M., Fediuk, R., Vatin, N.: Influence of aspect ratio on flexural behavior of reinforced self-compacting concrete beams. Case Stud. Constr. Mater. **17** (2022)
15. Zhang, Z., Abbas, E.M.A., Wang, Y.: Experimental study on flexural behavior of the BFRC-concrete composite beams. Case Stud. Constr. Mater. **15** (2021)
16. Abdullah, Q.N., Abdulla, A.I.: Flexural behavior of hollow self compacted mortar ferrocement beam reinforced by GFRP bars. Case Stud. Constr. Mater. **17** (2022)
17. Bureau of Indian Standards: Indian Standards (2016)
18. Garcia, A.L., Achaiah, A.T., Bello, J., Donovan, T.: Ferrock: a life cycle comparison to ordinary Portland cement. In: Industrial Ecology, California (2017)
19. Prashanth, M., Gokul, V., Shanmugasundaram, M.: Investigation on Ferrock Based Mortar an Environment Friendly Concrete. SSRN Electr. J. (2019)
20. Das, S., Hendrix, A., Stone, D., Neithalath, N.: Flexural fracture response of a novel iron carbonate matrix – glass fiber composite and its comparison to Portland cement-based composites. Construct. Build. Mater. **93**, 360–370 (2017)
21. Das, S., Stone, D., Convey, D., Neithalath, N.: Pore- and micro-structural characterization of a novel structural binder based on iron carbonation. Mater Charact **98**, 168–179 (2014)

22. Ge, W., et al.: Study on the workability, mechanical property and MMME 132–9 water. Case Stud. Constr. Mater. **18** (2023)
23. Huang, Y., Wang, Q., Shi, M.: Characteristics and reactivity of ferronickel slag powder. Constr. Build. Mater. **156**, 773–789 (2017)
24. Kim, H., Lee, C.H., Ann, K.Y.: Feasibility of ferronickel slag powder for cementitious binder in concrete mix. Constr. Build. Mater. **207**, 693–705 (2019)
25. Yang, H.J., Lee, C.H., Shim, S.H., Kim, J.H.J., Lee, H.J., Park, J.: Performance evaluation of cement paste incorporating ferro-nickel slag powder under. Case Stud. Constr. Mater. **15** (2021)
26. Egyptian Organization for Standards and Quality: ES 4756/1, Cement Part: 1 Composition. Specifications and Conformity Criteria for Common Cements, Cairo (2013)
27. The British Standard: EN 197–1, Cement - Part 1: Composition, Specifications and Conformity Criteria for Common Cements (2011)
28. ASTM International: ASTM C-494 Type A, Standard Specification for Chemical Admixtures for Concrete (2017)
29. The European Standard: EN 934–2, Admixtures for Concrete, Mortar and Grout - Part 2 (2009)
30. British Standard: BS 5075, Part 3, Concrete Admixtures Specification for Super-Plasticizing Admixtures (2002)

Best Practices for Use of Recycled Asphalt Pavement

Yara Chedid[1], Abhilash Reddy Gujjula[1], Shadi Saadeh[1(✉)], Ashraf Rahim[2], and John Harvey[3]

[1] California State University, Long Beach, CA, USA
{yara.chedid01,AbhilashReddy.Gujjula01}@student.csulb.edu,
shadi.saadeh@csulb.edu
[2] California Polytechnic State University, San Louis Obispo, CA, USA
arahim@calpoly.edu
[3] University of California, Davis, Davis, USA
jtharvey@ucdavis.edu

Abstract. The utilization of reclaimed asphalt pavement (RAP) for road construction has been a widely adopted practice due to its economic, environmental, and resource conservation benefits, which are benefits to humanity. This paper provides a comprehensive review and guidelines on using RAP in road construction. It discusses performance evaluation, typical distresses, and best management techniques for RAP. Performance tests show that increasing RAP percentages can decrease the chance of rutting, but the possibility of low-temperature cracking may be heightened. This study describes typical distresses, frequently ascribed to aged asphalt binder and including cracking, rutting, and moisture damage. Essential best practices for RAP processing, storage, and quality control are outlined. Particular focus is placed on the effects of milling equipment, depth, and speed on RAP quality and the critical function of moisture content in RAP storage. The paper concludes by emphasizing the need for appropriate rejuvenating agents in high RAP content mixtures and the essential role of effective RAP management practices. This paper is a valuable resource for specialists in the industry, offering insights that can be put into practice to maximize RAP utilization in sustainable road construction.

Keywords: RAP · Environment · Cost · Aggregates · Rejuvenator

1 Introduction

According to the U.S. Department of Transportation, 94% of the United States paved roads are asphalt pavements [1]. Approximately 420 million tons of asphalt mixes were produced out of the 3,600 asphalt production sites in the United States in 2019. That same year, 97 million tons of RAP were gathered for reuse, and $3.3 billion was saved by employing 90 million tons of recycled components in asphalt roads. Additionally, 5.5 million tons of RAP are replacing aggregates in road-related applications such as cold-mix and warm-mix.

A. S. Mosallam et al. (Eds.): ICATH 2023, SCI, pp. 77–91, 2024.
https://doi.org/10.1007/978-3-031-76557-5_7

The multiple asphalt recycling techniques fall into one of five groups: full-depth reclamation, cold planning, hot in-place recycling, cold in-place recycling, and hot recycling. Some of these recycling techniques might be combined and used for some projects.

Compared to new asphalt binders, binders, including RAP, are more viscous and extensively oxidized. Rejuvenating an old asphalt binder is a standard procedure to restore its properties to a state similar to the virgin binder. In fact, rejuvenator use is actually appropriate for recycling mixtures that are either severely oxidized or contain more than 25% RAP. To enhance the mixture's low-temperature properties without degrading those at higher temperatures, the proper rejuvenator type and dosage should be employed when building recycled pavements.

2 RAP Advantages

The two main benefits to humanity of using RAP are economic and environmental benefits. Construction costs will be substantially reduced by incorporating RAP materials, such as aggregates and binder, into new pavement. Moreover, using RAP contributes to more sustainable construction practices through its environmental benefits.

2.1 Environmental Benefits

Approximately 94% of the U.S. roadways are asphalt surfaced. Reusing RAP has three different environmental advantages, as shown in Fig. 1. By integrating them into new road-building projects, RAP reuse in recycled pavements is an alternate method of reducing waste from road construction materials and improving landfills management. As a result, we can save natural resources like aggregate and asphalt binder, which is essential for a sustainable environment for future generations. RAP material usage in building road pavements has been shown to minimize energy consumption and emissions from the transportation and extraction of virgin resources. Binder manufacturing consumes the most energy and produces the most emissions within the pavement sector. RAP reduces the quantity of binder needed, which eventually uses less energy, emits fewer emissions, leads to better air quality, and helps compact global warming. The reduction in landfill area is another advantage for the environment.

2.2 Cost Analysis

The overall cost of RAP-containing asphalt mixtures was compared to asphalt mixtures without RAP by Zaumanis et al. (2014) [2]. As shown in Fig. 2, various RAP material percentages, from 0% to 100%, were assessed in their investigation. At zero RAP, it costs around $40 per ton for asphalt binder in the overall mixture, while the binder is the costliest component of constructing a standard asphalt mixture. This is followed by the total cost of the aggregate of $20/ton. The use of 50% RAP materials decreased the cost of construction by over $20/ton to less than $50/ton from $70/ton. Cost categories for RAP processing, testing, rejuvenator, pollution control, and RAP material cost are assigned to recycled mixes, as shown in Fig. 2. Overall, using 100% RAP material lowered construction costs by 50–70%, which is a significant amount.

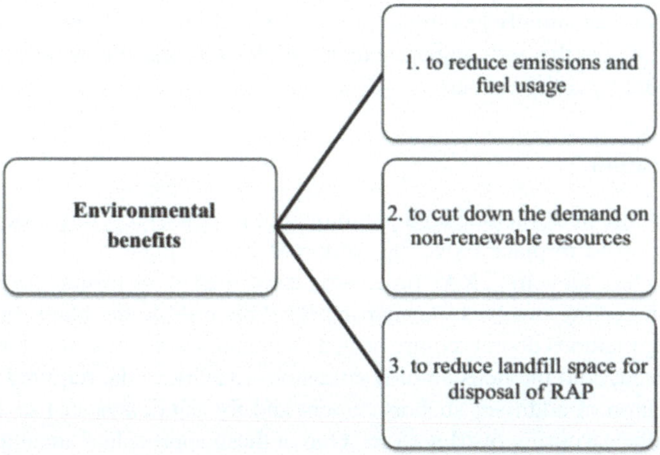

Fig. 1. Advantages for the environment of employing RAP in new pavement projects. Figure Courtesy T. Baghaee Moghaddam, H. Baaj/Construction and Building Materials 114 (2016) 805-816.

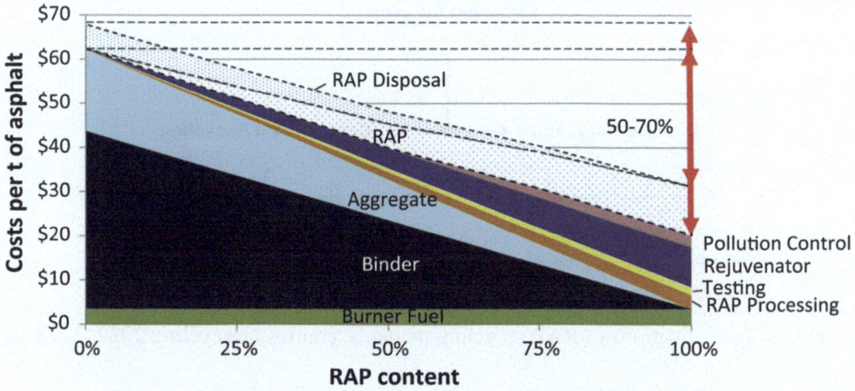

Fig. 2. Cost of material in hot mix recycling, Zaumanis et al 2014

3 RAP Disadvantages

As we already know, using RAP materials in pavements has many advantages and disadvantages due to the difficulty of RAP mix designs. A poorly designed recycled pavement reduces the pavement's service life, spikes maintenance costs, and raises emissions and energy consumption.

Due to RAP's stiff binder component, it has been determined that utilizing large quantities of RAP substance in road pavement raises the danger of early pavement distresses, including low-temperature and fatigue cracking, and makes the mixture more susceptible to moisture. In addition, the stiff binder may be a limiting factor affecting field compaction, eventually leading to premature field failure. However, several investigations have shown that utilizing more than 50% RAP is possible without affecting the mechanical qualities of the mixes.

3.1 Classification

RAP recycling methods include cold planning, hot in-place recycling, hot central plant recycling mix, cold in-place recycling, and cold central plant recycling (see Fig. 3). Regarding in-place recycling, RAP processing is finished at the project site. Conversely, central plant recycling involves transporting RAP material to the plant for processing. Cold recycling methods do not require heat, a characteristic feature of hot mix recycling methods. Instead, cold methods employ emulsions to achieve the required consistency. The incorporation of additives such as cement and fly ash delays surface raveling and gives the additives a curing overlay to set. One of these approaches' most notable benefits is their environmental impact. They support sustainable construction techniques by contributing to minimizing air pollution and the need for material hauling.

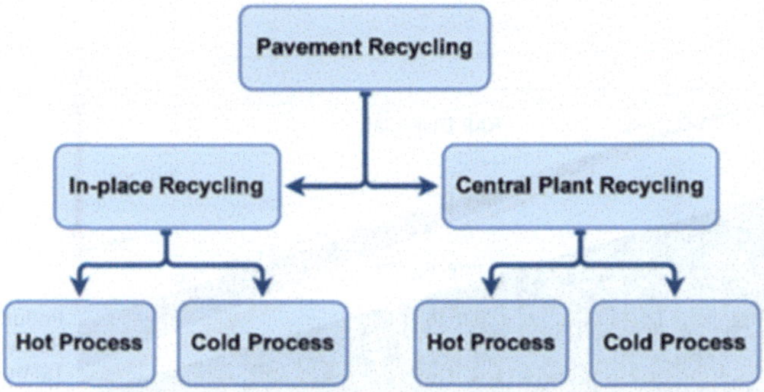

Fig. 3. Classifications of RAP recycling methods, courtesy Recycling 2022, 7, 35

Several engineering and logistical factors affect the selection of RAP recycling approach. The primary engineering considerations include pavement distress levels, structural strength, and resource availability. This selection is further influenced by other considerations such as equipment requirements, long-term maintenance expenses, initial cost, and the impact of construction on traffic flow. Moreover, additional criteria include the treatment's purpose, suitability for the required application, traffic conditions, climate, and expected performance life. Table 1 provides a summary of the characteristics associated with the different methods. A complete description is provided in the following subsections.

Among the different techniques of RAP recycling, the cold in-place recycling technique is frequently applied for construction and rehabilitation works. This method leads to lower energy consumption, fewer greenhouse gas emissions, and increased cost-effectiveness compared to other on-site recycling techniques.

Table 1. Asphalt pavement maintenance and rehabilitation using RAP, courtesy Abdalrhman Milad et al. / Jurnal Teknologi (Sciences & Engineering) 82:3 (2020) 35–44

Application Technique	Evaluation factor			
	Climate	Traffic	Condition Addressed	Contraindication
Cold In-Place Recycling (CIPR)	Remediation performs well in all weather conditions	Very successful in both high- and low-volume roadways	Reconstruction of old pavements	• Long remaining life, • Extend the service life of roadway pavements by 10–15 years
Hot mix asphalt recycling (HMAR)	Driedtreatment sealants perform better in warmer climates	Performance is not significantly affected by different ADT or truck levels	Reconstruction of old pavements	Extend service life of roads for over 12 years
Hot in-place recycling (HIR))	Driedtreatment sealants perform better in warmer climates	Performance is not significantly affected by different ADT or truck levels	Correct shallow-depth HMA surface distress.	Preservation treatment process is expected to extend pavement life by 10–12 years
Full-depth reclamation (FDR)	Remediation performs well in all climate conditions	Higher traffic bearing capacity	Stabilized base course	Extend the service life of roadway pavements by 10 years

4 Recycled Material Properties

4.1 Properties of Aggregate

RAP aggregate and virgin aggregate are differentiated by certain properties, including angularity, specific gravity, and abrasion. These properties significantly influence new mixes' volumetric characteristics and performance, specifically when a significant amount of RAP is utilized. Hence, evaluating RAP aggregates' physical and consensus properties is crucial before adding them to new pavement structures.

Aggregate gradation is usually performed after removing asphalt from recycled materials, employing an ignition furnace, or solvent extraction procedures. According to Roque et al. (2015) [10], the analysis of aggregate gradation yields more concise results when the ignition oven is used compared to aggregate gradation analysis following solvent extraction.

Bulk Specific Gravity of Aggregate

Bulk specific gravity (Gsb) of aggregate helps to find out the mixture's void content, an important consensus property for mix design.

$$Gse = \frac{100 - Pb}{\left(\frac{100}{Gmm}\right) - \left(\frac{Pb}{Gb}\right)} \tag{1}$$

where:

- G_{se} = Effective specific gravity
- P_b = Recycled material Asphalt content
- G_b = Recycled material binder specific gravity
- G_{mm} = Recycled material Theoretical maximum specific gravity

$$Gsb\ estimated = \frac{Gse}{\left(\frac{GsePba}{(100)(Gb)}\right) + 1} \tag{2}$$

- $G_{sbestimated}$ = Bulk Specific gravity
- P_{ba} = Percentage of asphalt absorbed

4.2 RAP Binder

The properties of the aged binder within the RAP and its quantity in the mixture play significant roles in defining the attributes of asphalt mixtures that incorporate RAP. Typically, in field applications, it is presumed that the RAP binder blends entirely with the virgin binder; however, this blending is not often fully achieved. When incorporating a small proportion of RAP, the existing aged binder in the RAP can follow for a decrease in the necessary quantity of virgin asphalt binder. Nevertheless, excessive RAP content or overly aged RAP can cause the particles to behave similarly to "black rock", impeding adequate mixing with the new binder and potentially yielding suboptimal outcomes.

It was hypothesized that complete blending of materials would result in the recovered RAP binder having qualities equivalent to or similar to the virgin binder. As blending becomes less complete, it is anticipated that the disparities in the characteristics of the binder from recovered RAP aggregate and that from newly sourced aggregate will become more pronounced. In scenarios where there is no blending, the characteristics of the binder obtained from recovered virgin aggregate should mirror those of a virgin binder. In contrast, the properties of the binder from recovered RAP aggregate should mirror a combination of RAP binder with a suitable amount of new binder enveloping the RAP elements.

Adhering to the approach outlined in the Bailey method (P. Shirodkar et al., 2010) [11], the estimations of the ratios of RAP binder to new binder surrounding the RAP aggregate are carried out, assuming no blending occurs. Two equations are proposed: the Blending ratio (3) and the Degree of partial blending (4).

$$Blending\ ratio = \frac{\left(\frac{G*}{Sin(\delta)}\right) blend\ binder\ virgin\ agg - \left(\frac{G*}{Sin(\delta)}\right) blend\ binder\ RAP\ agg}{\left(\frac{G*}{Sin(\delta)}\right)\ virgin\ binder - \left(\frac{G*}{Sin(\delta)}\right) RAP\ virgin\ binder\ zero\ blend} \tag{3}$$

Where:

- $(G^*/\sin(\delta))$ blend binder virgin agg = Rolling Thin-Film Oven (RFTO) $G^*/\sin(\delta)$of blended binder coating the coarse virgin aggregate
- $(G^*/\sin(\delta))$ blend binder RAP aggregate = RFTO $G^*/\sin(\delta)$of blended Binder coating the fine RAP aggregate
- $(G^*/\sin(\delta))$ Virgin binder = RFTO $G^*/\sin(\delta)$of virgin Binder, and
- $(G^*/\sin(\delta))$ RAP virgin binder 0 blend = RFTO $G^*/\sin(\delta)$ of RAP and virgin binder that coated RAP aggregate assuming 0% blending

$$Degree\ of\ partial\ blending(\%) = 100|1 - blending\ ratio| \tag{4}$$

5 RAP Mix Design

The Quality Improvement Series 124 from the National Asphalt Pavement Association (NAPA), titled "Designing HMA Mixtures with High RAP Content: A Practical Guide," is an invaluable guide on mix designs with significant RAP content. Many of the standards in this comprehensive guide align with the RAP mix requirements stipulated in AASHTO M 323, extending additional guidelines for defining RAP materials, sample preparation, mechanical property testing, and plant mix changes. For mixes with RAP levels of more than 25%, the guidelines recommend identifying RAP binder characteristics on a regional basis.

As shown in Fig. 4, the aggregates undergo rigorous property examinations before combining the selected RAP fractions in a preliminary mixed formulation. Following this, the binder is extracted from the mix for a detailed analysis of its characteristics, simplifying the process of choosing the suitable kind and amount of recycling agent. In the laboratory, the asphalt is blended, compacted, and assessed to determine the necessary volumetric and functional characteristics. The process is replicated with the adjustments needed if the mix fails to meet the required criteria at any level. If designing a mix with 100% RAP proves unachievable due to the qualities of RAP, especially the fine particles and binder content, the addition of virgin aggregates and binder might be considered (Gencer et al., 2012; Arnold et al., 2012) [12, 13]. However, extra caution is advised in order to ensure the effective mixing of virgin binders and RAP, in addition to the homogeneous coating of RAP and virgin aggregates.

5.1 RAP Binder

RAP aggregates must meet the exact requirements as virgin mineral aggregates to achieve a good pavement mix design using RAP. A sequence of testing is conducted to assess the aggregates' quality. These tests include evaluations of aggregate fracture faces, sand equivalent for fine aggregates, Los Angeles (L.A.) abrasion, and bulk specific gravity.

5.2 Asphalt Binder Content

The overall content of asphalt binder in a mixture is calculated by considering the quantities of new and salvageable asphalt originating from recycled materials. Three

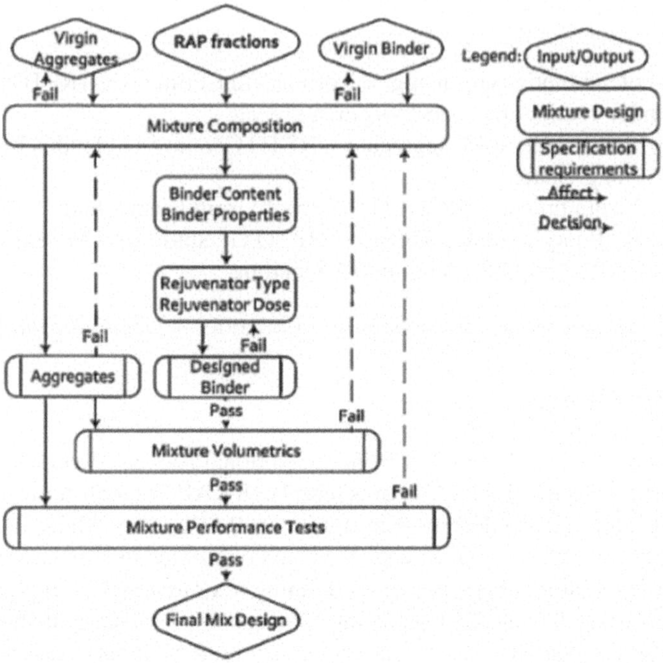

Fig. 4. Flow chart of Mix Design Procedure Figure Courtesy M. Zaumanis et al. / Resources, Conservation and Recycling.

methods may be utilized to ascertain the asphalt binder content of reused materials. The initial method supposes that the complete asphalt makeup of the recycled substance adds to the total content of the asphalt binder. Another method assumes that no recycled asphalt is usable, often called the "black rock" scenario. The third method recognizes that the situation's reality falls somewhere between these two extremes, although determining the exact proportion is challenging. Regardless of the method employed, the standard formula used to calculate the total asphalt content of an asphalt mix is:

$$TAC = \left[FRAP(RAP\ AC)(RAP\%) + FRAS(RAS\ AC)(RAS\%) + Virgin\ AC\%\right] \quad (5)$$

where:

- TAC = Total asphalt binder content
- FRAP, FRAS = Asphalt availability factors for RAP and/or RAS asphalt content.
- RAP AC = Asphalt content of RAP, decimal form.
- RAP% = Percentage of RAP in mixture, %.
- RAS AC = Asphalt content of Recycled Asphalt Shingle, decimal form.
- RAS% = Percentage of RAS in mixture, %

Factors influencing the binder content in a mix design encompass the RAP source, fines content, quantity, and quality of recycling agents, and virgin binders. In efforts to optimize the impact of these factors, they may be altered in combinations or individually.

The limit of RAP percentage in HMA is determined by three factors: the RAP material gradation, the stiffness of the RAP binder following its removal, and the design composition of the RAP material.

The RAP percentage can be calculated using the subsequent formula (McDaniel & Anderson, 2001).

$$RAP\ Percentage = (T\ blend - T\ virgin) / (TRAP - T\ virgin) \quad (6)$$

where:

- T_{virgin}: critical temperature of virgin binder
- T_{blend}: critical temperature of blend binder
- T_{RAP}: critical temperature of extracted RAP binder
- RAP: percentage of RAP expressed as a decimal.

The temperature where the smallest strain occurs for one million cycles in the asphalt mix is critical.

A specific methodology must be followed to design an optimal RAP-HMA mix (A Sharma et al., 2021). First, it is important to identify the RAP material's gradation. Next, the virgin aggregate's required amount is to be established around the final gradation of the mix and the RAP percentage to be incorporated. Following this, the bitumen content within the RAP material is determined using centrifugal methods, as per ASTM D2172, 2017 [13], and its absolute viscosity is determined at 60°C according to ASTM D2171, 2018 [14]. Subsequently, the combined aggregate's asphalt demand is calculated, keeping in mind the specific RAP replacement tier as demonstrated in Table 2 (Tier 1 is for less than 15% RAP, Tier 2 is for 15% to 25% RAP, and Tier 3 is for more than 25% RAP).

Table 2. Mix design guidelines (A Sharma et al. 2021 IOP Conf. Ser.: Mater. Sci. Eng. 1075 012025)

Tier	Amount of RAP	Recommendation
1	<15%	No changes in the mix design process
2	15% to 25%	One grade lower than conventional Performance Grade (PG)
3	>25%	Refer blending chart provided in ASTM D4887, 2016

Based on this, the target grade of the recycled binder is examined, and adjustments are made for the recycling agent dosage and RAP replacement level as per ASTM D2172, 2017 [13]. Finally, the ideal binder quantity for the recycled mix is established through the Marshall or Superpave Design approach, creating a corresponding Job Mix Formula.

5.3 Rejuvenator

Efficient utilization of rejuvenators should counter the aging effects on the RAP binder, refresh its qualities for extended utility, and ensure its adequacy for inclusion in new asphalt mixes. Refer to Table 2 of Baghaee and Baaj (2016) to understand how the recycling agent is selected considering its various characteristics.

Short-term, the rejuvenator should be able to interact with mixtures high in RAP content, swiftly permeating the RAP binder and activating the aged asphalt to yield uniformly coated blends. The rejuvenator ought to lessen the stiffness of the binder sufficiently to create a pliable, manageable mixture that facilitates straightforward paving and compaction to achieve the required density while also reducing the potential for harmful emissions. It is important to note that the process of diffusing ought to be completed prior to opening new pavement to traffic to mitigate issues such as reduced friction and greater rutting sensitivity.

Long-term, rejuvenating agents ought to restore and stabilize the chemical and physical properties of the old binder for an extended service life. The rheology of the binder needs to be carefully adjusted to diminish the likelihood of cracking from fatigue and low temperature without excessively softening the binder, leading to issues with rut formations. Adequate cohesiveness and adhesion must be present to avoid raveling and damage due to moisture. In recent years, the microcapsule concept has shown promise in asphalt recycling. When the microcapsule fractures, the escaping agent permeates the fine cracks, assisting in the healing process. Microcapsule shells are constructed from polymeric materials and provide adequate strength, toughness, and strong thermal stability at elevated temperatures. Fig. 5 illustrates the potential applications of rejuvenator-containing microcapsules, which have become a typical approach in recycling asphalt pavement.

6 RAP Performance

Table 3 summarizes examples of recent research findings for various performance testing for RAP asphalt mixes. Generally, increasing the percentage of RAP tends to reduce susceptibility to rutting while increasing material stiffness. However, as the proportion of recycled materials increases, there is a higher chance of low-temperature cracking, indicated by an increase in the critical low-temperature threshold. Findings regarding the potential for cracking at mid-range temperatures and susceptibility to moisture are inconclusive, varying within individual studies and across different investigations.

Common distresses in traditional asphalt mixes containing extremely high RAP content include cracking, rutting, water susceptibility, flushing, and raveling. These distresses often arise from asphalt binder that is aged and used in recycled material. Raveling of recycled asphalt pavement occurs due to insufficient blending between the two binders, leading to improper coating and suboptimal mixture. Rutting in recycled mixes can stem from overdosing with a binder that is softer or rejuvenating agents. Cracking is another prevalent distress in recycled pavement mixtures, allocated to the decreased ductility of aged binders, rendering the pavement more fracture susceptible. Moisture damage occurs when water undermines the binder and aggregate adhesion within the mix, potentially weakening the bond. Excessive or incompatible use of rejuvenators can lead to bleeding in the RAP mix, defined by the ascending movement of oil toward the top of the layer of asphalt.

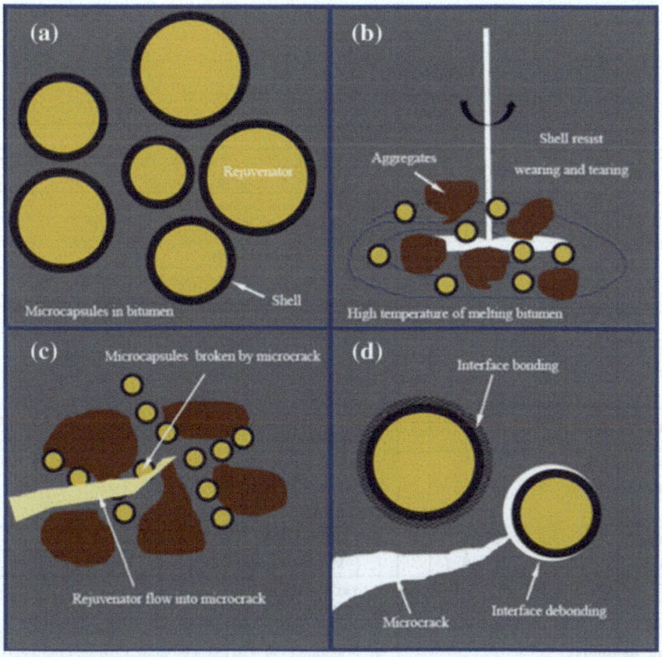

Fig. 5. Depiction of the potential conditions of microcapsules filled with rejuvenator within bitumen/asphalt binder, Baghaee Moghaddam, H. Baaj / Construction and Building Materials 114 (2016) 805–816 Figure Courtesy T. Baghaee Moghaddam, H. Baaj/Construction and Building Materials 114 (2016) 805-816

7 Best Practices for RAP Management

Optimizing the entire supply chain for RAP, which encompasses the stages of milling, processing, storing, and quality control, can significantly improve the quality of the end product. This section explores the optimal methods for managing RAP to attain outstanding outcomes.

7.1 RAP Milling and Processing

The quality of RAP can be significantly affected by the milling process, which can be either partial or complete. For road structures with multiple layers of varying aggregates or binders, partial milling is preferred to facilitate the strategic employment of RAP in higher-value layers later. Factors such as the type of milling equipment, depth of cut, and operation speed all affect the quality of RAP.

To minimize the content of undesirable fines, special care, and operational adjustments are advised. For instance, a slower advancement rate or faster rotation of the drum often results in the production of additional fine materials. According to the Smart Pave System creators, M. Zaumanis et al. (2014) [2], RAP obtained through upward-cut milling heads typically deviates no more than 10% of the original aggregate gradation.

Table 3. Summary of changes in RAP mixture properties, NCHRP 495 (2016)

Testing		Influence of Increasing RAP				References
Rutting						
Rutting (loaded wheel units)		S	2			Maupin (2008); Zaumanis et al. (2014); Willis et al. (2012); Watson et al. (2008)
Creep flow time					M	Daniel and Lachance (2005)
Creep stiffness				1		Abdulshafi et al. (2002)
Cracking (Intermediate Temperatures)						
Fatigue		S	1	2	M	Maupin et al. (2008); Abdulshafi et al. (2002); Zaumanis et al. (2014); Vukosavlievic (2006); Watson et al. (2008); McDaniel et al. (2012)
Reflective cracking (Overlay Tester)				1		Willis et al. (2012)
Dissipated energy				1		Vukosavlievic (2006)
Fracture energy			1			Vukosavlievic (2006)
SBC	**Fracture energy**		2	1		Lee et al. (2016); Willis et al. (2012); Johnson et al. (2013)
	Tensile strength			2		Vukosavlievic (2006); Johnson et al. (2013)
	Fracture work				M	Lee et al. (2015)
Moisture Sensitivity						
Moisture sensitivity		S		1		Olard (2010)
Toughness index			1			Vukosavlievic (2006)
Differences over Range of Temperatures						
Stiffness (dynamic modulus)				2	M	Daniel and Lachance (2005); Abdulshafi et al. (2002); Roque et al. (2015); Lee et al. (2015); Olard (2010). McDaniel et al. (2012)
Phase angle, mix			1			Abdulshafi et al. (2003); Vukosavlievic (2006)
Fracture toughness		S			M	Lee et al. (2015); Johnson et al. (2013)
Low Temperature Testing						
Indirect tensile strength					M	Roque et al. (2015)
Indirect tensile creep compliance		S	1	2		Zaumanis et al. (2014); Roque et al. (2015); Johnson et al. (2013); Watson et al. (2008)
Thermal cracking				2		Zaumanis et al. (2014)
Critical cracking temperature				2		Vukosavlievic (2006); McDaniel et al. (2012)
Notched fracture energy			1			Swiertz et al. (2011)
Failure strain			1			Roque et al. (2015)
Energy Ratio			2			Roque et al. (2015); Willis et al. (2012)

S – Denotes consistent findings in the specified research; M – signifies varied outcomes in the specified research.
1 – Represents a decrease in a certain characteristic or test outcome as the proportion of recycled material rises in the study.
2 – Signifies an increase in a certain characteristic or test outcome with a higher percentage of recycled material in the study.

An overabundance of fine particles can significantly impact the design of RAP mixtures due to not meeting specified requirements for aggregate size distribution, the ratio of dust to binder, air void content, and voids in mineral aggregate (VMA).

7.2 Storage of RAP

To reduce contamination and ensure the uniformity of materials, the management of stockpiles of RAP ought to resemble those of virgin aggregate. Startup waste should be strictly segregated from RAP to maintain its quality. When RAP from various origins is combined in one pile, it is recommended to mix it prior to processing or introduce it uniformly into the cold feed to guarantee consistency. The level of moisture in RAP is a pivotal aspect that limits the maximum amount of RAP that can be utilized and influences subsequent production stages. For every 1% increase in moisture, drying, and heating expenses are elevated, overall plant production goes down, and emissions are increased by 10%. To mitigate these challenges, the following steps can be taken to reduce moisture content: (1) store stockpiles under a roof or covered area, (2) use paved, sloping storage space for water runoff, (3) utilize tall conical stockpiles, and (4) perform small-scale crushing and screening of RAP on the day of usage.

7.3 Quality Control and Variability Assessment of RAP

According to some 1980s and 1990s studies, the composition of RAP was considered to vary significantly. However, recent findings suggest that RAP material from a single project tends to be more consistent.

RAP must be comprehensively characterized for accurate formulation of the mix and assurance of quality. To accomplish this, it is necessary to gather samples from different locations within the RAP stockpile using the back-dragging approach. This aids in understanding the variations and material properties. While evaluating the binder concentration and aggregate gradation might suffice for mixes with low RAP content, a more detailed analysis for both aggregate and binder characteristics becomes more essential when dealing with mixes with high RAP content.

8 Summary

This investigation encompasses an examination of RAP utilization in road construction. The ensuing inferences can be made from the documented research:

(1) Much research has been conducted to integrate RAP into newly constructed pavement structures.
(2) Employing RAP substance in constructing newly constructed pavement structures can save considerable cost.
(3) As discussed in this technical brief, using RAP has various environmental and health benefits due to reduced greenhouse gas emissions.
(4) The performance of recycled mixtures can be compromised by the aggregate and binder from RAP. Consequently, it is essential to determine the properties of both RAP aggregate and RAP binder.
(5) When constructing asphalt pavement with substantial RAP content, it is common practice to use a rejuvenating substance. The selection and quantity of the rejuvenator are particularly critical.
(6) RAP management is the most important step to be followed for effective use of RAP.

References

1. U.S. Department of Transportation, Federal Highway Administration: 2018, August 23 Table HM-12 - Highway Statistics 2017 (2017). https://www.fhwa.dot.gov/policyinformation/statistics/2017/hm12.cfm
2. Zaumanis, M., Mallick, R.B., Frank, R.: 100% recycled hot mix asphalt: a review and analysis. Res. Conserv. Recycl. **92**, 230–245 (2014). https://doi.org/10.1016/j.resconrec.2014.07.007
3. Shen, J., Amirkhanian, S., Aune Miller, J.: Effects of rejuvenating agents on superpave mixtures containing reclaimed asphalt pavement. J. Mater. Civ. Eng. **19**(5), 376–384 (2007). https://doi.org/10.1061/(ASCE)0899-1561(2007)19:5(376)
4. Bennert, T., Ericson, C., Pezeshki, D.: Rejuvenating agents with RAP in hot mix asphalt (HMA). FHWA-NJ-2015-008 (2015)
5. National Academies of Sciences, Engineering, and Medicine: Use of Reclaimed Asphalt Pavement and Recycled Asphalt Shingles in Asphalt Mixtures. The National Academies Press, Washington, DC (2016). https://doi.org/10.17226/23641
6. National Academies of Sciences, Engineering, and Medicine: Improved Mix Design, Evaluation, and Materials Management Practices for Hot Mix Asphalt with High Reclaimed Asphalt Pavement Content. The National Academies Press, Washington, DC. https://doi.org/10.17226/22554
7. Federal Highway Administration Publications Research: 97148 User Guidelines for Waste and Byproduct Materials in Pavement Construction, FHWA-RD-97-148 (2008)
8. Ncube, A.T., Bobet, A.: Use of recycled asphalt FHWA/IN/JTRP-2021/14 (2021). https://doi.org/10.5703/1288284317316
9. Commonwealth of Pennsylvania: PennDOT recycling material fact sheet reclaimed asphalt pavement (RAP) (2020)
10. Roque, R., Yan, Y., Cocconcelli, C., Lopp, G.: Perform an investigation of the effects of increased reclaimed asphalt pavement (RAP) levels in dense graded friction courses. Florida Department of Transportation Research (2015)
11. Shirodkar, P., et al.: A study to determine the degree of partial blending of reclaimed asphalt pavement (RAP) binder for high RAP hot mix asphalt. Constr. Build. Mater. **25**(1), 150–155 (2010). https://doi.org/10.1016/j.conbuildmat.2010.06.045
12. Gencer, M.N., Gencer, S.E., Karapahin, M., Memis, M., Kunduraciodlu, G.: A new method for hot recycling of asphalt. In: 5th Euroasphalt & Eurobitume Congress (2012)
13. Arnold, J.C., Nölting, M., Riebesehl, G., Denck, C.: Unlocking the full potential of reclaimed asphalt pavement (RAP)–high quality asphalt courses incorporating more than 90% RAP, a case study, paper No. O5EE-187. In: 5th Eurasphalt & Eurobitume Congress, Istanbul (2012)
14. AASHTO M 323: Standard Specification for Superpave Volumetric Mix Design (2004)
15. ASTM D4887: Standard Practice for Preparation of Viscosity Blends for Hot Recycled Asphalt Materials BT - Standard Practice for Preparation of Viscosity Blends for Hot Recycled Asphalt Materials (2016). https://doi.org/10.1520/D4887_D4887M-11R21E01
16. ASTM D2172: Standard test methods for quantitative extraction of asphalt binder from asphalt mixtures (2017). https://doi.org/10.1520/D2172_D2172M-17E01
17. ASTM D2171: Standard test method for viscosity of asphalts by vacuum capillary viscometer (2018). https://doi.org/10.1520/D2171-07E01
18. EPA: Chapter 11: Mineral Products Industry, AP 42, Fifth Edition, Volume I (2004). https://www3.epa.gov/ttn/chief/ap42/ch11/index.html
19. Williams, B.A., Willis, J.R., Shacat, J.: Annual Asphalt Pavement Industry Survey on Recycled Materials and Warm-Mix Asphalt Usage: 2018, 9th Annual Survey (IS 138). National Asphalt Pavement Association, Greenbelt, Maryland (2019). https://doi.org/10.13140/RG.2.2.21946.82888

20. West, R.C.: Best Practices for RAP and RAS Management, QIP 129 (2015)
21. Dughaishi, H.A.: Encouraging sustainable use of RAP materials for pavement construction in Oman: a review. Recycling **7**(3), 35 (2022). https://doi.org/10.3390/recycling7030035
22. Milad, A., Taib, A.M., Ahmeda, A.G., Solla, M., Yusoff, N.I.M.: A review of the use of reclaimed asphalt pavement for road paving applications. J. Teknologi **82**(3) (2020). https://doi.org/10.11113/jt.v82.14320
23. Sharma, A., Ransinchung, G.D., Kumar, P., Kumar, D.S.N.V.A.: Design mixture of RAP-HMA pavements: a review. In: IOP Conference Series: Materials Science and Engineering, vol. 1075 (2021). https://doi.org/10.1088/1757-899X/1075/1/012025
24. Kuang, D., Yu, J., Chen, H., Feng, Z., Li, R., Yang, H.: Effect of rejuvenators on performance and microstructure of aged asphalt. J. Wuhan Univ. Technol. **29**(2), 341–345 (2014). https://doi.org/10.1007/s11595-014-0918-3
25. Shen, J., Amirkhanian, S., Tang, B.: Effects of rejuvenator on performance-based properties of rejuvenated asphalt binder and mixtures. Constr. Build. Mater. **21**(5), 958–964 (2007). https://doi.org/10.1016/jconbuildmat.2006.03.006
26. Mogawer, W.S., Booshehrian, A., Vahidi, S., Austerman, A.J: Evaluating the effect of rejuvenators on the degree of blending and performance of high RAP, RAS, and RAP/RAS mixtures. Road Mater. Pavement Des. **14**(sup2), 193–213 (2013). https://doi.org/10.1080/14680629.2013.812836
27. Zaumanis, M., Mallick, R.B., Frank, R.: Evaluation of different recycling agents for restoring aged asphalt binder and performance of 100% recycled asphalt. Mater. Struct. **48**, 2475–2488 (2015). https://doi.org/10.1617/s11527-014-0332-5
28. Chen, M., Xiao, F., Putman, B., Leng, B., Wu, S.: High temperature properties of rejuvenating recovered binder with rejuvenator, waste cooking and cotton seed oils. Constr. Build. Mater. **59** (2014). https://doi.org/10.1016/j.conbuildmat.2014.02.032
29. Im, S., Zhou, F., Lee, R., Scullion, T.: Impacts of rejuvenators on performance and engineering properties of asphalt mixtures containing recycled materials. Constr. Build. Mater. **53** (2014). https://doi.org/10.1016/j.conbuildmat.2013.12.025
30. Nayak, P., Sahoo, U.C.: A rheological study on aged binder rejuvenated with Pongamia oil and composite castor oil. Int. J. Pavement Eng. **53**, 596–603. https://doi.org/10.1080/10298436.2015.1103851
31. Mogawer, W.S., Austerman, A., Roque, R., Underwood, S., Mohammad, L., Zou, J.: Ageing and rejuvenators: evaluating their impact on high RAP mixtures fatigue cracking characteristics using advanced mechanistic models and testing methods. Road Mater. Pavement Des. **16**(sup2), 1–28. https://doi.org/10.1080/14680629.2015.1076996

Construction Materials Durability: A Review

Yasser Tajmout[1,2(✉)], Safae Merzouk[2], Brahim Elbhiri[2], and Aniss Moumen[1]

[1] Laboratory of Engineering Sciences, National School of Applied Sciences, Ibn Tofaïl University, Kenitra, Morocco
tajmout.yasser@gmail.com
[2] SMARTiLab, Moroccan School of Engineering Sciences (EMSI), Corner of Zerhouni and Ejoukak - Aid Brahim Souissi, Rabat, Morocco

Abstract. The study, which is presented in this paper, uses bibliographic analysis of a number of indexed documents to look at a variety of factors, such as the number of publications, authors, nations, institutions, and journals involved in research on the durability of building materials. Since 2012, there have been more pertinent documents reviewed, which is a reflection of the growing significance and interest in this crucial area of study around construction materials durability. It also lists the authors, nations, institutions, and publications that have been referenced the most in the area. The evaluation also looks at the most popular research subjects as well as the keywords that are utilized in searches. Citation analysis seeks to discover major contributors to this literature, including notable researchers and materials. Finding the leading fields in this investigation is the goal of author co-citation analysis. This bibliometric analysis covers both the quantitative and qualitative components of research output, deepening our understanding of the building materials durability research landscape and its changing dynamics.

Keywords: Durability · Materials · Bibliometric Study

1 Introduction

The area of building materials durability has received substantial attention in recent decades because of the vital role in assuring the lifespan and performance of constructed structures [1]. Understanding the trends, accomplishments, and future directions of research in this topic is critical for encouraging innovation, overcoming obstacles, and furthering the discipline. This work gives a thorough bibliometric evaluation of research on the durability of construction materials over the last three decades, with the goal of documenting and summarizing the important results and advancements during the last three decades.

The major goal of this study is to assess and synthesize current literature, shed light on the growing research environment, and identify significant authors, nations, institutions, and journals participating in this topic. This study uses bibliographical analysis of a large number of indexed articles to identify patterns, trends, and variations in research interest across time [1, 2].

A. S. Mosallam et al. (Eds.): ICATH 2023, SCI, pp. 92–100, 2024.
https://doi.org/10.1007/978-3-031-76557-5_8

To accomplish this goal, the review goes into numerous aspects of the study domain. It evaluates publication output, charting the growth and distribution of research through time, and identifies prominent authors, nations, organizations, and journals that have made substantial contributions to the area. Furthermore, the study intends to find relevant papers and authors who have played essential roles in creating the discourse on building material durability using citation analysis [3, 5].

This review examines the qualitative components of the literature as well as the quantitative aspects. It aims to discover the primary areas of investigation within the field of building materials durability by investigating the most commonly used keywords and recurrent research themes. This study will give significant insights into the present emphasis and research goals, assisting academics and practitioners in understanding the most pressing concerns as well as prospective routes for future research.

Finally, this extensive bibliometric study provides an overview of the research environment on building materials durability, capturing the advances made over the previous years. This study intends to enlighten academics, policymakers, and industry experts alike by integrating current information and identifying research gaps, allowing evidence-based decision-making, and paving the way for future research initiatives in this critical subject.

2 Methodology

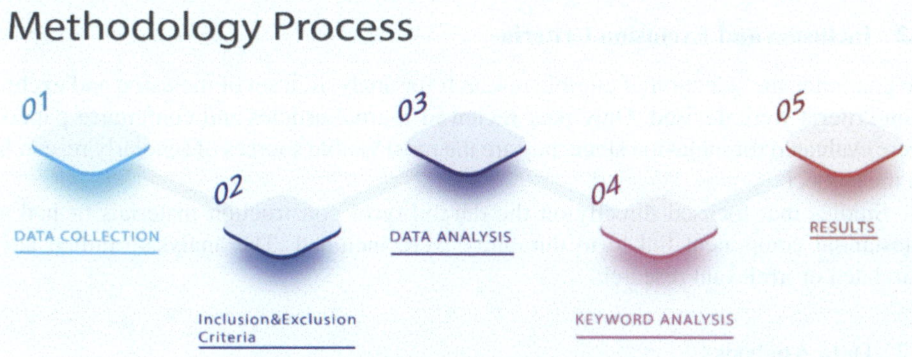

Fig. 1. Methodology process chart flow

2.1 Data Collection

The initial stage in conducting this bibliometric study was to obtain pertinent data from multiple databases i.e., Scopus, Web of Science, ScienceDirect and Springer, which are well-known for their extensive coverage of scientific publications. They were chosen as the major data sources because of their vast indexing of scholarly publications and conference proceedings in a variety of areas, including the durability of construction materials (Fig. 1).

The search was carried out using appropriate keywords linked to the durability of construction materials, assuring the inclusion of relevant research published between 1990 and 2023. The search approach was intended to return a complete set of papers that covered the scope of field study. This paper selected the previously mentioned databases, to download bibliographic records for analysis. They were chosen as one of the largest citation and abstract databases of peer-reviewed journals.

The following terms were entered to search the databases: "construction material", durability, material, and sustainability. In addition to the terms, an advanced search was set through a set of rules, regarding subject area, type of document, etc. The query used for Scopus was: TITLE-ABS-KEY ("construction material" AND durability AND material) AND (PUBYEAR > 1989) AND (LIMIT-TO (SUBJAREA, "ENGI") OR LIMIT-TO (SUBJAREA, "MATE") OR LIMIT-TO (SUBJAREA, "ENVI")) AND (LIMIT-TO (DOCTYPE, "ar") OR LIMIT-TO (DOCTYPE, "cp") OR LIMIT-TO (DOCTYPE, "re") OR LIMIT-TO (DOCTYPE, "ch") OR LIMIT-TO (DOCTYPE, "bk")).

Conducted on March 8, 2023, the search collected 2483 initial records from Scopus, 473 initial records from ScienceDirect, around 70 initial records from Springer, and around 226 initial records from Web of Science. Some of the initially collected records were published in irrelevant disciplines, such as computer science or social sciences. A screening permitted to filter these records. Finally, 1998 articles were selected covering a period of 33 years; i.e., the oldest record dated 1990. Each bibliographic record contained the metadata of the article, i.e., title, authors, abstract, …, list of references and DOI.

2.2 Inclusion and Exclusion Criteria

To guarantee the selection of eligible research for analysis, a set of inclusion and exclusion criteria were devised. Only peer-reviewed journal articles and conference papers were evaluated for inclusion since they are the most visible sources of scholarly research in the subject.

Studies that focused directly on the durability of construction materials or had a substantial component linked to durability were included. The analysis omitted any unrelated or irrelevant research.

2.3 Data Analysis

Quantitative Analysis: The selected documents were analyzed quantitatively to identify patterns and trends in research output. Various bibliometric indicators, such as publication counts, publication growth rates over time, authorship patterns, institutional affiliations, and countries of origin, were examined. The analysis aimed to provide a comprehensive overview of the publication landscape in the field of construction materials durability.

Citation Analysis: Citation analysis was conducted to identify highly cited documents and influential authors. The number of citations received by individual documents and authors was examined to gauge their impact on the field. Co-citation analysis was employed to identify clusters of related studies and prominent research themes within the construction materials durability literature.

2.4 Keyword Analysis

A keyword analysis was conducted to identify the most frequently used terms within the selected documents. This analysis aimed to identify the core concepts and research topics that have received significant attention in the field. The frequency and distribution of keywords were examined to gain insights into the prevailing research interests and priorities.

2.5 Research Trends and Future Directions

The final step of the methodology involved analyzing the research trends observed over the 30-year period. By synthesizing the findings from the quantitative and qualitative analyses, the study aimed to identify key developments, research gaps, and emerging areas of interest within the field of construction materials durability. Based on these findings, the paper proposes future research directions and areas that warrant further investigation.

3 Bibliometric Techniques and Tools

The downloaded dataset was analyzed using five bibliometric techniques: First is the study of the number of publications, which examines how the number of publications has changed through time; Second is the keyword co-occurrence analysis essentially identifies keywords in documents based on their co-occurrences; Third is the co-citation patterns across documents are produced via bibliographic coupling analysis; Fourth is the direct citation analysis, specifically focusing on citations between journals; And lastly co-authorship study, which reveals the trends in author cooperation across nations and regions.

The first technique is used to identify trends in the emergence of publications on construction material durability; the second and third techniques are used to illustrate the body of knowledge surrounding the durability of construction materials; the fourth technique is used to determine the importance of journals for publications on construction material durability; and the fifth technique is used to identify the collaboration network in the community of researchers studying the durability of construction materials.

VOSviewer is a commonly used tool to build networks of articles, journals, researchers, organizations, nations, and keywords. VOSviewer was deemed an appropriate tool for this investigation as a result.

4 Results and Discussion

Figure 2 shows the number of publications in construction materials durability studies over time. The first papers were published in 1990 by Lasisi in the journal of Building and Environment. From Fig. 2. We documented a minor growth in publications between 1990 to 2009, followed by second phase of variant publication numbers between 2010 to 2015, and this was followed by an important expansion of research in subsequent years starting from 2016.

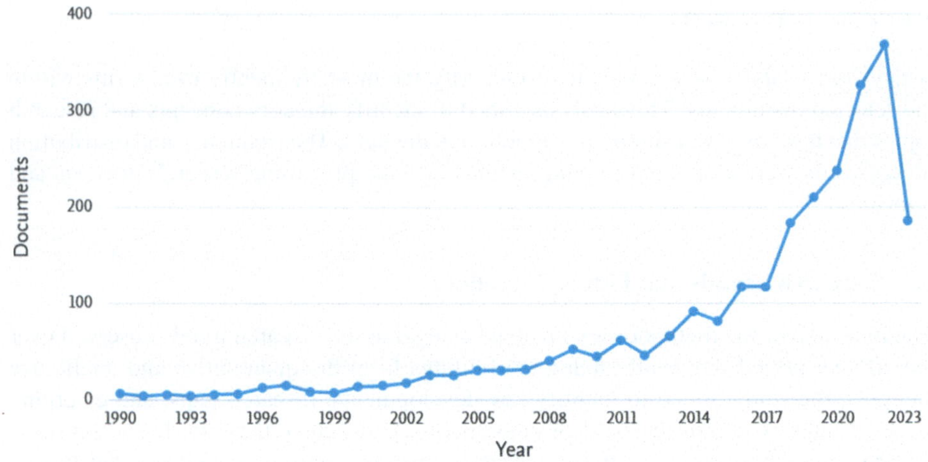

Fig. 2. Number of publications in construction materials durability (1990–2023)

Main Research Themes (Keyword Co-Occurrence Analysis). Authors in scientific research typically employ keywords to succinctly explain their primary focuses and areas. The keywords may be mapped in terms of occurrence patterns and connections

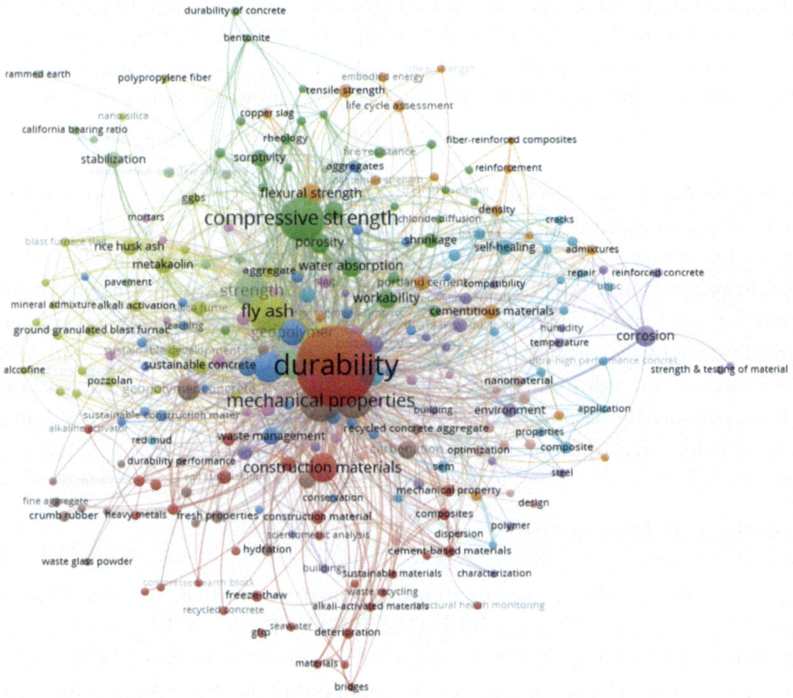

Fig. 3. Mapping of Keyword's co-occurrence analysis

utilizing keyword co-occurrence analysis. In an effort to identify the key subject matter in building material durability studies, this approach was adopted.

Figure 3 shows the visualization of keyword's co-occurrence analysis which contains 11 main clusters, with 2290 links and a link strength of 4141.

From Fig. 3. We found that the durability aspect and the properties of construction materials are tightly linked, so in terms of working on durability we need to focus our review on these properties and their effect on the materials used in construction today.

Construction material durability keyword co-occurrence analysis offered us important insights into the elements and ideas surrounding this subject. We may better comprehend the main issues and arguments around the durability of construction materials by recognizing the frequent use of terminology and their links. A study of the mapping is given below:

Construction materials are the primary focus of the analysis. It serves as the cornerstone for the keyword co-occurrence mapping. The focus will mostly be on various sorts of building materials.

The main characteristic under examination is durability. It speaks to the capacity of building materials to tolerate deterioration, weathering, and wear over time. This term stands for the analysis primary goal. Durability & strength are strongly related, as materials with higher strengths typically have longer lifespans.

Resistance: The ability of a material to survive a certain factor, such as moisture, corrosion, fire, impact, or chemical exposure, is referred to as resistance.

Maintenance: Maintenance entails tasks necessary to sustain and prolong the life of building materials. It implies that using good maintenance techniques might increase durability.

Life Cycle: The processes that a building material goes through, from extraction to disposal, are referred to as the life cycle.

Examining the co-occurrence of durability and a term can reveal sustainable and durable products. Testing: Testing is the process of determining if a building material will hold up over time using either laboratory testing or realistic simulations. It stresses how crucial it is to consider durability before use.

Design: When assuring the endurance of construction materials, design factors are important. Structural integrity, load-bearing capability, and material selection are just a few design-related buzzwords.

Climate: Keywords that refer to the climate draw attention to how the environment affects the durability of materials. Temperature, humidity, UV exposure, and extreme weather occurrences are examples of these concepts.

The interdependencies and contextual importance of the co-occurring keywords may be discovered by analyzing their connections. For instance:

Strength and durability: Stronger materials are expected to have more durability, according to the link between strength and durability.

Resistance and durability: Materials with higher levels of resistance are often more durable since they can withstand external pressures and degradation processes.

Maintenance and robustness: The inclusion of maintenance in the mapping shows that appropriate maintenance and repair can favorably affect the robustness and lifetime of building components.

Durability and life cycle: Analyzing the life cycle of a material can reveal information about its durability and long-term sustainability.

Testing and long-term performance: The existence of testing shows how crucial it is to assess materials' long-term performance before construction.

Design and durability: Taking durability into account during the design process may help choose the right materials and structural components to increase lifetime.

Climate and durability: The link between climate-related variables and robustness emphasizes the need of taking into consideration local environmental characteristics when choosing materials for certain places.

In conclusion, comprehending the essential ideas and elements relating to the durability of building materials may be fully grasped by examining the mapping of keyword co-occurrence. It demonstrates the significance of considering and obtaining lasting building materials in terms of strength, resistance, maintenance, life cycle, testing, design, and climate.

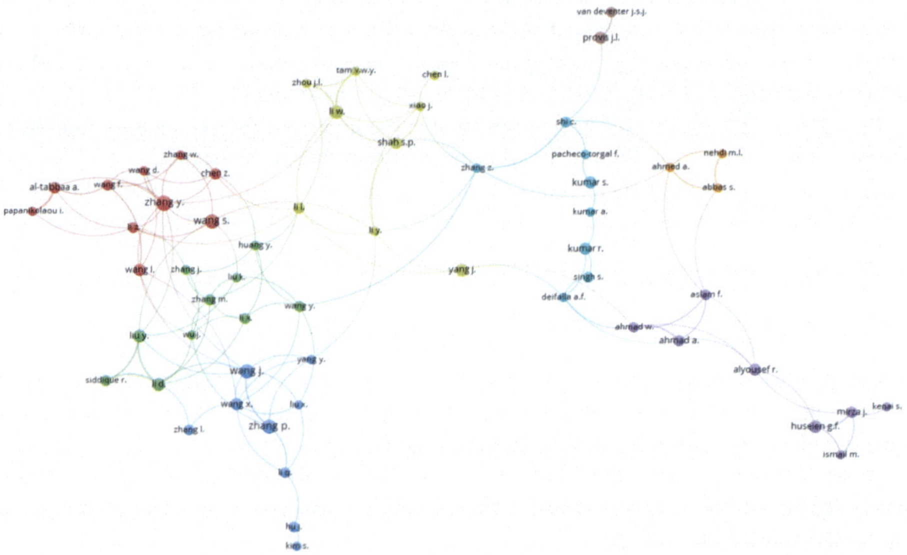

Fig. 4. Mapping of co-authorship analysis

Figure 4 shows the visualization of co-authorships analysis which contains 8 main clusters, with 126 links and a link strength of 175.

Figure 5 shows the visualization of co-occurrence analysis by country which contains 6 main clusters, with 482 links and a link strength of 1045. The countries showing the highest interest in the subject matter are the United States, India and China. Morocco is linked to Spain, Portugal, France and Australia, producing 7 documents of research on the subject.

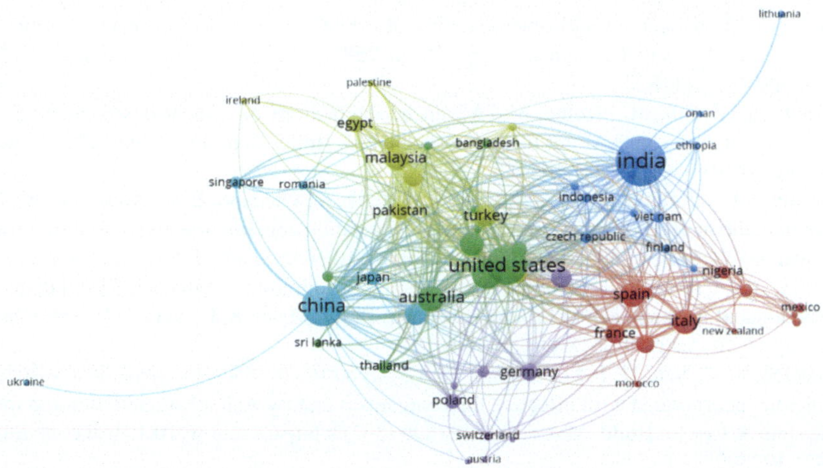

Fig. 5. Country co-occurrence map

5 Conclusion

This bibliometric evaluation offers a thorough overview of the body of knowledge on the durability of construction materials. This report highlights the major issues, trends, and developments in the subject through a methodical examination of several research publications. Insights into the most significant authors, journals, and academic institutions may be gained through the synthesis of bibliometric data, which also highlights the networks of collaboration and the flow of information within the field.

Additionally, the assessment suggests new fields of study that point the way for future research into things like cutting-edge testing techniques, sustainable practices, and novel materials. Overall, this bibliometric study is a useful tool for encouraging a greater understanding of the durability of building materials and directing the course of future research projects in this important area.

References

1. Zheng, L., Chen, K., Lu, W.: Bibliometric analysis of construction education research from 1982 to 2017. J. Prof. Issues Eng. Educ. Pract. **145**(3) (2019). https://doi.org/10.1061/(asce)ei.1943-5541.0000412
2. Udomsap, A.D., Hallinger, P.: A bibliometric review of research on sustainable construction, 1994–2018. J. Cleaner Prod. **254**, 120073 (2020). https://doi.org/10.1016/j.jclepro.2020.120073
3. Omrany, H., Chang, R., Soebarto, V., Zhang, Y., Ghaffarianhoseini, A., Zuo, J.: A bibliometric review of net zero energy building research 1995–2022. Energy Build. **262**, 111996 (2022). https://doi.org/10.1016/j.enbuild.2022.111996
4. Soliman, A.M., et al.: Innovative construction material technologies for sustainable and resilient civil infrastructure. Mater. Today Proc. **60**, 365–372 (2022). https://doi.org/10.1016/j.matpr.2022.01.248

5. Li, Y., Li, M., Sang, P.: A bibliometric review of studies on construction and demolition waste management by using CiteSpace. Energy Build. **258**, 111822 (2022). https://doi.org/10.1016/j.enbuild.2021.111822
6. Makul, N., Fediuk, R., Szeląg, M.: Advanced interactions of cement-based materials with microorganisms: a review and future perspective. J. Build. Eng. **45**, 103458 (2022). https://doi.org/10.1016/j.jobe.2021.103458
7. Ibrahim, M., Alimi, W., Assaggaf, R.A., Salami, B. A., Oladapo, E.A.: An overview of factors influencing the properties of concrete incorporating construction and demolition wastes. Constr. Build. Mater. **367**, 130307. https://doi.org/10.1016/j.conbuildmat.2023.130307
8. Lewry, A.J., Crewdson, L.: Approaches to testing the durability of materials used in the construction and maintenance of buildings. Constr. Build. Mater. **8**(4), 211–222 (1994). https://doi.org/10.1016/s0950-0618(09)90004-6
9. Huseien, G.F., Sam, A.R.M., Algaifi, H.A., Alyousef, R.: Development of a sustainable concrete incorporated with effective microorganism and fly Ash: characteristics and modeling studies. Constr. Build. Mater. **285**, 122899 (2021). https://doi.org/10.1016/j.conbuildmat.2021.122899
10. Kim, I.S., Choi, S.Y., Yang, E.I.: Evaluation of durability of concrete substituted heavyweight waste glass as fine aggregate. Constr. Build. Mater. **184**, 269–277 (2018). https://doi.org/10.1016/j.conbuildmat.2018.06.221
11. Korenkova, S., Sidorenko, Y.: Improving durability of cement composite materials. Procedia Eng. **111**, 420–424 (2015). https://doi.org/10.1016/j.proeng.2015.07.110
12. Osareh, F.: Bibliometrics, Citation Analysis and Co-Citation Analysis: A Review of Literature I. Libri **46**(3) (1996). https://doi.org/10.1515/libr.1996.46.3.149
13. Robichaud, L.B., Anantatmula, V.S.: Greening project management practices for sustainable construction. J. Manag. Eng. **27**(1), 48–57. https://doi.org/10.1061/(asce)me.1943-5479.0000030
14. Santos, R., Costa, A.A., Grilo, A.: Bibliometric analysis and review of building information modelling literature published between 2005 and 2015. Autom. Constr. **80**, 118–136 (2017). https://doi.org/10.1016/j.autcon.2017.03.005
15. Tan, Y., Shen, L., Yao, H.: Sustainable construction practice and contractors' competitiveness: a preliminary study. Habitat Int. **35**(2), 225–230. https://doi.org/10.1016/j.habitatint.2010.09.008
16. Imani, M., Donn, M., Balador, Z.: Bio-inspired materials: Contribution of biology to energy efficiency of buildings. In: Martínez, L., Kharissova, O., Kharisov, B. (eds) Handbook of Ecomaterials. Springer, Cham (2019). https://doi.org/10.1007/978-3-319-68255-6_136
17. Ball, R., Tunger, D.: Bibliometric analysis - a new business area for information professionals in libraries? Scientometrics **66**(3), 561–577 (2006). https://doi.org/10.1007/s11192-006-0041-0
18. Hosseini, M.R., Martek, I., Zavadskas, E.K., Aibinu, A.A., Arashpour, M., Chileshe, N.: Critical evaluation of off-site construction research: a Scientometric analysis. Autom. Constr. **87**, 235–247 (2018). https://doi.org/10.1016/j.autcon.2017.12.002
19. Rawat, P., et al.: A state-of-the-art review on mechanical performance characterization and modelling of high-performance textile reinforced concretes. Constr. Build. Mater. **347**, 128521 (2022). https://doi.org/10.1016/j.conbuildmat.2022.128521

Review in the Behavior of Ferro-Cement in Construction Technology

Taha Elsayed[1](✉) and Ahmed F. Deifalla[2](✉)

[1] Department of Structural Engineering, Benha University's Shoubra Faculty of Engineering, Cairo, Egypt
Taha.ibrahim@feng.bu.edu.eg

[2] Department of Structural Engineering and Construction Management, Future University in Egypt, New Cairo City, Egypt
ahmed.deifalla@fue.edu.eg

Abstract. Ferro cement is a crucial enhancement in civil engineering field. According to its light weight, few loads will be added upon the structures. This paper will present various properties of Ferro cement such as: impact resistance, cracking, tensile behavior, compression, and fire resistance. Based on previous performed tests, different types of applications in addition to its benefits will be specified sequentially.

Keywords: Ferro cement · cement · tensile behavior · repairing materials · light weight concrete

1 Introduction

Ferro cement was improved by European since 1848, Joseph Louis Lambot created various rowing boats, seats, plant pots and other items [1]. Ferro cement is a complex of ferro (iron) and cement (cement mortar). It can be utilized as a kind of thin-walled RC construction in which massive numbers of wire mesh which uniformly distributed through the cross section instead of the reinforced bars as shown in Fig. 1 [2]. Sustainable materials or steel can be used to made the mesh used as reinforcement. Thus, the result of the thin composite material, higher fraction will be existed of steel rather than the normal reinforced concrete. The maximum cover for this reinforcing within 2 mm to 5 mm [3]. Recently, its applications can be found around the world for example, in Malaysia, low-cost houses were constructed by prefabricated ferro cement wall panels. In Jordon and Indonesia, mosques domes where established. In addition to unique buildings are found with this unique technology in India. Ferro cement material demonstrates to have high qualities of impact resistance, crack control and toughness as well, because of the short spacing and the uniform distribution of reinforcement. It can be designed with a high quality, cost, and properties according to the customer's budget. Ferro cement consists of mortar matrix which has been identified by IFS committee 10, like a homogeneous mixture of cement, fine aggregate, and water. Some additives can be added as an example, silica ash, fly ash and super plasticizer to improve its properties [4]. There are some

A. S. Mosallam et al. (Eds.): ICATH 2023, SCI, pp. 101–118, 2024.
https://doi.org/10.1007/978-3-031-76557-5_9

recommendations should be followed for mix properties [5]. For wire mesh, there are manufacturing process produce according to both ASTM C 847 and ASTM A 185 [6].

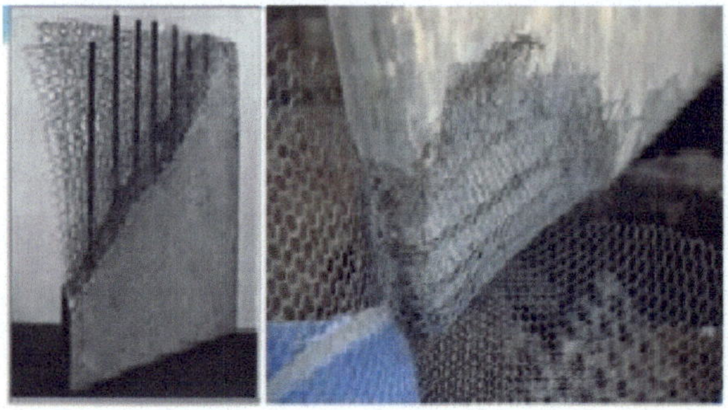

Fig. 1. Cross section of ferrocement

2 Axial Compression Behavior

An experimental program created for nine squared columns (200 mm x 200 mm x 1500 mm) under compression load with 25 mm eccentricity from the center in direction x. After 28 days, the target strength used to design the mix of the concrete was 35 MPa. Specimens classified as shown in both Table 1 and Figs. 2 & 3. The results as shown in Table 2, in addition to a comparison between the experiment load as shown in Fig. 4. Furthermore, there is an extended data about the proportion of enhancement in load

Table 1. Types of tested specimens [7]

Groups	Specimen Id	Description
Group I	Cc	Control specimen reinforced by 4 φ 12 longitudinal bars and 11 φ 8 stirrups
Group II	C1-E	One layer expanded + 4 12 + 6 φ 6 stirrups
	C2-E	Two layers expanded + 4 12 without stirrups
	C3-E	Three layers expanded + 4 12 without stirrups
Group III	C4-W	One layer welded + 4 12 + 6 φ 6 stirrups
	C5-W	Two layers welded + 4 φ 12 without stirrups
	C6-W	Three layers welded + 4 φ 12 without stirrups
Group IV	C7-ME	One layer of modified and galvanized expanded mesh + 4 φ 12 without stirrups
	C8-MW	One layer of modified and galvanized welded mesh + 4 φ 12 without stirrups

and the first crack load in the Figs. 7 and 8 respectively [7]. Another experiment using silica fume, ground granulated blast furnace (GGBS), metakaolin, fly ash and other cementitious materials, is created. It used to minimize the cost by utilizing such materials with low cost. Some different materials of mesh where used. Nine columns of geopolymer (150 mm × 150 mm × 1600 mm) with various volume-fraction and layers. The properties of the mesh as shown in Table 3. The designed mix of (HSC) was used after 28 days to target 60 MPa as follow per m3, 400 kg RSA, 1150 kg coarse aggregate, 650 kg sand, 47 kg water, 50 kg sodium hydroxide (NaOH), and 150 kg sodium meta- silicate (Na_2 SiO3). The whole samples described as in table 5. And the details of the samples in Fig. 7. The test is sat up by using a compression test machine within capacity up to 5000 kN as shown in Fig. 8. The measurements of deflection were taken by using LVDT which tested till failure. The results can be shown in Table 5.

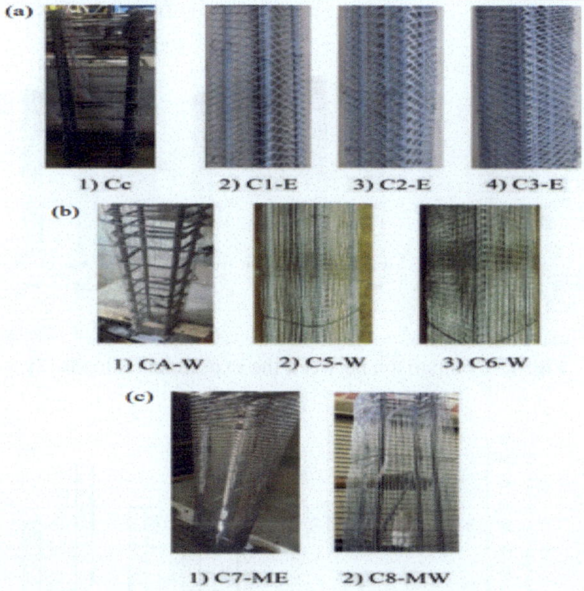

Fig. 2. Experimental specimens before casting [7]

3 Shear Behavior

The crack pattern in ferro-cement depends on the kind of its bond between matrix wire and mortar. First cracking is the most important factor in ferro-cement. The wide of the first crack within 0.005 mm until 0.1 mm depends on the method of noticed. The RC specific surface area (SSA) increase leading to crack increases [9]. Figure 9 shows the dimensions and reinforcement details of 7 tested beams in this program. The control specimen was box cross section with 2Ø 12 for the tension side and 2Ø 10 for the compression in addition to, 13Ø 6 as stirrups. As for the other 6 beams were without

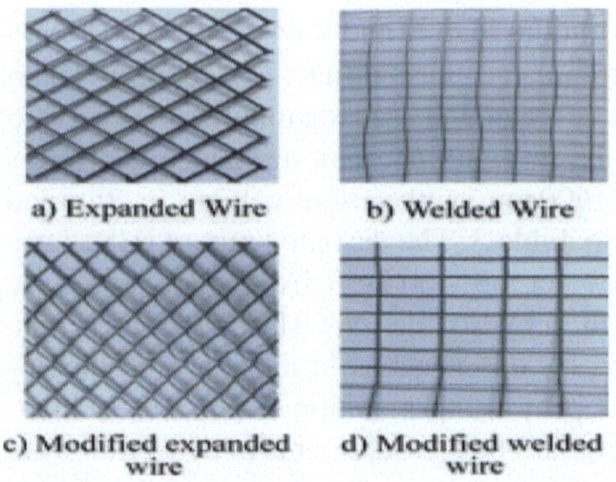

Fig. 3. Configurations of composite materials [7]

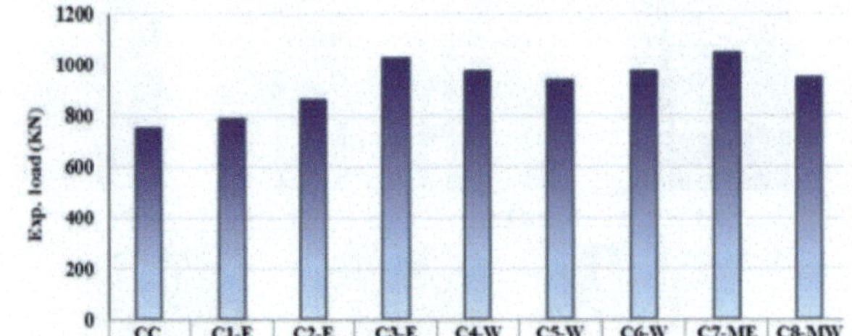

	CC	C1-E	C2-E	C3-E	C4-W	C5-W	C6-W	C7-ME	C8-MW
Exp. load	753.8	787.5	860	1030	975	939.9	975	1048	950

Fig. 4. Comparison between the experimental loads [7]

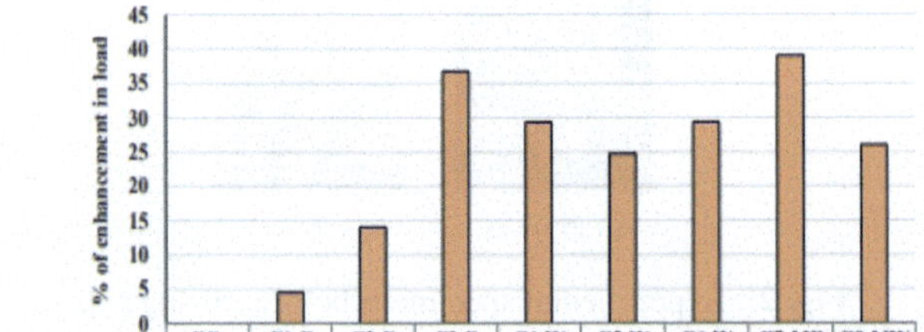

	CC	C1-E	C2-E	C3-E	C4-W	C5-W	C6-W	C7-ME	C8-MW
%enhancement		4.5	14	36.7	29.3	24.7	29.3	39	26

Fig. 5. Comparison between the percentage of experimental failure load enhancement [7]

stirrups, but glass fibers used instead of using stirrups. The experiment is divided into two groups. For the first one, different numbers of layers were used as one, two, three layers for beam 1, beam 2, and beam 3 respectively. While the next group was used tensar wire mish as stirrups and using one layer, two and three for beams 1,2, and 3 as shown in table 6. The results were as shown in Table 7 and in Fig. 10.

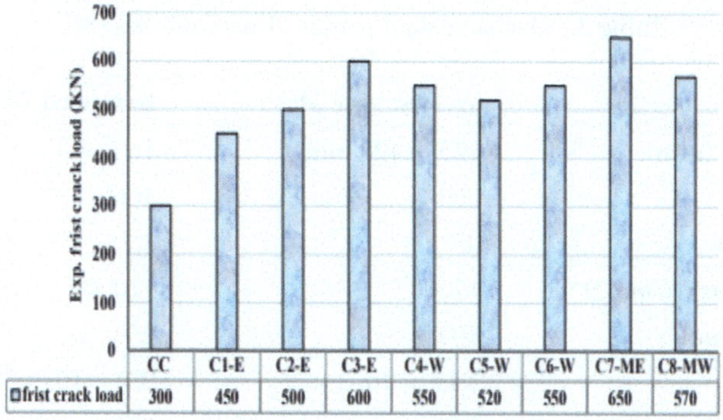

Fig. 6. Comparison between the first crack loads in each group [7]

Table 2. Types of tested specimens [7]

Groups	Specimen Id	Experimental failure L.d (kN)	Experimental produced moment (kN.m)	Load- carrying capacity enhancement (%)	Experimental moment enhancement (%)
Group I	Cc	733.8	18.9	-	-
Group II	C1-E	787.5	19.7	4.5	4.2
	C2-E	860	21.5	14	13.8
	C3-E	1030	25.8	36.7	36.5
Group III	C4-W	975	24.4	29.3	29.1
	C5-W	939.9	23.5	24.7	24.3
	C6-W	975	24.4	29.3	29.1
Group IV	C7-ME	1048	26.2	39	38.6
	C8-MW	950	23.8	26	26

4 Durability

The durability of each material relies on the kid of surrounded environment. The best performance of each material based on its durability as the same as strength properties. More invincible the mortar layer of ferro-cement is more durable in comparison with normal concrete cover. However, ferro-cement seems a thin RC, which durability is affected by the following factors [10].

- Because the cover in ferro-cement is exceptionally low, corrosive chemicals can easily penetrate the reinforcement.
- In comparison to reinforced concrete, ferro-cement has a larger volume of tiny diameter wire mesh. The SSA of the wire mesh is particularly big, offering a broad area of contact for corrosion to occur, perhaps boosting the rate of corrosion in it.

Table 3. Mechanical properties of wire-meshes [8]

Properties	Welded Wire Mesh	Expanded Wire Mesh
Dimensions size (mm)	125 × 125 mm	16.5 × 3 mm
weight (gm/m^2)	600	1660
Thickness (mm)	0.7	1.25
Young's Modulus (N/mm2)	17000	12000
Yield stress (N/mm^2)	400	250
Yield Strain	1.17 × 10–3	9.7 × 10–3
Ultimate Strength	600	380
Ultimate Strain	0.0585	0.0592

Table 4. Samples description [8]

Series	Sample ID	Sample Description	Volume of fraction	RFT.	Stirrups
Control	C1	Control	-	4 φ 12	6 φ 8
Group A: Welded wire-mesh	C1-A	1-Layer welded	0.0027	5 φ 12	_
	C2-A	2-layers welded	0.0054	6 φ 12	_
	C3-A	3-layers welded	0.0081	7 φ 12	_
Group B: Expanded wire-mesh	C4-B	1-Layer expanded	0.00753	8 φ 12	_
	C5-B	2-Layers expanded	0.0151	9 φ 12	_
Group C: Tensar-mesh	C6-C	1-Layer Tensar	0.0204	10 φ 12	_
Group D: Fiber glass-mesh	C7-D	1-Layer fiber glass	0.00535	11 φ 12	_
	C8-D	2-Layers fiber glass	0.0107	12 φ 12	_

- Although the majority of reinforcing meshes used in galvanized ferro-cement are corrosion resistant; however, the zinc coating could cause negative consequences due to gas bubble production.

Using low water to cement ratio in the mortar of the ferrocement, could minimize probability of corrosion in and increase durability. The mortar's permeability and porosity can be decreased. The mortar compaction of should be well executed. Mineral additives and admixtures should also be used in Ferro-cement. A coating layer should be provided according to the condition [10].

Table 5. Experimental results [8]

Column ID	First crack load (kN)	Serviceability Load (kN)	Ultimate Load (kN)	Def. at first crack. Load (mm)	Def. at Ult. Load (mm)	Ductility Ratio	Energy Absorption (kN.mm)
C1	296	460.59	738.7	3.8	12.02	3.16	6320.08
C1-A	325	481.27	771.8	4.3	12.72	2.96	9201.48
C2-A	380	525.46	842.5	4.7	14.36	3.06	9353.37
C3-A	455	596.15	955.6	5.2	14.5	2.79	10380.91
C4-B	443	507.46	813.7	5.9	15.45	2.62	10122.72
C5-B	449	574.96	921.7	6.3	16.96	2.69	12008.52
C6-C	515	640.9	1027	4.4	13.55	3.08	10414.93
C7-D	391	496.65	796.4	3.9	12.08	3.1	6983.16
C8-D	378	524.71	841.3	4.5	14.07	3.13	9179.36

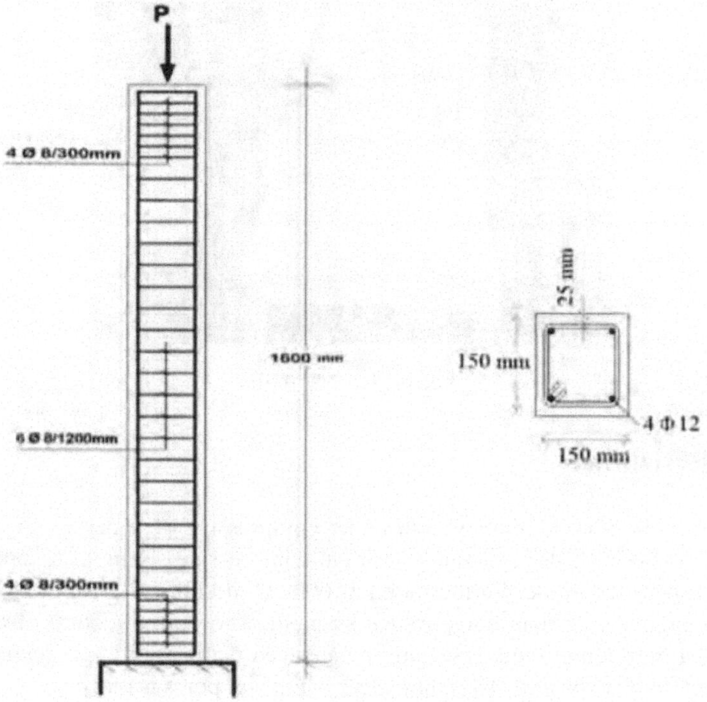

Fig. 7. Cross section Details of samples [8]

Fig. 8. Test setup [8]

5 Fire Resistance

When exposed to 600 °C, the strength under compressive of concrete might drop by up to half. At 800 °C, the concrete's residual compressive strength is approximately one fifth of its initial value. Furthermore, at 600 °C, traditional concrete loses tensile splitting strength faster than compressive strength. The pore structure of concrete is altered under high temprature, resulting in enhanced permeability but decreased durability. After 600 °C exposure, The cumulative volume of pores is twice that of traditional concrete. Pore structure coarsening and Extreme C-S-H gel dehydration lead to concrete strength reduction at temperatures above 600 °C. Fires in concrete structures seldom cause significant major structural damage, and the 95% of damaged structures may be

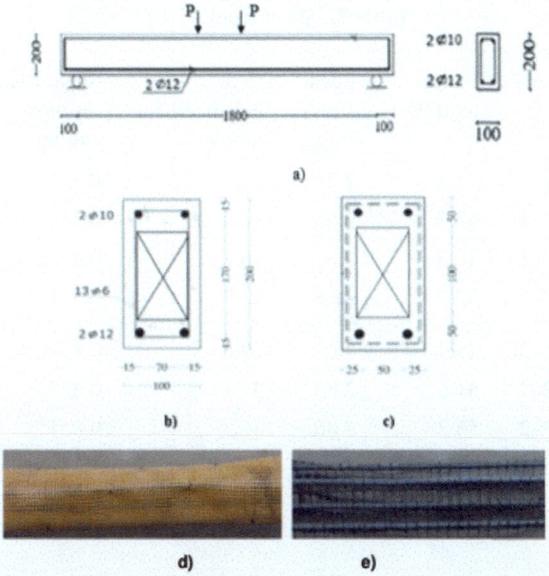

Fig. 9. Beams geometric shape and reinforcement details [9]

Table 6. Box beams specimen's descriptions [10]

Series	Specimen no.	Specimen description	Reinf. Tension	Reinf. Compression	Stirrups
Control	BOX 1	Control specimen	2 φ 12	2 φ 10	13 φ 6
Group 1 "Glass fiber wire mesh"	BOX 1–1	One-layer glass fiber	2 φ 12	2 φ 10	-
	BOX 2–1	Two-layer glass fiber	3 φ 12	2 φ 10	-
	BOX 3–1	Three-layer glass fiber	2 φ 13	2 φ 10	-
Group 2 "Tensar wire mesh"	BOX 1–2	One-layer tensar	4 φ 12	2 φ 10	-
	BOX 2–2	Tow-layers tensar	2 φ14	2 φ 10	-
	BOX 3–2	Three-layer tensar	5 φ 12	2 φ 10	-

successfully repaired. As a result, repairing concrete structures damaged due to fire is a more practical and cost-effective choice than demolishing and reconstructing [11].

Table 7. Shear results [10]

Series	Specimen no.	Failure load (kn)	Shear stress (mpa)	% of enhancement in load	Shear load ultimate strength	Deflection at failure load (mm)
Control	BOX 1	40.5	2.25	–	0.833	0.4
Group 1 "Glass fiber wire mesh"	BOX 1–1	45.7	2.53	12.8	0.83	0.29
	BOX 2–1	47.3	2.62	16.8	0.83	0.278
	BOX 3–1	50.2	2.78	23.9	0.831	0.25
Group 2 "Tensar wire mesh"	BOX 1–2	48.4	2.69	19.6	0.834	0.27
	BOX 2–2	51.6	2.86	27.4	0.832	0.25
	BOX 3–2	55.2	3.06	36.3	0.831	0.23

Table 8. Load capacity of beams [12]

Name of specimen	Ultimate load (kN)	Improvement over fire damage specimen (%)	Decline over control specimen (%)
Control beam	114	–	–
Fire damaged specimen	61	–	46
single layer ferrocement strengthened	89	46	22
Double layer ferrocement strengthened	105	72	8

The resistance of elements with concrete to fire is determined by parameters such as fire severity, element materials, geometry, and so on. The capacity of structural components to resist fire can be classified as fire resistance. Different results were achieved based on numerous experiments conducted on ferrocement by various researchers on its resistant potential under fire. The majority of the researchers felt that adding a jacket with Ferrocement on constructions increased resistance due to fire. Which is due to its slightly larger specific heat capacity than that of concrete cover, allowing it to absorb more heat than concrete cover? When exposed to extreme temperatures, the structural members get weakened, leading the structures to collapse. The experiments discovered that applying a thin coating of Ferrocement jacket reinforcing the components minimized cover spalling caused by mesh of wire used in reinforcing. Under normal conditions, increasing the Ferro-cement wire mesh content increases the toughness and the flexural. However, this effect is only before fire, while following fire exposure, it is no effect. Furthermore, when the wire mesh content increases, the insulating quality degrades,

Table 9. Pipes Descriptions [13]

Pipe ID	Pipe Type	Layers Number
P1	Control	-
P2	Control + Poly.fiber	-
P3	welded mesh	1
P4		2
P5		3
P6		4
P7	expanded mesh	1
P8		2
P9		3
P10	Fiber glass-mesh	1
P11		2
P12		3

Table 10. Pipes Descriptions [14]

	Load		Displacement			
Pipe Id	1st crack load pcr (n)	Ultimate load (n)	1st crack displacement δcr (mm)	Ultimate displacement δu (mm)	Ductility δu/δcr	Energy absorption (kn.mm)
P1	9	11.35	0.13	0.5	3.85	2.84
P2	9	14.58	0.06	0.56	9.33	4.08
P3	9	23.04	0.19	0.64	3.37	7.37
P4	9	24.3	0.03	0.72	24	8.75
P5	9	24.72	0.01	0.86	86	10.63
P6	9	28.44	0.02	1.02	51	14.5
P7	9	12.29	0.07	0.47	6.71	2.89
P8	9	16.07	0.15	0.59	3.93	4.74
P9	9	19.28	0.14	0.74	5.29	7.13
P10	9	27.74	0.02	0.65	32.5	9.02
P11	9	28.8	0.03	0.74	24.67	10.66
P12	9	31	0.02	0.79	39.5	12.25

owing to a decrease in the specific heat capacity of the ferrocement. Increased mortar coating. On the contrary, it resulted in enhanced insulating ability. Ferrocement behavior under fire is an ongoing area of research, and further trials are needed before concluding ferrocement qualities pertaining to fire resistance [10].

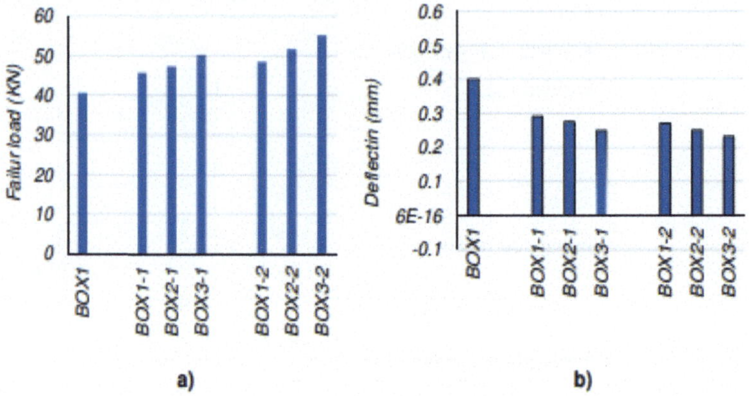

Fig. 10. A comparison between the results [10]

According to Al Amin et al., his experiment was casting rectangular beams 200 mm x 150 mm x 1200 mm, after 7 and 28 days of curing, the specimens exposed to fire and the results was as shown in Fig. 11. Plain concrete's flexural strength is nearly entirely based on its tensile strength. However, experiments reveal that the strength in tension is significantly outmatched by the rupture modulus. Concrete pavement design must take into account flexural strength. For the test of this specimen's flexural strength, a third point loading technique was used. A chamber is constructed using steel reinforcements, CI sheets, and bricks to heat the beams. Thermocouple and a digital meter are used to measure the temperature. The temperature in the building peaked at 805 °C during the fire, fluctuated between 700 and 800 °C for 2 h, and then fell to normal temp.. There are minor fractures and concrete spalling on the beams once they have cooled down. Due to the weakness of concrete under strain, fractures and spalling appeared in the the beams tensile region of that were being burned, rendering the concrete unusable. Within one inch, the fractures' depth is visible in the beam. As a result, the clear cover of the tension area) is removed, wire-brushed clean, and free of loose objects. After damaging the beams, the results were as shown in Table 8 [12].

The deformations started in the layers of ferro cement because of the strengthening action. The core failure happens after the layer failure. For concrete beams strengthened using double- and single-layer ferrocement, the fracture begins at the top and the base of the beam simultaneously. With an increase in load, fractures spread after being initiated. The presence of densely populated wire mesh prevents fractures from spreading quickly. In comparison to the fire-damaged specimens, both examples demonstrated a significant improvement in ultimate load bearing capability and delayed fracture development [12].

Fire resistance is the ability of a structural element's to resist fire. It is influenced by many parameters, including material, support condition, geometry and fire severity. Some of the most common fire resistance test standards such as ISO TR 834–3, BS 476–10, and ASTM E119–12 [23, 24]. Many countries use ISO TR 834–3:2012 [25],

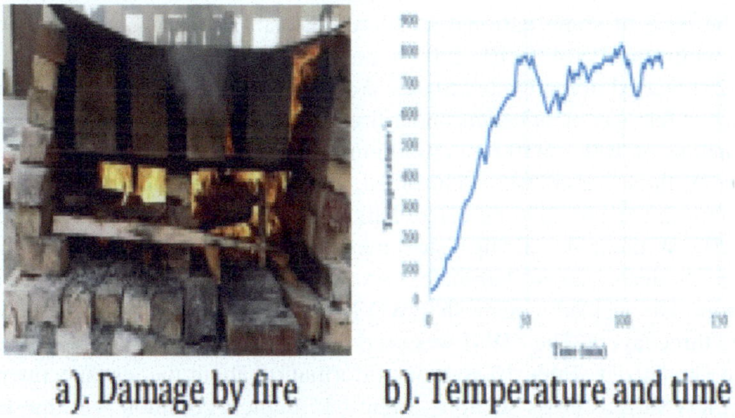

Fig. 11. The Process of beam damage [13].

Fig. 12. Cracking patterns for all pipes [14].

and it is the base of some national standards. Most European countries use similar standards, while the BS 476–10, 2009 standard is used by the British [24]. ASTM E119–12 [23] is used in the United States. According to Yousry et al., the strength and the behavior of ferro- cement beams under fire. To accomplish this goal, eighteen beams with dimensions of 100 mm100 mm1000 mm were tested after casting. As a standard reinforcement, three beams were reinforced. The control beam flexure reinforcement were 2D8 mm in tension, 2D6 mm steel bars in compression, and shear reinforcement of 6mm@200. Without any stirrups, steel mesh were added to the ferro-cement beams. To reinforce the ferro cement laminate, two types of wire mesh: expanded wire mesh (EWM) and square welded wire mesh (SWWM) were used. SWWM was used in single, double, and three layers. The EWM was used in single and double layers. In general, all meshes are galvanized. Table 10 contains information about the test specimens, while Fig. 14 depicts various cross sections. Figure 15 depicts common wire mesh used in ferro-cement applications [26].

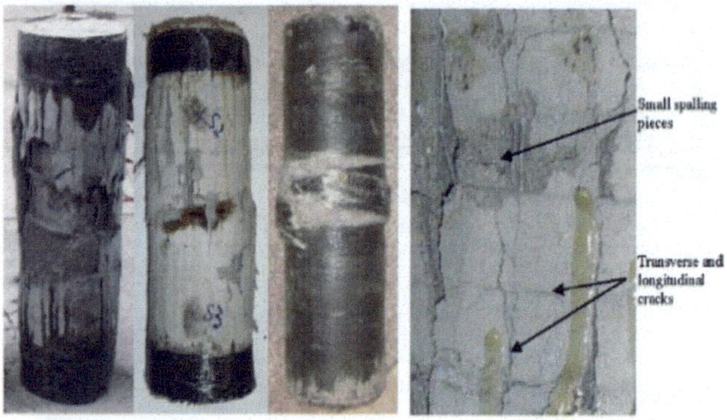

Fig. 13. ferrocement and FRP-strengthened circular columns [17].

6 Impact Resistance

During impact loading, the amount of absorbed energy is defined as impact resistance. In comparison to traditional reinforced concrete, ferro-cement is better at absorbing impact energy, making it more able to withstand impacts. Ferro-high cement's specific surface area and substantial volume of reinforcing account for its high impact energy. Energy absorption capacity due to impact rises with the addition of silica fume and fly ash to the mix. Impact resistance is also influenced by the ferro-thickness. Cement's Impact damage often only affects the area of contact and is localized. The impact may result in internal mortar layer spalling and mesh layer delamination. Due to its extensive use of wire mesh reinforcing and high impact energy, ferro-cement is susceptible to punching failure.

According to an experiment for Atta Shekh, A special steel frame was constructed in order to drop onto the simply supported panels, a load from a specified height. Inside

a- Effect of drooping load b- Place of panels inside the testing frame c- Impact testing frame

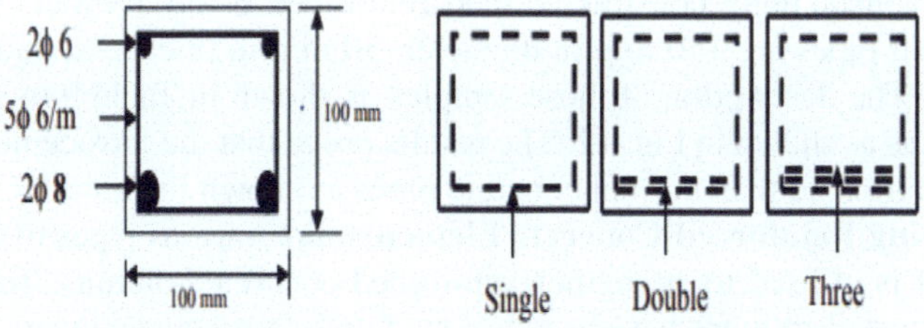

a) Control beams b) Beams with ferrocement form reinforced with single or double layers or three of steel mesh

Fig. 14. ferrocement and FRP-strengthened circular columns [17].

the impact frame, the ferro-cement specimen was placed on the supports. In addition, the steel ball was left to hit the surface of the panels vertically (by gravity), as shown in Fig. 14. After 28 days, all sandwich ferro-cement panels were tested and compared to control panels. After a 3.0 kg ball was dropped from a height of 1.30 m, the cracks were measured onto the center of the top surface of the panel. Repeating several times for the procedure resulted in the appearance of more new cracks on the surface, splitting of the layers of the panels, spalling of mortar and eventally collapse or failure [22].

7 Applications of Ferro-Cement

Ferro-cement is used in a broad variety of building projects. Due to the materials that make up the product being readily available and its incredibly adaptable character. It can be built with very little machinery and very little expert labour. Some of the uses for ferro-cement include such as: ferro-cement duct, ferro cement building, rehabilitation of structures, floor riles, containers and tanks, fire resistance structures, rural applications, floors & roofs, and Marine resistance structures [10].

- In 1975, MERI introduced the ferrocement subject. MERI officers at the time had already conducted some research. MERI has conducted numerous ferrocement tests and experiments. A water tank and partition wall was built out of ferrocement and are still on display. MERI used ferrocement to build two boats. The boats are still on display for viewing. Rainwater is collected in one of the boats, but there is no evidence of leakage. In Canal Lining, ferrocement lining was used as an experiment in 2014–15. Tapi Irrigation Development Corporation has completed the construction of a 100-m-long Punand Right Bank Canal near Kalwan. In this section of the canal, there was a lot of leakage. The leakage was eliminated after the ferrocement lining was stoppedcompletely [14].
- **Replacing steel water pipes by using Ferro-cement ones**. According to an experiment for Taha A. El-Sayed et al., the fundamental purpose of this research is to investigate the application of ferro-cement technology in the construction of water pipelines with qualities such as high deformation behavior, high strength, fracture resistance, high ductility, and elastic stiffness. Twelve ferro-cement pipes were manufactured, painstakingly constructed, and reinforced using a variety of composite materials. All pipes are 150 mm in diameter, 50 mm in thickness, and 1000 mm in length [13]. The description of these samples as shown in Table 9 in addition to its crack patterns as shown in Fig. 12. The results prove that the ferrocement pipes have a higher failure load than normal concrete pipes as shown in Table 15.
- **Strengthening Reinforced Concrete Elements**: For various types of inadequacies, ferrocement is utilised to strengthen reinforced concrete columns. To further limit the section, tested columns have been jacketed with ferrocement layers. Ferrocement typically ranges in thickness from 1.5 to 3 cm. To prevent opening of the mesh reinforcement, wire mesh is typically given with overlap. The wire mesh is then plastered with cement, let to cure, and tested. To increase the axial capacity of columns, better confinement of the current element is the mechanism used. According to studies by Kaishet al., Mourad et al., Kaish et al., Xiong et al., and, Kondraivendhan et al. This material has been shown to successfully increase the strength, stiffness, ductility, and cracking behaviour of deficient columns. Ferrocement acts as an external confinement element, preventing lateral expansion of the column, increasing axial capacity, and providing a more uniform stress distribution across the section. In contrast to the brittle failure of concrete columns, the failure mode is ductile. Many hair-like cracks appear prior to failure as shown in Fig. 13 [16–21].

8 Conclusion

The strength under compressive of Ferrocement hollow column was 13.3% higher than that of RCC column. Increasing the layers of weld mesh increased columns strength. By contrasting the ferro-cement solid column with the ferro-cement hollow column. The hollow ferro-cement column will have a higher strength under compressive and less deflection. When compared to conventional columns, ferro-cement hollow columns save 34.66% on costs. The flexural strength of ferro-cement concrete beam is 30 to 40% higher than that of conventional concrete beam.

In the case of ferro-cement pipe, the ferro-cement pipes fail at higher load than conventional concrete pipes. Glass fiber mesh reinforced pipes crack and fail loads at higher loads. Among the other models, the pipes reinforced with a mix of expanded and welded wire meshes have the highest cost (strength) and the lowest failure load. Using pipes fabricated with only the ferro-cement mixture increased the pipe's strength by 35%, while using both the ferro-cement mixture and the wire mesh increased the pipe's strength by 60%.

The axial load of under-reinforced columns is reduced more that of over reinforced columns under heating.

References

1. Sasiekalaa, K., Malathy, R.: A review report on mechanical properties of ferrocement with cementitious materials, Int. J. Eng. Res. Technol. **1**(9), ISSN: 2278–0181 (2012)
2. Omnia, M.: Structural Behavior of Ferrocement Concrete Plates Subjected to Flexural and Dynamic Loadings. MSC thesis Faculty of Engineering Menoufia university (2021)
3. Batson, G.B., et al.: Guide for the design, construction, and repair of ferrocement. ACI Struct. J. **85**(3), 325–351 (1988)
4. Code, F.M.: Building Code Recommendations for Ferrocement (IFS-Committee 10–2001). International Ferrocement Society, Asian Institute of Technology, Thailand (2001)
5. Naamam, A.E.: Performance Criteria for Ferrocement. Journal of Ferrocement **09**(02), 75–91 (1979)
6. Hanson, J.M., Shah, S.P., Harmon, J.D.: Use of expanded metal as reinforcement in concrete slabs ACI J. Proc. **73** 2 (1976). https://doi.org/10.14359/11058
7. Erfan, A.M., Ahmed, H.H., Mina, B.A., El-Sayed, T.A.: Structural performance of eccentric ferrocement reinforced concrete columns. Nanosci. Nanotechnol. Lett. **11**(9), 1213–1225 (2019)
8. El-Sayed, T.A.: Axial compression behavior of ferrocement geopolymer hsc columns. Polymers **13**(21), 3789 (2021)
9. Arif, M., Kaushik, S.K.: Mechanical behaviour of ferrocement composites: an experimental investigation. Cement Concr. Compos. **21**(4), 301–312 (1999)
10. Erfan, A.M., El-Sayed, T.A.: Structural shear behavior of composite box beams using advanced innovated materials. J. Eng. Res. Reports **5**, 1–14 (2019)
11. Zahid, M.M., Bakar, B.A., Nazri, F.M., Ahmad, M.M., Muhamad, K.: Review of repair materials for fire-damaged reinforced concrete structures. In: IOP Conference Series: Materials Science and Engineering, vol. 318, No. 1, p. 012023. IOP Publishing (2018)
12. Al, A.M.İN., TAMAL, S., Bari, A.F., Mazumder, M., Hasan, M.A.: Strengthening of fire damaged reinforced beams by using ferro cement. Turkish J. Eng. **6**(3), 206–210 (2022)
13. Abdullah, G.M., Alshaikh, I.M., Zeyad, A.M., Magbool, H.M., Bakar, B.A.: The effect of openings on the performance of self-compacting concrete with volcanic pumice powder and different steel fibers. Case Stud. Const. Materials **17**, e01148 (2022)
14. More, K.K., Ghane, V.R., Parkhe, D.D.: applications of ferrocement in construction. Int. J. Adv. Mech. Civil Eng. ISSN: 2394–2827 **6**(5) (2019)
15. Hardik, S.: A Review study of Application of Ferro-Cement", All content following this page was uploaded by Hardik Saxena on 12 February 2020
16. Xiong, G.J., Wu, X.Y., Li, F.F., Yan, Z.: Load carrying capacity and ductility of circular concrete columns confined by ferrocement including steel bars. Constr. Build. Mater. **25**(5), 2263–2268 (2011)
17. Kaish, A.B.M.A., Alam, M.R., Jamil, M., Zain, M.F.M., Wahed, M.A.: Improved ferrocement jacketing for restrengthening of square RC short column. Constr. Build. Mater. **36**, 228–237 (2012)

18. Mourad, S.M., Shannag, M.J.: Repair and strengthening of reinforced concrete square columns using ferrocement jackets. Cement Concr. Compos. **34**(2), 288–294 (2012)
19. Kaish, A.A., Alam, M.R., Jamil, M., Wahed, M.A.: Ferrocement jacketing for restrengthening of square reinforced concrete column under concentric compressive load. Proc. Eng. **54**, 720–728 (2013)
20. Kondraivendhan, B., Pradhan, B.: Effect of ferrocement confinement on behavior of concrete. Constr. Build. Mater. **23**(3), 1218–1222 (2009)
21. Masood, A., Arif, M., Akhtar, S., Haquie, M.: Performance of ferrocement panels in different environments. Cem. Concr. Res. **33**(4), 555–562 (2003)
22. Abdullah, A.S.K.: Structural Behavior of Ferrocement Panels Exposed to Fire. Sulaimani J. Eng. Sci **5**, 7–17 (2018)
23. Standard Test Methods for Fire Tests of Building Construction and Materials" Annual Book of ASTM Standards, Vol 04.07. Designation ASTM E119–12
24. Fire tests on building materials and structures. Guide to the principles, selection, role and application of fire testing and their outputs BS 476–10:, 2009, December 2008
25. Fire-resistance tests-Elements of building construction; Part 3: Commentary on test method and guide to the application of the outputs from the fire-resistance test ISO/TR 834–3:2012
26. Shaheen, Y.B., Etman, Z.A., Ramadan, A.G.: Pre and post-fire strength assessment of Ferro cement beams. Concr. Res. Lett.. Res. Lett. **3**(3), 459–476 (2012)

Influence of Pozzolanic Material in Cement Concrete Mixes Contain Cement Kiln Dust

Rafik K. Abdel Wahab[1](✉), M. Hassan[2], and Ahmed Deifalla[3]

[1] Civil Engineering Department, Faculty of Engineering – Qena Branch, Al-Azhar University, Cairo, Egypt
dr.rkhairy@gmail.com, mqader81@yahoo.com

[2] Basic Sciences Department, High Institute of Engineering and Technology, 21 k Belbies-Cairo Road, Cairo, Egypt

[3] Future University in Egypt, New Cairo, Egypt
ahmed.deifalla@fue.edu.eg

Abstract. This study aimed to examine the effects of pozzolanic materials as an additive on the compressive, tensile, and flexural strengths of concrete mixes containing different percentages of cement kiln dust. To determine the effectiveness of cement kiln dust (CKD) as a weight-for-weight replacement of cement, There were 60 samples manufactured of five different concrete mixtures, including CKD. The percentage of replacement was 0% in the control mix and 40% in the other mixes. The mixes (G3, G4, and G5) also contain additions of 5%, 10%, and 15% of fly ash, respectively. As a result of using cement kiln dust as a replacement for cement, the results demonstrate that all concrete strengths decreased. When fly ash was added as a pozzolanic material to mixes containing cement and cement kiln dust, all of the concrete's strengths increased. However, all the results were below those of the control specimens: fly ash, CKD.

Keywords: Cement Kiln Dust · tensile strength

1 Introduction

Cement kiln dust production was estimated to account for 15–20% of global cement production, which ranges from 420 to 560 million tons annually. Due to the high concentration of chlorides, sulfates, and alkalis in this industrial waste and its high softness (20 to 100 microns), it is dangerous to human health and the environment [5]. The use of CKD in concrete and mortar samples at different proportions was investigated as a replacement for Portland cement. It was found that when adding low percentages of CKD, the strength properties of these samples did not change [2]. According to studies on the use of CKD on the corrosion behavior of embedded reinforcements and the compressive strength of cement paste, using it in place of up to 5% by weight of cement did not affect the strength of the cement paste or the passivity of the reinforcement [4, 7]. When powdered coal is burned in power plants to generate electricity, a fine powder called fly ash is created. A pozzolanic is a substance with aluminous and siliceous material that forms cement when water is present. Fly ash is one type of pozzolanic. Fly ash creates a compound that is similar to Portland cement when combined with lime and water.

A. S. Mosallam et al. (Eds.): ICATH 2023, SCI, pp. 119–128, 2024.
https://doi.org/10.1007/978-3-031-76557-5_10

Utilization of Fly Ash Because it is not considered a hazardous waste, numerous studies have been conducted over the past 30 years with the intention of increasing its use in various industries [14, 18].

Because it is an oxide-rich material, fly ash is suitable for use as a raw material in numerous industries and construction [18]. Currently, fly ash is used to improve construction materials and is excellent for agricultural applications to improve soil properties [14]. The fly ash can also be used to make ceramics, bricks, concrete, build roads, and do other things. The production of concrete, either in conventional or geopolymer form [22], is the most significant application of fly ash. The use of fly ash in cement, concrete, and other cementitious composites has been the subject of numerous studies.

Utilization of fly Ash in building construction although fly ash is not a brand-new technology, it is gaining popularity as a means of enhancing the quality of both construction and the surrounding environment. There are financial, environmental, and technical advantages to including fly ash in concrete [13–25]. Fly ash is currently utilized by the cement industry as a pozzolanic material for the production of Portland Pozzolana Cement due to its similar SiO_2 and Al_2SiO_3 content [17]. Compounds with cementitious properties are produced when it chemically reacts with calcium hydroxide in the presence of moisture and at room temperature [16–26]. Technically, using Portland cement and fly ash together contributes to the consumption of $Ca(OH)_2$, which is produced when cement hydrates and results in the formation of cementitious products [24–27]. Fly ash's glassy phase and the calcium hydroxide it produces produce more CSH gel, which increases density and strength [25]. When compared to earlier ages, the compressive strength increased at later ages [27]. This green innovation improves the solidness and administration life of substantial designs [15–28]. Fly ash can be added to the concrete mix as an admixture or as a cement weight replacement [19]. In the production of cement, the incorporation of utilization fly ash as a component of blended cement can result in significant cost and energy savings [15–21]. Fly ash can replace up to 75% of the cement weight with fly ash, but its use is strictly regulated by strict standards like ASTM C 618 and EN 450–1[20] from the American Society for Testing Materials. Geopolymer concrete containing fly ash is the new cement substitute for building and construction materials [29, 30]. This concrete mixture produced a concrete with low shrinkage, good acid resistance, and high compressive strength [16].

Fly ash is usually used in blended cement to make in-place concrete, but it can also be used to make high-strength precast and prestressed concrete. The use of fly ash in high-strength precast and prestressed concrete has been restricted due to its slow early strength development. The fact that super plasticized fly ash concrete with a low water-to-cement ratio can be proportioned to meet the very early age strength and other requirements for precast or prestressed concrete products has somehow been confirmed by recent studies [15].

2 Materials and Methodology

2.1 Cement

Ordinary Portland cement from Nahda Cement Factory in Qena governorate was used in this study. Chemical composites mentioned in Table 1.

The Cement compressive strength test had been done at 3and 7 days the results were (185–280) kg/cm^2 for 3 and 7 days respectively.

2.2 Cement Kiln Dust (by Pass) (CKD)

Cement Kiln Dust by pass (CKD) is a fine, highly alkaline powder. Chemical composition of CKD used in this study mentioned in Table 1.

Table 1. The Chemical Composites of Ordinary Portland Cement and by pass (CKD)

Component	SiO_2	AL_2O_3	Fe_2O_3	Cao	MgO	SO_3	K_2O	Na_2O	CL^-
OPC	20.6	4.5	3.6	62.5	2.6	2.7	0.5	0.2	0.01
CKD	15.8	3.6	2.8	63.8	1.9	1.7	3.0	0.3	1.1

2.3 Fly Ash

Fly ash used in this article was produced by Sika Egypt Company. Its chemical composition mentioned in Table 2.

Table 2. Chemical Composites of Fly ash

Component	SiO_2	AL_2O_3	Fe_2O_3	Cao	MgO	SO_3
Fly ash	55	26	7	9	2	1

2.4 Coarse Aggregate

Coarse aggregate used was Dolomite crushed stone. Its MNS was 20 mm, volumetric weight was 1650 kg/ m^3. The grading curves for used crushed dolomite are presented in Fig. 1.

2.5 Fine Aggregate

Natural sand used in the concrete as a fine aggregate. Its fineness modulus Obtained by sieve analysis was 2.8. The grading curves for used sand are presented in Fig. 2.

2.6 Mix Design

ACI mix design method had been used to select mix proportions of normal concrete ACI manual of concrete practice has been used to provide the all-concrete Mixes [1]. The mix ingredients are shown in Table 3.

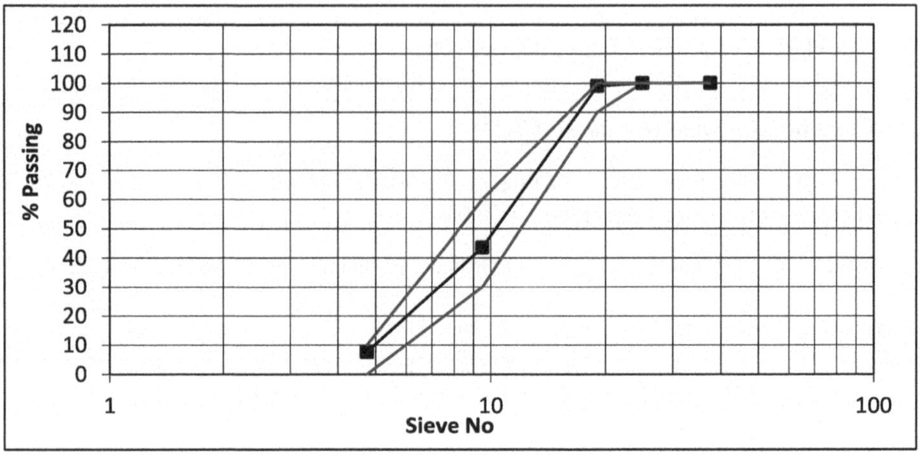

Fig. 1. The Grading Curve for Crushed Dolomite

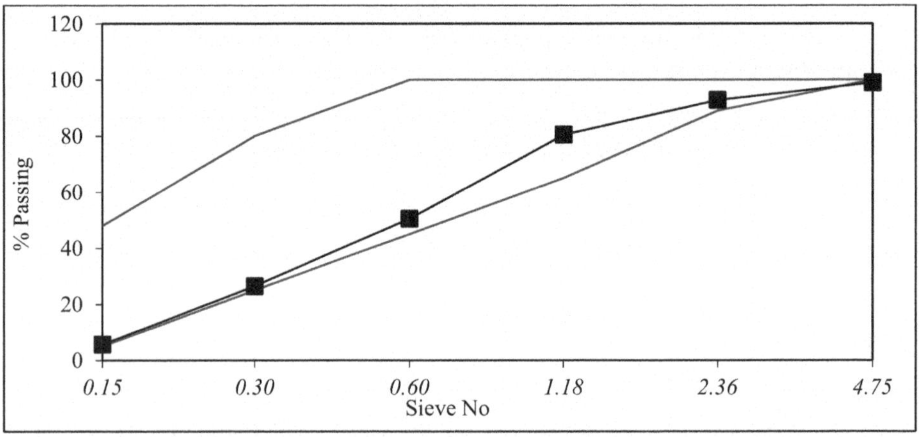

Fig. 2. The Grading Curve for Fine Sand

Table 3. Concrete Mix Ingredients

Water (kg/m^3)	Cement (kg/m^3)	Coarse (Kg/m^3)	Fine Agg. (kg/m^3)
205	380	1023	737

2.7 Mix Design Criteria

- Characteristic Compressive strength (F_{cu}) 300 kg / cm^2
- Slump (75 – 100) mm
- MNS for Coarse Agg. 19 mm
- FM for sand 2.8
- ACI Design mix method according to British method [3]

3 Experimental Program

Six groups were casted, each group had six (100*100*100) cubes, three prisms (150*150*70) and three cylinders (150 × 300), for compressive, flexural and splitting tensile strength. As mentioned in the Table 4.

Table 4. Mix proportioning

Group No	Cement (kg/m^3)	Fly ash (kg/m^3)	CKD (kg/m^3)	Coarse agg. (kg/m^3)	Fine agg. (kg/m^3)	Water (kg/m^3)
(G1) Pure cement (C) (control)	380	0	0	1023	737	205
(G2) Cement + 10% fly ash	380	38	0	1023	737	215
(G3) CM (60%C + 40%CKD)	228	0	152	1023	737	215
(G4) (95%CM) + (5%FA)	217	19	144	1023	737	220
(G5) (90%CM) + (10%FA)	205	38	137	1023	737	225
(G6) (85%CM) + (15% FA)	194	57	129	1023	737	230

Prepare six cubes, three prisms, and three cylinders for each mix. Curing all samples by submerging them in a water tank until the test date. The compressive strength of three of the six cubes was determined after seven days, while the remaining cubes were tested after 28 days. After 28 days, the three samples were tested for flexural and tensile strength, and the average was taken. The relationship was plotted.

3.1 Test of Compressive Strength

Pressure test was led by pounding six shapes, three after 7 days and the rest following 28 days All examples had been tried under pressure static burden by utilizing 2000 kN pressure testing machine with rate (0.6 N/mm^2) until disappointment. As indicated by the Egyptian code C203 [8].

3.2 Parting Rigidity Following 28 days

The Egyptian code for this is C203.[8] 4–3 Flexural strength for 28 days. Using an X-ray diffraction (XRD, PANalytical Empyrean), the Egyptian code C203.[8] requires XRD analysis of raw materials.

The Portland cement chemical composition that was used in this work is listed in Table 1, and Fig. 1 depicts the XRD pattern. Aluminates are only found in trace amounts, but common calcium silicates like alite and belite are present.

The CKD XRD patterns reveal low levels of calcite and high levels of free lime, indicating that there will be more free calcium available for a pozzolanic reaction. Dolomite and significant amounts of anhydrite ($CaSO_4$) are present in CKD. Lastly, there are also significant amounts of quartz and traces of sylvite (KCl). The amorphous component is based on an aluminosilicateglass phase, and the mixed crystalline phases of quartz and Mullite are visible in the fly ash XRD results.

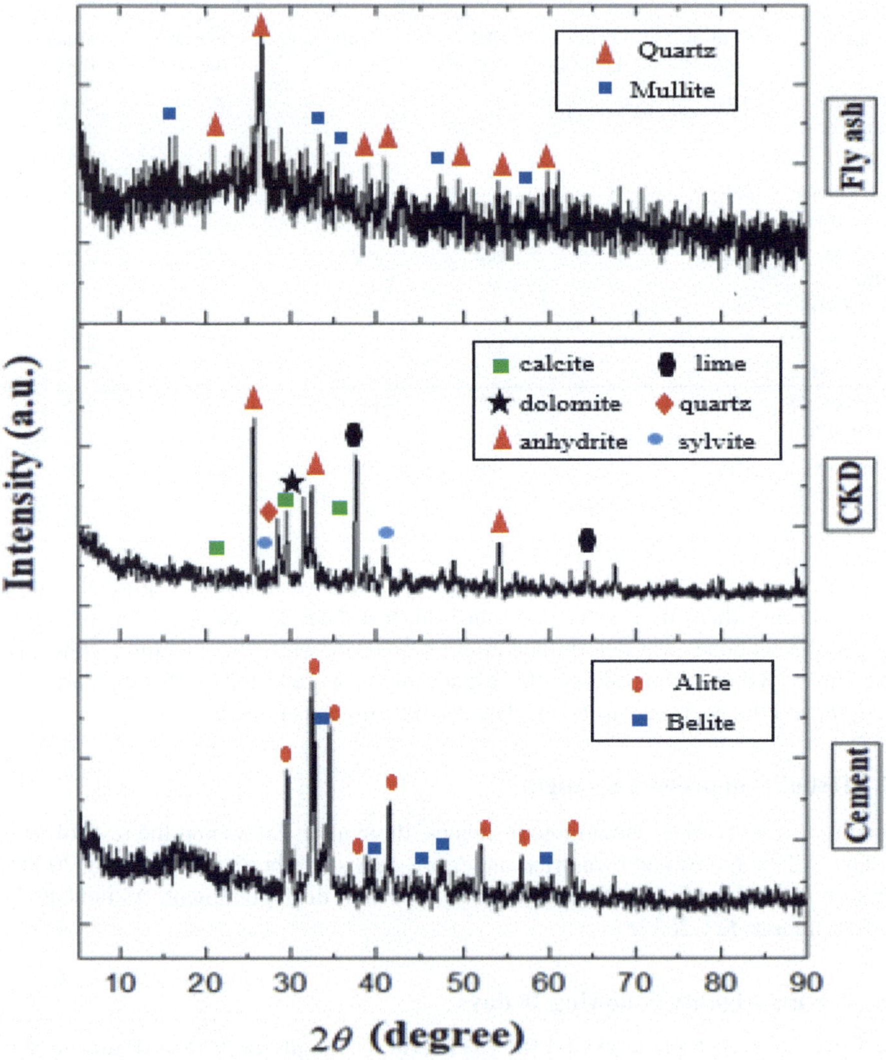

Fig. 3. XRD patterns of Cement, Cement Kiln Dust and Fly ash

The average compressive strength for the control group (G1) was (26.22–36.67(N/mm2) for 7 and 28 days, respectively, as shown in Table 5. The weight addition

Table 5. Results of Compressive, Splitting and Flexural strength

Group No.	Compressive Strength (N/mm^2)		Splitting Strength (N/mm^2) 28 days	Flexural Strength (N/mm^2) 28 days
	After 7 days	After 28 days		
(G1) Pure cement (C) (control)	26.22	36.67	3.67	5.18
(G2) Cement + 10% FA	28.1	39.23	3.93	5.54
(G3) CM (60%C + 40%CKD)	9.07	12.84	1.3	1.8
(G4) (95%CM) + (5%FA)	10.61	16.08	1.61	2.25
(G5) (90%CM) + (10%FA)	15.33	23.87	2.41	3.34
(G6) (85%CM) + (15%FA)	18.79	29.89	2.95	4.18

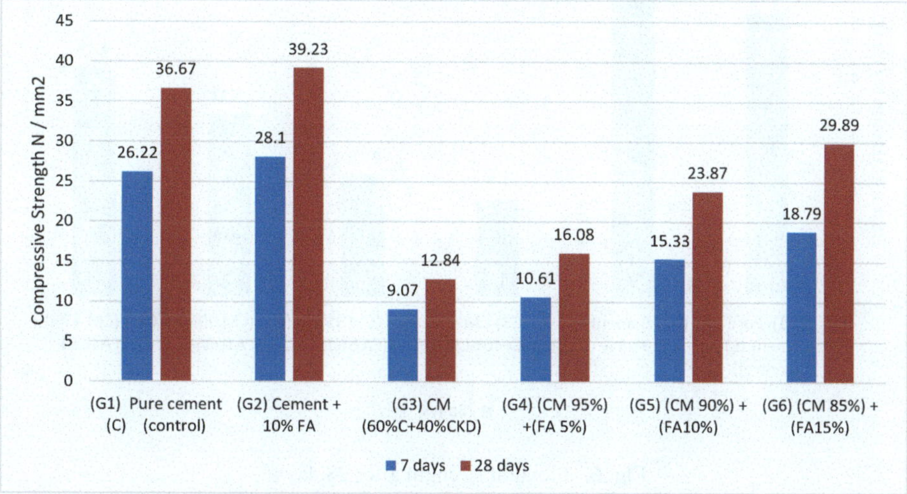

Fig. 4. Compressive strength after 7 and 28 days

of 40% cement kiln dust to the mixture results in a 65 percent decrease in strength. When fly ash is added as a replacement for a percentage of cementitious material, the compressive strength rises (Fig. 3, Fig. 4, Fig. 5, Fig. 6).

As can be seen in Figure, there was a direct correlation between the percentage of Fly ash added and the strength increasing. For 5% Fly Ash Added (G4), the percentage of strength increasing increased by approximately 125% compared to Group3's results, but it remained 56% below the control value. For seven days and 28 days, the percentage

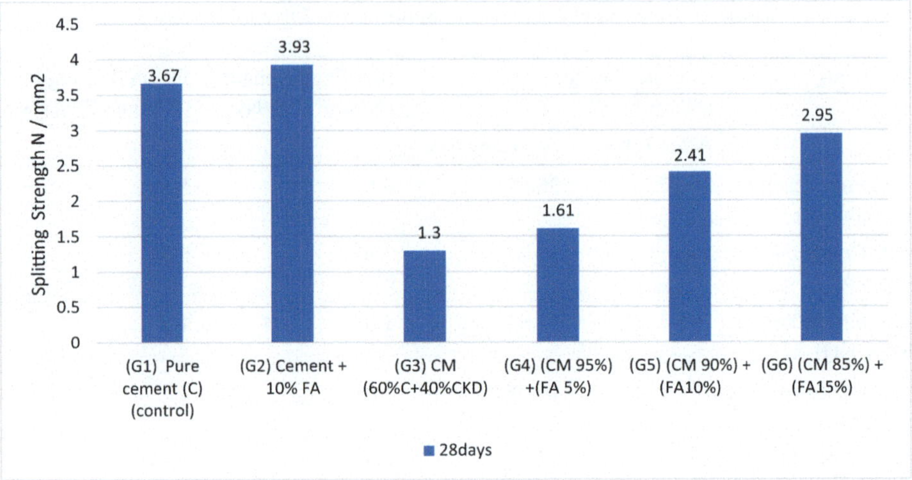

Fig. 5. Splitting strength after 28 days

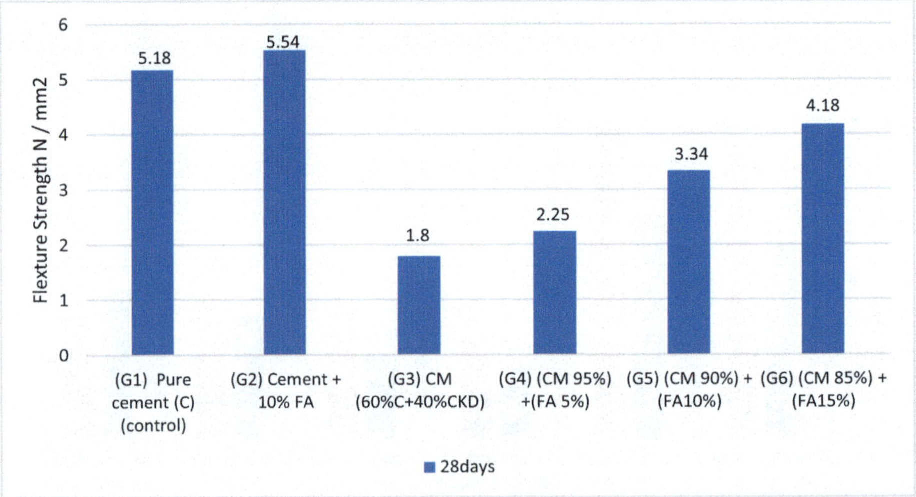

Fig. 6. Flexural strength after 28 days

of strength increase for 10% Fly Ash Added (G5) was approximately 50% higher than that of Group 4 (G4), but it remained approximately 35 percent below the control value. There is a slight increase in compressive strength between the results of (G6) and (G5) when 15% fly ash is added (G6). For seven days and 28 days, the strength increased by about 25%. For both the 7-day and 28-day compressive strengths, the compressive strength of the 15% added Fly Ash was still approximately 19% lower than that of the control group.

4 Conclusion

The following conclusions could be drawn from the results of the experiments:

- Adding fly ash by 10% increases the compressive strength by about 7%
- The compressive strength of concrete is decreased when CKD or CKD and fly ash are used to partially replace cement.
- A 40 percent increase in CKD results in a roughly 65% decrease in compressive, splitting, and flexural strength over the course of 28 days, respectively.
- In the cementitious materials (CM) of Group (G4), the replacement of 5% fly ash by weight increased the compressive strength after 28 days while remaining under control by approximately 56 percent.
- In the cementitious materials (CM) of Group (G5), substituting 10% fly ash by weight increased the compressive strength after 28 days while remaining under control by approximately 35%.
- In Group (G6), the cementitious materials (CM) containing 15% fly ash by weight increased the compressive strength after 28 days while remaining under control by approximately 19%.
- In order to reach the control strength, a partial replacement of CKD could be reduced by adding fly ash in a ratio of 15%, which is the most effective ratio.

References

1. American Concrete Institute (ACI): ACI Manual of Concrete Practice, 1997, Part 1. American Concrete Institute (ACI) (1997)
2. Al-Harthy, A.S., Taha, R., Al-Maamary, F.: Effect of cement Kiln dust (CKD) on mortar and concrete mixtures. Constr. Build. Mater. **17**(5), 353–360 (2003)
3. Neville, A.M., Brooks, J.J.: Concrete Technology. Prentice Hall (1987)
4. Batis, G, Rakanta, E., Sideri, E., Chaniotakis, E., Papageorgiou, A.: Advantage of simultaneous use of cement kiln dust and blast furnace slag. In: Proceeding of the International Conference of the Challenges of Concrete Construction, Dundee, UK, University of Dundee (2002)
5. Adaska, W.S., Taubert, D.H.: Beneficial uses of cement KILN dust. In: 2008 IEEE/PCA 50th Cement Industry Technical Conference, Miami, FL, May 19–22 (2008)
6. Bhatty, J.I.: Alternative Uses of Cement Kiln Dust, RP 327. Portland Cement Association, Skokie, Illinois, USA (1995)
7. El-Sayed, H.A., Gabr, N.A., Hanafi, S., Mohran, M.A.: Re-utilization of by-pass kiln dust in cement manufacture. In: International Conference on Blended Cement in Construction, Sheffield, UK (1991)
8. Egyptian code for design of reinforced concrete
9. Maslehuddin, M., Al-Amoudi, O.S.B., Rahman, M.K., Ali, M.R., Barry, M.S.: Properties of cement kiln dust concrete. Constr. Build. Mater. **23**(6), 2357–2361 (2009)
10. McKay, M., Emery, J.: Stabilization/solidification of contaminated soils and sludges using cementitious systems. In: Proceeding of the Cement Industry Solutions to Waste Management, Calgary, Alberta, Canada, October 7–9, pp.135–1511992 (1992)
11. Bhatty, M.S.Y.: Properties of blended cement made with Portland cement, cement kiln dust, fly ash and slag. In: Proceeding of the 8th International Congress on the Chemistry of Cement, Rio de Janeiro, Brazil, September 22–28, pp. 118–127 (1986)

12. Siddique, R.: Utilization of waste materials and by-products in producing controlled low-strength materials. Resour. Conserv. Recycl. **54**(1), 1–8 (2009)
13. Hsu, S., Chi, M., Huang, R.: Effect of fineness and replacement ratio of ground fly ash on properties of blended cement mortar. Constr. Build. Mater. **176**, 250–258 (2018)
14. Dwivedi, A., Jain, M.K.: Fly ash – waste management and overview: a review. Recent Res. Sci. Technol. **6**, 30–35 (2014)
15. Naik, T.R., Singh, S.S.: Fly ash generation and utilization - an overview (1993). http://www4test.uwm.edu/cbu/Papers/1993%20CBU%20Reports/REP-190.pdf
16. Nordin, N., Abdullah, M.M.A.B., Tahir, M.F.M., Sandu, A.V., Hussin, K.: Utilization of fly ash waste as construction material. Int. J. Conserv. Sci. **7**, 161–166 (2016)
17. Bahu, B.K., Jayaram, S.: Environmental impacts of fly ash. Botswana J. Technol. **9**(1), 9–13 (2000)
18. Tiwari, M.K.: Fly ash utilization a brief review in Indian context fly ash utilization : a brief review in Indian context. Int. Res. J. Eng. Technol. **3**, 949–956 (2016)
19. Openshaw, S.C.: Utilization of coal fly ash master dissertation, Ph. D. thesis, University of Florida, Florida (1992)
20. Blissett, R.S., Rowson, N.A.: A review of the multi-component utilisation of coal fly ash. Fuel **97**, 1–23 (2012)
21. Sett, R.: Flyash : characteristics, problems and possible utilization. Adv. Appl. Sci. Res. **8**, 32–50 (2017)
22. Malhotra, V.M.: Durability of concrete incorporating high-volume of low-calcium (ASTM Class F) fly ash. Cem. Concrete Compos. **12**, 271–277 (1990)
23. Al-Kutti, W., Nasir, M., Johari, M.A.M., Islam, A.S., Manda, A.A., Blaisi, N.I.: An overview and experimental study on hybrid binders containing date palm ash, fly ash, OPC and activator composites. Constr. Build. Mater. **159**, 567–577 (2018)
24. Hemalatha, T., Ramaswamy, A.: A review on fly ash characteristics – towards promoting high volume utilization in developing sustainable concrete. J. Clean. Prod. **147**, 546–559 (2017)
25. Liu, B., Xie, Y., Zhou, S., Yuan, Q.: Influence of ultrafine fly ash composite on the fluidity and compressive strength of concrete. Cem. Concr. Res. **30**, 1489–1493 (2000)
26. Fauzi, A., Nuruddin, M.F., Malkawi, A.B., Abdullah, M.M.A.B.l: Study of fly ash characterization as a cementitious material. Procedia Eng. **148**, 487–493 (2016)
27. Kocak, Y., Nas, S.: The effect of using fly ash on the strength and hydration characteristics of blended cements. Constr. Build. Mater. **73**, 25–32 (2014)
28. Bhanumathidas, N., Mehta, P.K.: Concrete mixtures made with ternary blended cements containing fly ash and rice-husk ash. Am. Concr. Inst. **199**, 379–392 (2001)
29. Palomo, A., Grutzeck, M.W., Blanco, M.T.: Alkali-activated fly ashes: a cement for the future. Cem. Concr. Res. **29**, 1323–1329 (1992)
30. Ghazali, N., et al.: Utilization of fly ash in construction. IOP Conf. Ser. Mater. Sci. Eng. **601**, 012023 (2019). https://doi.org/10.1088/1757-899X/601/1/012023
31. Temuujin, J., Van Riessen, A., MacKenzie, K.J.D.: Preparation and characterisation of fly ash based geopolymer mortars. Constr. Build. Mater. **24**, 1906–1910 (2010)

A Review of the Most Important Parameters that Influence the Reinforcement of Structures by Using the FRP Materials

Yasmina ED-Dariy[1](✉) and Brahim El Bhiri[2]

[1] National School of Architecture, Fez, Morocco
yasminadariy@gmail.com, yasminaeddariy@research.emi.ac.ma
[2] SMARTiLab /Moroccan School of Engineering Sciences (EMSI), 10000 Rabat, Morocco

Abstract. According to the literature, several methods have been developed to reinforce structures in Civil Engineering while using fiber-reinforced polymers (FRP) materials. This paper presents the most parameters that can influence the reinforcement of structures such as orientation, the number of layers, the width of the composite, the thickness, surface treatment and the environment. It was shown by some researchers that the number of layers increases the strength and improves the mechanical properties of reinforced elements up to a certain level of layers. Otherwise, it was confirmed by some others that the surface preparation not only improved the ultimate load capacity but also improved the adhesion between the concrete and the CFRP sheets. Other studies concluded that the influence of curing temperature could make the element more ductile at high temperatures as well as it can make the element weak at low temperatures.

Keywords: Reinforcement · Composite materials · Concrete · FRP

1 Introduction

The increase in loads that stress the structure like the increase in the level of activity in the structure, change of use of the structure, or defects in the project or in the execution, insufficient or poorly placed reinforcement, insufficient dimensions of structural elements are the causes which lead to structural reinforcement. According to the literature, one of the methods developed and which has met with great success in the field of upgrading reinforced concrete structures, is the exterior bonding of composite materials based on carbon fibers (or glass). (Wong and Vecchio) [1] showed that reinforcement with carbon fibers increase the ultimate load capacity. (Wu, 2004) [2] confirmed that bonding the fabric increases the ultimate strength of reinforced beams by reducing the deflection of the structures. As it increases their stiffness, it also limits crack propagation. (Correia et al., 2015) [3] were also able to show that the reinforcement increased the rigidity of the slabs reinforced with carbon fiber reinforcement and consequently reduced the deflection for a specific load level.

The effectiveness of a repair depends, largely, on the load transfer in the concrete-fiber interface. Several researches have confirmed that the polymerization conditions

A. S. Mosallam et al. (Eds.): ICATH 2023, SCI, pp. 129–138, 2024.
https://doi.org/10.1007/978-3-031-76557-5_11

directly affect the adhesion and more precisely the influence of the curing temperature. It has confirmed that the resistance obtained after 1 h of curing at 60 °C is greater than that obtained after 10 to 20 days of curing [4]. Ed-Dariy et al. concluded that the ductile failure mode was observed in the case of samples cured at 30 °C and 50 °C compared to all the samples cured at ambient conditions [5]. (Van de Weyenberg et al., 2006) [6] showed that the longitudinal strength and stiffness were increased following the improvement of the interface rather than the fiber properties. (ESPCI, 2005) [7] noticed that the transition zone created by the penetration of resin within the porous network of the concrete is more important in the case where the total porosity of the substrate is higher. The role of porosity present within cementitious materials is a factor which has been highlighted in the literature by (Sasse and Fiebrich, 1983; Toutanji and Ortiz, 2001) [8, 9]. It is, in fact, an essential parameter since it governs the impregnation of the surface by the resin. This adhesion depends on several parameters, which are presents in this paper like the surface treatment, fiber orientation the dimensions of the reinforcement.

On the other hand, the durability of the reinforcement depends also on certain factors, which can influence the proper functioning of the reinforcement system in the long term. This paper presents some of those.

2 Composite Materials

A composite material is a combination of at least two materials, which must be both immiscible and have a high adhesion capacity. The assembly of these materials will give a new product with different properties from its constituents. One, called reinforcement, is the basis of the material thus formed and ensures mechanical strength. The other, called matrix (most often a plastic material), ensures cohesion between the fibers. The matrix is the bond of the fiber-reinforced polymer (FRP) and plays several important roles.

The polymer matrix is an organic compound made up of long molecular chains formed by repetitive assembly of smaller units called monomers. Matrix materials can be grouped into two broad categories: thermosetting resins and thermoplastic resins. The reinforcement can be by using Carbon, Glass, or Aramid fibers.

2.1 Comparison Between Fibers

- Glass fibers have a low modulus and lack resistance to saline and/or basic solutions and which are likely to lose their resistance over time;
- Aramid fibers, which have a low modulus, limited fatigue resistance in an aqueous environment and limited thermal resistance;
- Carbon fibers have a set of physical or mechanical properties, which make them a top quality reinforcing material in certain applications linked to buildings [10] (Fig. 1).

HR: High Strength , HM: High Modulus

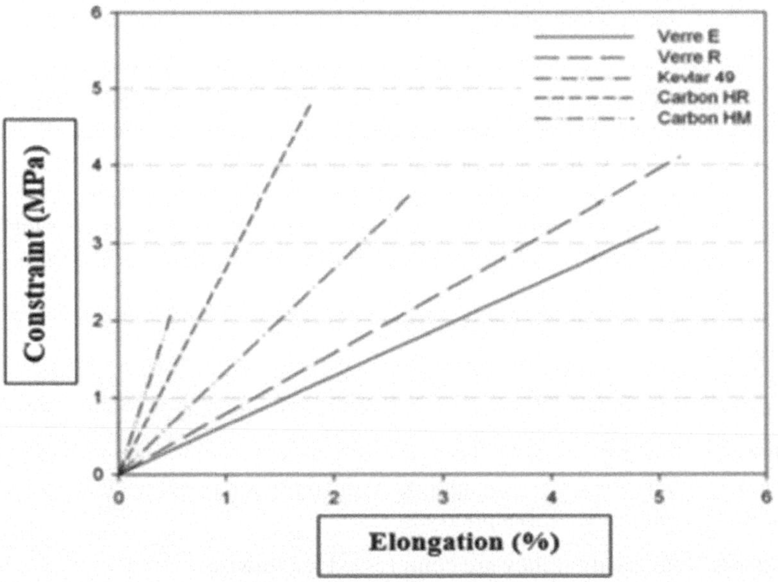

Fig. 1. Mechanical characteristics of carbon fibers [10]

2.2 Fiber Orientation

The fibers have excellent properties like very high tensile and compressive strength and in order to ensure good sizing there is a factor which remains the most important and which must be controlled. This is indeed the factor "Orientation». Is essential that all illustrations are clear and legible.

(Grace et al., 1999) [11] reported that vertically oriented fibers with a U-shape around the sample cross-section, significantly reduce sample deviations and increase capacitance loading resistance of the beam.

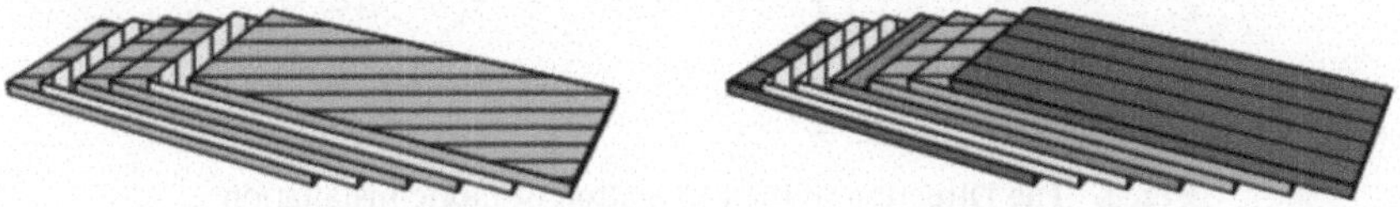

Fig. 2. Laminate configurations [12]

2.3 Continuous Fibers

In the case of high-performance composites, which are generally made up of continuous fibers and a thermosetting matrix, it is relatively easy to orient a fabric of fibers in a preferred and chosen direction. These fibers often exist in the form of fabrics or plates (lamellae) which are characterized by the intersection at right angles of two yarn systems (warp and weft). Most of these reinforcements work well in tension, but offer less good performance in compression and shear. There are different geometries and textures of reinforcements:

- **Unidirectional (UD):** in a layer (UD), the fibers are assembled parallel to each other using a very light weave. Resistance and elasticity will then be concentrated along a single axis. This type is intended for reinforcement of beams, pillars, walls and floor slabs;
- **Laminates**: the laminated composite is composed of several plies of varied orientations, each ply is a unidirectional composite (UD) in order to obtain the desired mechanical properties for the final structure [12] (Fig. 2).

2.4 Direction of Fibers

The fibers are strong in one direction only. For traction, there are two directions:

- **Direction L:** (direction of the fibers) in this direction the composite works optimally;
- **Direction T:** where the forces are applied in the direction perpendicular to the fibers; in this sense the composite works poorly.

The fibers break easily if they are compressed lengthwise.

In the case of reinforcing shear forces it is preferable to use a fabric inclined at 45°. This inclination makes it possible to sew stress cracks also inclined at 45°.

Furthermore, the use of bidirectional composites (composed of fibers oriented in both directions) can be an advantageous solution to improve both axial and transverse behavior.

Furthermore, identifying the preferred direction is important, this identification is done simply by indicating the direction of installation of the composite (see Fig. 3): the example of a bidirectional fabric with preferred direction of resistance in the warp, made up of 70% fibers in the warp direction and 30% fiber in the warp direction frame).

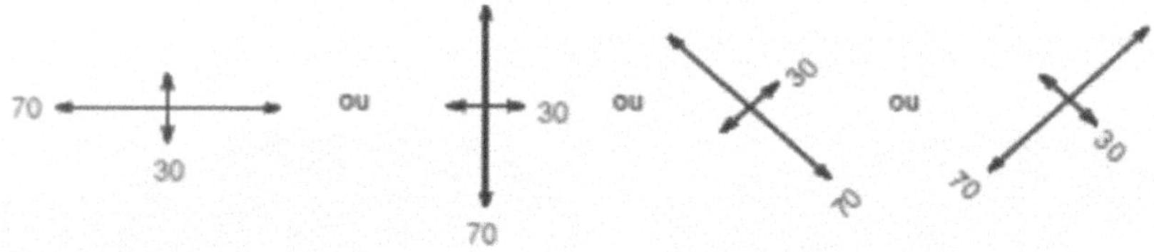

Fig. 3. The Direction of the orientation of fabric installation

2.5 The Influence of the Environment

The mechanical properties of engineering materials are the most important parameters for structural designers. The environmental considerations and durability are also important for the evaluation of materials used in reinforcement of structures.

Exposure to adverse conditions can degrade the mechanical properties of FRP materials because almost all engineering materials are sensitive to aggressive environments.

- **Humidity**
  The mechanical performance of the composite depends on the moisture absorption rate; almost all polymers are very negatively influenced by the presence of humidity. Moisture penetration is mainly concentrated at the matrix level. This penetration leads to a weakening in the development of internal bonds between the fiber and the resin. This leads to a degradation of the mechanical properties of the PRF. When the polymer matrix reaches its saturation point of moisture absorption no reduction in properties appears afterward.
- **Temperature**
  FRP-based composite materials are very influenced by temperature; they display a significant change in mechanical properties when temperatures are very high this change results in a drop in their properties. Long-term durability is influenced by increasing temperatures. On the other hand, low temperatures only have a negative impact on durability in cases where they lead to material weakening.
- **Fire**
  Fire can constitute a major risk on the behavior of the polymer resin from a mechanical point of view as well as on the environmental side because they display a large quantities of toxic fumes.
- **Ultraviolet rays**
  Ultraviolet (UV) rays can damage FRP materials. Aramid fibers are known to be particularly sensitive to UV rays, while carbon fibers and glass fibers are resistant to UV light. Furthermore, due to UV the majority of polymers degrade slightly. Degradation from UV light can be surmount by using various additives in the matrix, or by applying an opaque paint to the FRP.

2.6 The Influence of Number of Layers

David et al., (2001) noted for beams with the following nomenclature P1 (unreinforced beam), P2 (Beam reinforced by a single layer) P5 (Beam reinforced by 2 layers):

- A gain in flexion of 51% and 73% of P2 and P5 respectively (see Fig. 4) compared to P1;
- An increase in rigidity;
- A reduction in mid-span deflection of 18 mm and 16 mm for P2 and P5 compared to 50 mm of the control beam.

Wu, (2004) [2] confirms that flexural failure of reinforced beams can occur when the reinforced concrete element is reinforced by too much FRP fabric as he adds that for a poorly reinforced beam at the shear force (reinforcement rate <0.2%), increasing the quantity of longitudinal fabric only slightly improves the ultimate strength.

2.7 The Influence of the Width of the Reinforcement

Better performance was observed with the widest reinforcement, according to an experimental study carried out by Correia et al., (2015) [3] because this type of reinforcement displayed greater rigidity and an increase in ultimate capacity by 23%.

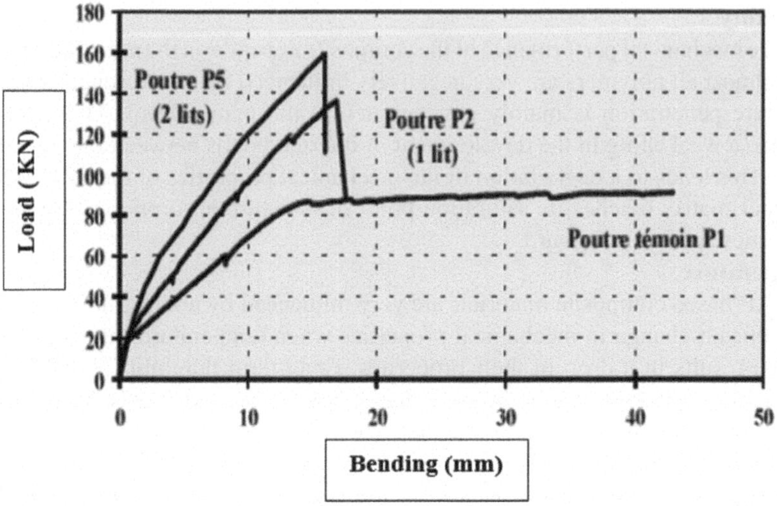

Fig. 4. Comparison between a beam reinforced by a one CFRP layer and by two layers [13]

According to Neto et al., (2016) [14] if the concrete width is greater than the adhesive width, a non-uniform shear stress distribution develops in both the inner and outer region of the carbon fiber width.

Khshain et al., (2015) [15] observed that increasing the CFRP strip width could increase the lifespan of the reinforced element.

2.8 Influence of Thickness

The studies of Quantrill et al., (1996) [16] indicated that the value of the shear stresses of beams reinforced by FRP with an adhesive layer of 1 mm thick is greater than those reinforced by FRP with an adhesive layer of 2 mm thick.

On the other hand and according to the experimental and numerical results of Ungureanu et al., (2018) [17], it was noted the variation of the thickness of the adhesive layer does not influence the stress-strain distributions and the load breakage. Correia et al., (2015) [3] also observed the same behavior of two samples of different thicknesses (1.4 and 1.2 mm).

2.9 Influence of Surface Treatment

According to Chicoine, (1998) [18] the bonding surface must be perfectly flat to avoid the risk of detachment.

Toutanji and Ortiz, (2001) [9] studied the influence of surface preparation on the bond strength and load capacity of beams reinforced by the EBR method.

The main aim of the surface preparation was to remove the outer weak layer, as well as the impurities hidden in the pores of the surface concrete, which prevent good attachment of the CFRP sheets. The results revealed that the surface preparation not only improved the ultimate load capacity, but also delayed the delamination of the CFRP sheets on the concrete surface by providing sufficient bonding interface between the CFRP sheets and the concrete.

Mostofinejad and Mahmoudabadi, (2010) [19] found that substrate preparation methods can improve the failure behavior of beams. This repair of the surface can also increase the ultimate load capacity of reinforced beams compared to those lacking surface preparation. (Correia et al., 2015) [3] concluded that the sandblasted samples carried higher ultimate loads than the loads of the reference samples, because the sandblasted samples obtained higher adhesion compared to the others ones.

Iovinella et al., (2013) [20] concluded that bush hammering and sandblasting were the most effective methods knowing that the most commonly used methods in surface preparation are: grinding, brushing, steel shot blasting, sandblasting and bush hammering (Fig. 5) [5]. And according to Wu, (2004) [4] a correct bonding between the composite and the concrete is based on the sanding of the faces of the element to be reinforced.

2.10 Influence of Temperature

Many researchers have shown that higher temperature accelerates the degradation process [21–23].

Tadeu and Branco, (2000) [24] studied the effect of temperature on reinforced concrete samples reinforced with CFRP strips using the EBR technique for four temperature levels (30 °C, 60 °C, 90 °C and 120 °C). Trial results showed a reduction in the breaking load for high temperature values. This reduction in strength was linked to the variation in the mechanical properties of the adhesive at high temperature. For the tests carried out at 30 °C, the rupture was initiated at the level of the concrete surface. For the tests at temperatures above 60 °C the breakage took place at the level of the adhesive and the tests carried out at 120 °C showed a break at the tape-epoxy interface.

Di Tommaso et al., (2001) [25] carried out three bending tests under very low (–100 °C), low (−30 °C), moderate (20 °C) and high (40 °C) temperatures. For values with high temperature (40 °C), the rupture is noticed at the level of the adhesive while for moderate temperatures (20 °C), the rupture is produced at the level of the concrete near the interface. For low temperature values (−100 °C and −30 °C), the failure of the beams is due to the delamination of the CFRP strip after the opening of a crack in the concrete.

The results showed that the carrying capacity decreased when the temperature increased from 20 °C to 40 °C. As they concluded that the mode of failure changes with temperature.

The behavior of the reinforced element becomes more ductile at high temperatures and more fragile at low temperatures.

Wu et al., (2005) [26] varied the temperature levels (20 °C, 30 °C, 40 °C, 50 °C, 60 °C), the results confirm that increasing the temperature leads to a reduction in breaking load. Wang et al., (2011) [27] found that at 300 °C the ultimate tensile strength of the carbon fiber reinforced material is equal to approximately 50% of that at room temperature and at 700 °C it only retains approximately 5% of its initial value (that of ambient temperature) (Fig. 6).

In the field of reinforcement with composite materials, the most critical point is the concrete/composite interface, which is why there is a lot of research that has studied the influence of very high temperature on the behavior of the system. Generally, very high temperatures significantly affect the mechanical properties and this when the temperature is close enough to the glass transition temperature (Tg) of the matrix.

On the other hand, certain fibers are intrinsically resistant to high temperatures. As the case of carbon fibers, its Young's modulus begins to decrease only when the temperature is above 500 °C and the tensile strength only begins to decrease slightly from 400 °C [1].

Nguyen, (2015) [28] summarizes the most important results relating to the mechanical behavior of an FRP composite exposed to high temperatures, which are:

- Reduction in strength and stiffness;
- Variation in behavior;
- Decrease of Young's modulus.

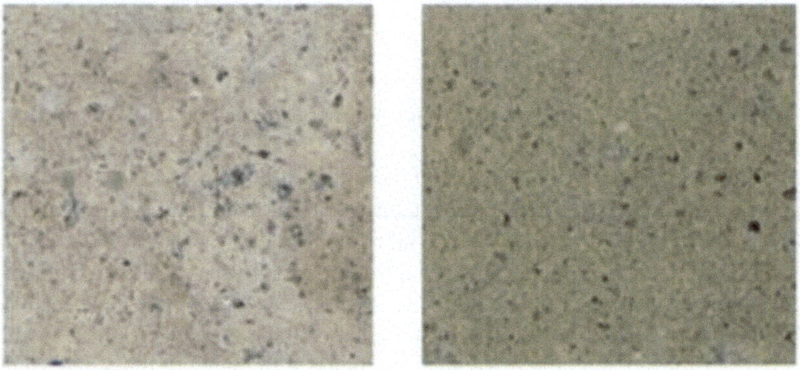

a) *b)*

Fig. 5. Different degrees of roughness of concrete surfaces **a)** grinding treatment **b)** sandblasting treatment [3]

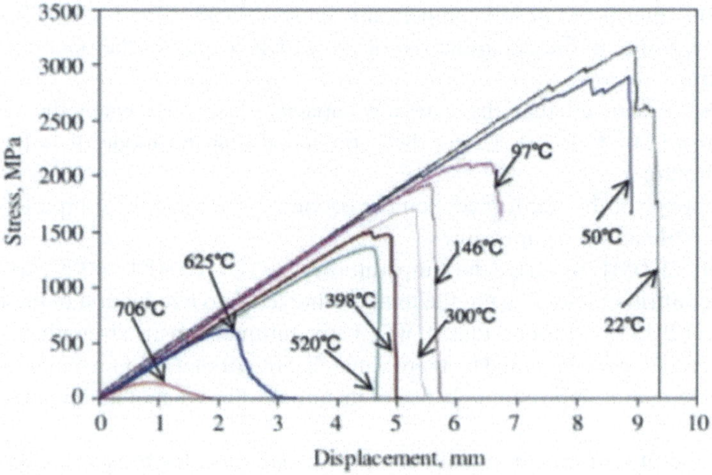

Fig. 6. Stress-displacement of heated-cooled carbon/epoxy laminated specimens [26]

3 Conclusion

It could be concluded from this review that:

For effective reinforcement while using composite materials using carbon fibers or glass fibers, it is necessary to take into consideration the orientation of the fibers because the fibers are strong in one direction only.

The reinforcement depend of numbers of layers because it has noted increases of strength and mechanical properties.

On the other hand improving the concrete interface can be done through surface treatment, which can improve the adhesion between the composite and the concrete, which results in an improvement in the behavior of the reinforced element.

Finally, to succeed in the long-term reinforcement system, temperature, fire and ultraviolet rays must be taken into consideration.

References

1. Wong, R.S., Vecchio, F.J.: Towards modeling of reinforced concrete members with externally bonded fiber-reinforced polymer composites. ACI Struct. J. **100**, 47–55 (2003)
2. Wu, Z.Y.: Etude expérimentale du comportement des poutres courtes en béton armé pré-fissurées et renforcées par matériaux composites sous chargement statique et de fatigue. Ecole des Ponts ParisTech (2004)
3. Correia, L., Teixeira, T., Michels, J., Almeida, J.A., Sena-Cruz, J.: Flexural behaviour of RC slabs strengthened with prestressed CFRP strips using different anchorage systems. Compos. Part B Eng. **81**, 158–170 (2015)
4. Dodiuk, H., Kenig, S.: Low temperature curing epoxies for structural repair. Prog. Polym. Sci. **19**(3), 439–467 (1994)
5. Ed-Dariy, Y., Lamdouar, N., Cherrad, T., Rotaru, A., Barbuta, M., Mihai, P.: The influence of the curing conditions on the behavior of jute fibers reinforced concrete cylinders. Periodica Polytechnica Civ. Eng. **65**(4), 1162–1173 (2021)
6. Van de Weyenberg, I., Truong, T.C., Vangrimde, B., Verpoest, I.: Improving the properties of UD flax fibre reinforced composites by applying an alkaline fibre treatment. Compos. Part Appl. Sci. Manuf. **37**, 1368–1376 (2006)
7. ESPCI, C.A.: Etude des mécanismes d'adhésion à l'interface résine/ciment en vue de la réparation des ouvrages de génie civil. INSA, Lyon (2005)
8. Sasse, H.R., Fiebrich, M.: Bonding of polymer materials to concrete. Mater. Struct. **16**, 293–301 (1983)
9. Toutanji, H., Ortiz, G.: The effect of surface preparation on the bond interface between FRP sheets and concrete members. Compos. Struct. **53**, 457–462 (2001)
10. Perrot, Y.: Influence des propriétés de la matrice sur le comportement mécanique de matériaux composites verre/polyester utilisés en construction navale de plaisance-Cas des résines polyester limitant les émissions de styrène. Université de Bretagne Sud (2006)
11. Grace, N.F., Sayed, G.A., Soliman, A.K., Saleh, K.R.: Strengthening reinforced concrete beams using fiber reinforced polymer (FRP) laminates. ACI Struct. J.-Am. Concr. Inst. **96**, 865–874 (1999)
12. Gornet, L.: Généralités sur les matériaux composites (2008)
13. David, E., Ragneau, E., Buyle-Bodin, F.: Etude expérimentale et modélisation du comportement en flexion de poutres BA renforcées par collage de composites. Rev. Fr. Génie Civ. **5**, 1181–1195 (2001)

14. Neto, P., Alfaiate, J., Vinagre, J.: Assessment of the dependence of CFRP-concrete behaviour on the width of the bonded materials. Compos. Part B Eng. **91**, 448–457 (2016)
15. Khshain, N.T., Al-Mahaidi, R., Abdouka, K.: Bond behaviour between NSM CFRP strips and concrete substrate using single-lap shear testing with epoxy adhesive. Compos. Struct. **132**, 205–214 (2015)
16. Quantrill, R.J., Hollaway, L.C., Thorne, A.M.: Experimental and analytical investigation of FRP strengthened beam response: Part I. Mag. Concr. Res. **48**, 331–342 (1996)
17. Ungureanu, D., Ţăranu, N., Lupăşteanu, V., Isopescu, D.N., Oprişan, G., Mihai, P.: Experimental and numerical investigation of adhesively bonded single lap and thick adherents joints between pultruded GFRP composite profiles. Compos. Part B Eng. **146**, 49–59 (2018)
18. Chicoine, T.: Analyse et conception d'ancrages sur les poutres renforcees en flexion a l'aide de materiaux composites (1998)
19. Mostofinejad, D., Mahmoudabadi, E.: Grooving as alternative method of surface preparation to postpone debonding of FRP laminates in concrete beams. J. Compos. Constr. **14**, 804–811 (2010)
20. Iovinella, I., Prota, A., Mazzotti, C.: Influence of surface roughness on the bond of FRP laminates to concrete. Constr. Build. Mater. **40**, 533–542 (2013)
21. Dawood, M., Rizkalla, S.: Environmental durability of a CFRP system for strengthening steel structures. Constr. Build. Mater. **24**, 1682–1689 (2010)
22. Nguyen, T.-C., Bai, Y., Zhao, X.-L., Al-Mahaidi, R.: Curing effects on steel/CFRP double strap joints under combined mechanical load, temperature and humidity. Constr. Build. Mater. **40**, 899–907 (2013)
23. Nguyen, T.-C., Bai, Y., Zhao, X.-L., Al-Mahaidi, R.: Durability of steel/CFRP double strap joints exposed to sea water, cyclic temperature and humidity. Compos. Struct. **94**, 1834–1845 (2012)
24. Tadeu, A.J., Branco, F.J.: Shear tests of steel plates epoxy-bonded to concrete under temperature. J. Mater. Civ. Eng. **12**, 74–80 (2000)
25. Di Tommaso, A., Neubauer, U., Pantuso, A., Rostasy, F.S.: Behavior of adhesively bonded concrete-CFRP joints at low and high temperatures. Mech. Compos. Mater. **37**, 327–338 (2001)
26. Wu, Z.S., Iwashita, K., Yagashiro, S., Ishikawa, T., Hamaguchi, Y.: Temperature effect on bonding and debonding behavior between FRP sheets and concrete. J. Soc. Mater. Sci. **54**, 474–480 (2005)
27. Wang, W., Sain, M., Cooper, P.A.: Study of moisture absorption in natural fiber plastic composites. Compos. Sci. Technol. **66**, 379–386 (2006). https://doi.org/10.1016/j.compscitech.2005.07.027
28. Nguyen, T.H.: Contribution à l'étude du comportement thermomécanique à très haute température des matériaux composites pour la réparation et/ou le renforcement des structures de Génie Civil. Université Claude Bernard-Ly (2015)

Emerging Trends in Construction Dispute Resolution: Integrating Technology and Collaboration for the Future

Elsayed Salem[1(✉)], Emad Elwakil[1], and Mohamed Hegab[2]

[1] School of Construction Management, Purdue University, West Lafayette, IN, USA
{esalem,eelwakil}@purdue.edu

[2] Department of Civil Engineering and Construction Management, California State University, Northridge, USA
Mhegab@csun.edu

Abstract. Construction disputes have historically caused massive delays and cost overruns, often stemming from adversarial relationships between stakeholders. However, the industry is transforming towards more collaborative, technology-enabled dispute prevention and resolution approaches. This paper explores these emerging trends by extensively analyzing current literature and industry outlooks. It examines innovations in digital documentation, data analytics, virtual collaboration, blockchain transactions, integrated project delivery, and relational contracting frameworks reshaping dispute cultures. The study aims to investigate the ability top integrate new technologies for instance BIM and fuzzy logic systems to reduce disputes in construction projects. The analysis also underscores the need for construction professionals to continually expand their expertise in the latest dispute management best practices as projects become more complex. It concludes that an amalgamation of technological tools, collaborative mindsets, proactive planning, and continuous knowledge building is crucial in developing efficient construction dispute management mechanisms worldwide. Also, the full sharing of information and collaboration between project parties is the effective way to reduce conflicts and disputes at early stages.

Keywords: construction disputes · delays · technology-enabled dispute resolution · digital documentation · data analytics · virtual collaboration · blockchain transactions · integrated project delivery · relational contracting frameworks · dispute cultures · dispute management · knowledge building · construction professionals

1 Introduction and Background

Construction projects are complicated, and the level of collaboration is continuously increasing, which increases disputes probability at any project stage due to the complications of contracts. As a result, construction project disputes are almost unavoidable. The owners' initial selection of delivery methods and contract types influence the seriousness of project disputes. On the other hand, the contractual parties' relationship is

A. S. Mosallam et al. (Eds.): ICATH 2023, SCI, pp. 139–150, 2024.
https://doi.org/10.1007/978-3-031-76557-5_12

complicated, contributing to conflicts. The conflicts directly become disputes if they are not managed [1]. 10–30% of projects in the construction industry experience serious disputes. Usually, from four disputes, one is filed to a claim. Litigation is the most traditional form for dispute resolution. In addition to conventional dispute methods, Alternative Dispute Resolution (ADR) methods can be used to resolve construction project Disputes. For instance, Mediation and Arbitration are popular ADR methods used to resolve disputes [1]. Many studies attempted to introduce a solution for construction disputes using fuzzy logic systems and BIM [2–10]. In construction projects, many parties are involved in the contract. Conflicts between parties are almost unavoidable due to the complexity of the construction work. Therefore, conflicts may arise, transforming into disputes if they are not managed. Disputes must be resolved, and the process may be transferred to third-party intervention. Disputes usually lead to claims if they are not resolved [1].

There is a lack in literature on integrating new technologies to solve construction disputes. Therefore, this study aims to investigate the available effective methods, tools, and technologies used to prevent and resolve construction project disputes. The study covers a timeline of 10 years, starting 2015 till 2024.

2 Research Methodology and Data Collection

The methodology applied in this study is based on an extensive review of the studies related to construction dispute resolution, as illustrated in Fig. 1.

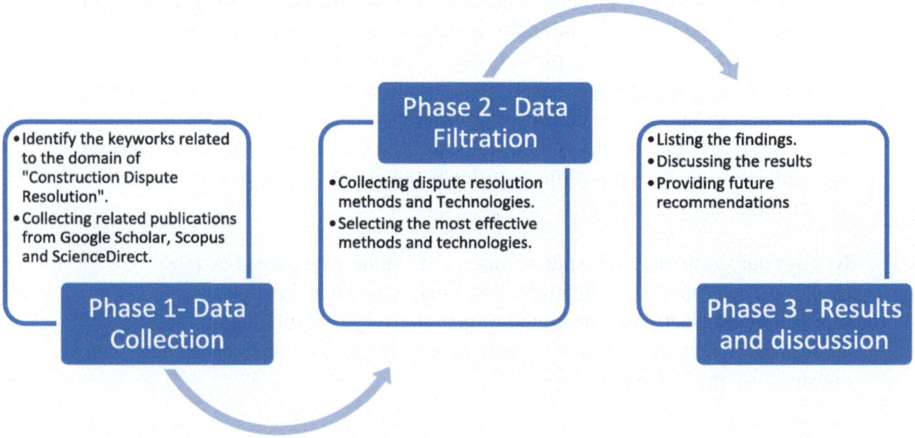

Fig. 1. Study Methodology Flowchart

3 Phase 1: Data Collection

3.1 Identifying the Keyworks Related to the Domain of "Construction Dispute Resolution"

To determine the research problem's importance, the keywords combination "Construction and Disputes and Resolution" was used to collect the related published research using the Scopus database. The publications were first classified by subject area. The results are introduced in Table 1 and plotted in Fig. 2.

Table 1. Top 10 Subject Areas Related to Construction Disputes Resolution

Rank	Subject area	Number of publications
1	Engineering	751
2	Social Sciences	320
3	Business, Management and Accounting	294
4	Environmental Science	75
5	Computer Science	75
6	Earth and Planetary Sciences	57
7	Arts and Humanities	49
8	Decision Sciences	45
9	Economics, Econometrics and Finance	31
10	Materials Science	24

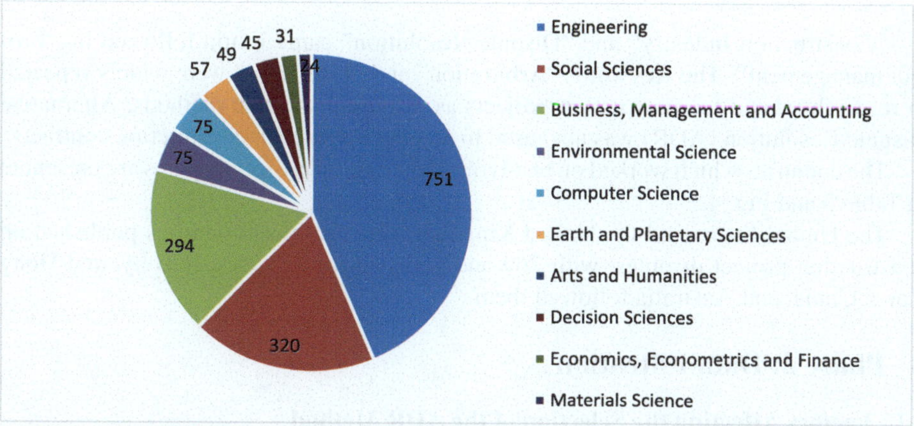

Fig. 2. Top 10 Subject areas related to Construction Disputes Resolution (Source: SCOPUS data base)

Engineering was the dominant subject area, with 751 publications. This was followed by social, business management, environmental, and computer science.

The top 15 keywords filtered the most used to show the essential keywords most related to the study problem. The resulting top 15 keywords are shown in Table 2 and plotted Fig. 3.

Table 2. Top 15 keywords related to Construction Disputes Resolution

Rank	Keywords	Number of publications
1	Construction Industry	351
2	Dispute Resolution	335
3	Project Management	223
4	Laws And Legislation	171
5	Contracts	137
6	Construction Projects	123
7	Construction	113
8	Alternative Dispute Resolution	107
9	Construction Dispute	73
10	Decision Making	63
11	Contractors	58
12	Arbitration	58
13	Costs	56
14	Construction Management	53
15	Litigation	52

"Construction Industry" and "Dispute Resolution" came in top followed by "Project management". The Keywords "Arbitration" and "Litigation" were widely repeated in research related to construction projects as dispute resolution methods. Alternative Dispute Resolution (ADR) was also used to resolve disputes in construction contracts.

The countries which worked on resolving construction project disputes are presented in Table 3 and Fig. 4.

The United States and the United Kingdom were the most countries published on construction project disputes, with 269 and 166 publications, respectively, and Hong Kong, China, and Australia followed them.

4 Phase 2: Data Filtration

4.1 Factors Affecting the Selection of the ADR Method

The construction projects are unique in nature, resources, specifications, contracts, and involved parties. Therefore, selecting a suitable ADR method is not easy as it depends on many factors. Table 4 summarizes the potential factors according to [11].

Most of the factors affecting the selection of the ADR method are related to the level of complexity of the dispute and the flexibility of the contract.

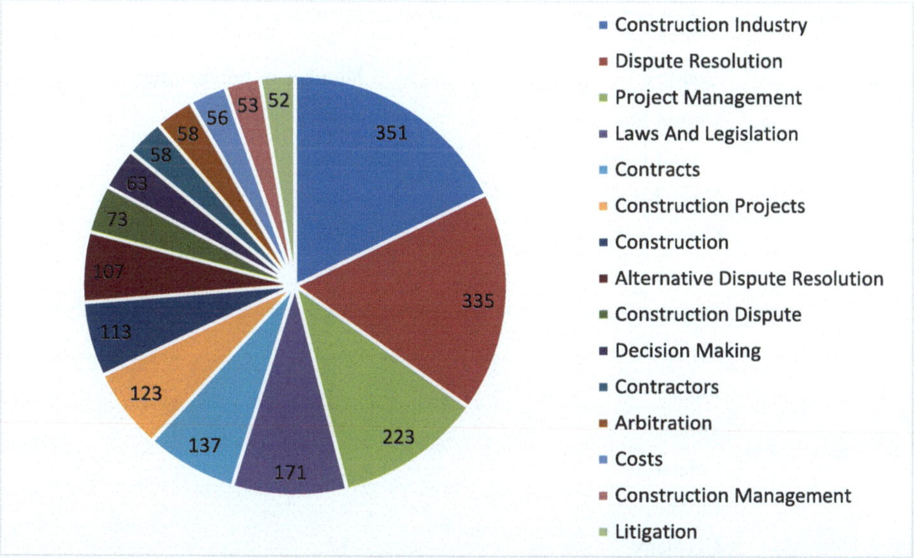

Fig. 3. Top 15 keywords related to Construction Disputes Resolution.

4.2 Causes of Disputes

Construction disputes should be controlled in earlier project stages. Therefore, identifying the causes of disputes is crucial. Table 5 summarizes the potential causes of disputes in construction projects [11].

Additionally, the probability of disputes depends on the project type. As shown in Fig. 5, the commercial projects recorded 60% of the total claims in 3 different project categories in KSA [12].

5 Phase 3: Results and Discussion

5.1 Early Dispute Resolution and Avoidance

Most research on construction project disputes focuses on what happens after disputes have occurred. However, there are advantages to reducing the causes and likelihood of disputes at early stages [13].

5.2 Traditional Method for Dispute Resolution: Litigation

Litigation is the most traditional method for disputes resolution. In construction, Litigation is defined as requesting a legal action in court. The court will determine or enforce the obligations and rights of one party. While there are Alternative Dispute Resolution Methods (ADR), Litigation is considered the last resort solution, sometimes un avoidable [1].

Table 3. Top 20 Countries with publications on Construction Disputes Resolution.

Rank	Country	Number of publications
1	United States	269
2	United Kingdom	166
3	Hong Kong	86
4	China	57
5	Australia	57
6	Canada	53
7	Malaysia	42
8	India	35
9	Turkey	30
10	Lebanon	23
11	Indonesia	21
12	Germany	17
13	New Zealand	16
14	Taiwan	15
15	South Africa	14
16	South Korea	13
17	Egypt	13
18	United Arab Emirates	11
19	Japan	11
20	Iran	10

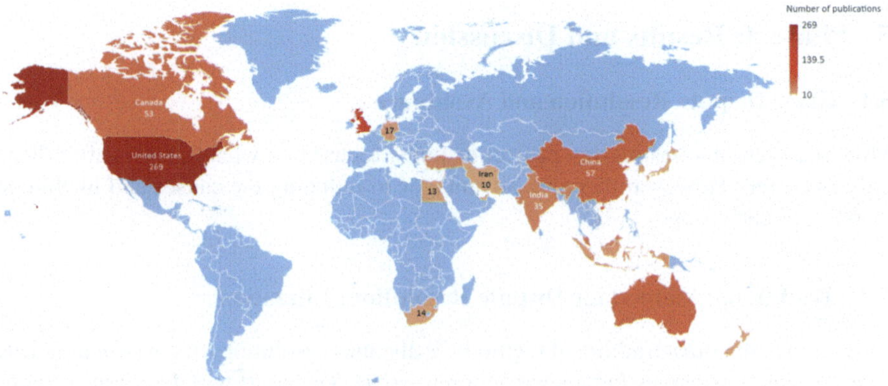

Fig. 4. Top 20 Countries with Publications on Construction Disputes Resolution

Table 4. Factors affecting the selection of the ADR method

Number	Factors affecting the selection of the ADR method
1	Confidentiality
2	Degree of control by parties
3	Third-party control on the process
4	Flexibility in implementation
5	Enforceability of the method
6	Bindingness of the decision
7	Ability to appeal if not satisfied with the result
8	Degree of formality
9	Cost of implementing the methods
10	Time to reach a settlement
11	Type of contract
12	Preservation of relationship between parties
13	Suitability in local law system
14	Complexity of the dispute

5.3 Effective Alternative Dispute Resolution Methods

ADR methods are more effective than traditional methods due to their flexibility. The flexibility of involving third party adviser and unformal way of negotiation make it easier to reach the final dispute solution. Figure 6 introduces a comparison between the most effective ADR methods.

5.4 Effectiveness and Ranking of ADR Methods

ADR methods methods are effective in solving construction disputes than traditional methods; however, each ADR methods has advantages and drawbacks. The most effective ADR methods are ranked in Table 6 and Fig. 7 based on the results adapted from [11].

Negotiation and Arbitration came on top of most effective ADR methods with RII score of 0.729 and 0.8 respectively. Mediation, Adjudication, and Conciliation recorded approximately the same lower RII score in range (0.614–0.643).

5.5 Tools and Technologies

- **Benefits of using BIM for reducing claims in the construction industry**

Building Information Modeling (BIM) is a computer-based technology that aims to manage construction projects effectively. However, BIM is a relatively new technology,

Table 5. Causes of disputes in construction projects. Adopted from [11]

Number	Causes
1	Failure to properly administer the contract
2	Error and/or omissions in contract documents
3	Incomplete design information or Employer requirement
4	Failure to understand and/or comply with its contractual obligations by either party
5	Poorly managed construction process leading to shortage of resources and quality issues
6	Diverse interpretation of contract terms
7	Inadequate risk identification/allocation
8	Lack of cooperation and trust among parties
9	Opportunistic behavior of project parties
10	The reluctance of project participants to deal with changes
11	Conflicting goals and objectives of project parties
12	Lack of experience in construction practices and management
13	Lack of interpersonal skills among professionals
14	External uncertain factors such as weather conditions or environmental regulations
15	External changes, such as changes in market conditions and environmental regulations

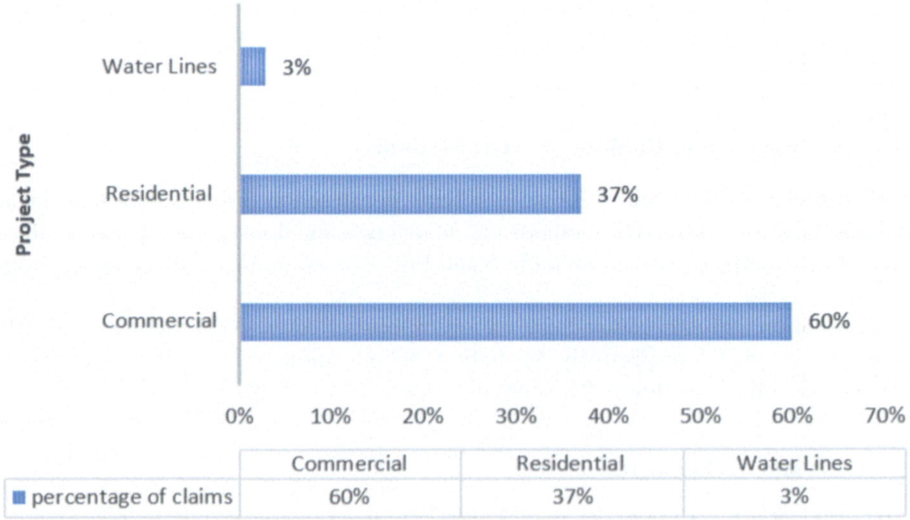

Fig. 5. Claims percentages for different types of projects. Adopted from [12]

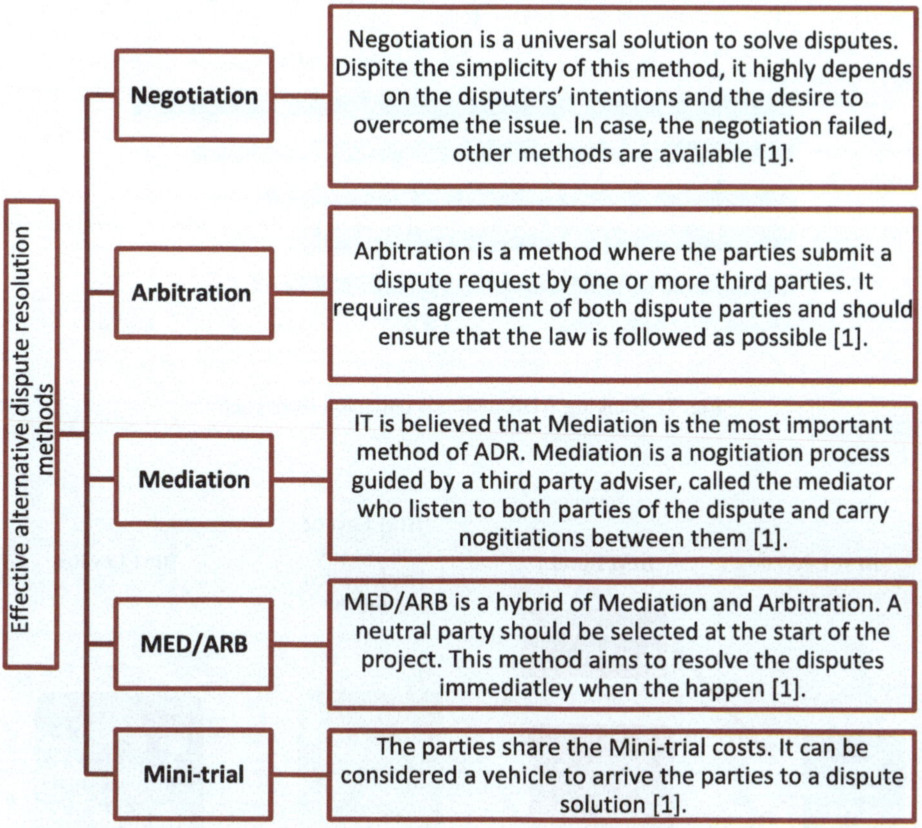

Fig. 6. Effective alternative dispute resolution methods

Table 6. Ranking ADR methods (Adapted from [11]).

ADR Method	RII	Rank
Negotiation	0.800	1
Arbitration	0.729	2
Mediation	0.643	3
Adjudication	0.614	4
Conciliation	0.614	5

and it rapidly gained acceptance in the construction field. BIM effectively reduces disputes by offering a way to try more options and analyze them in real time [12]. Figure 8 illustrates the development of BIM in 4 levels to reduce conflicts between construction project stakeholders. As a result, the disputes will be reduced as the conflicts are the leading cause of disputes.

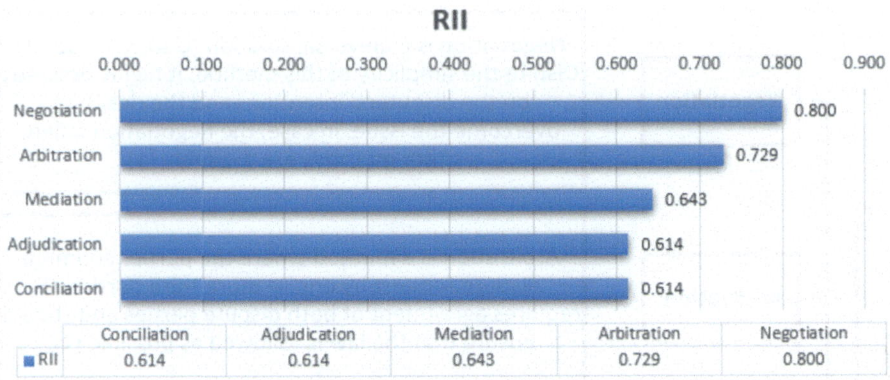

Fig. 7. Ranking ADR methods (Adapted from [11]).

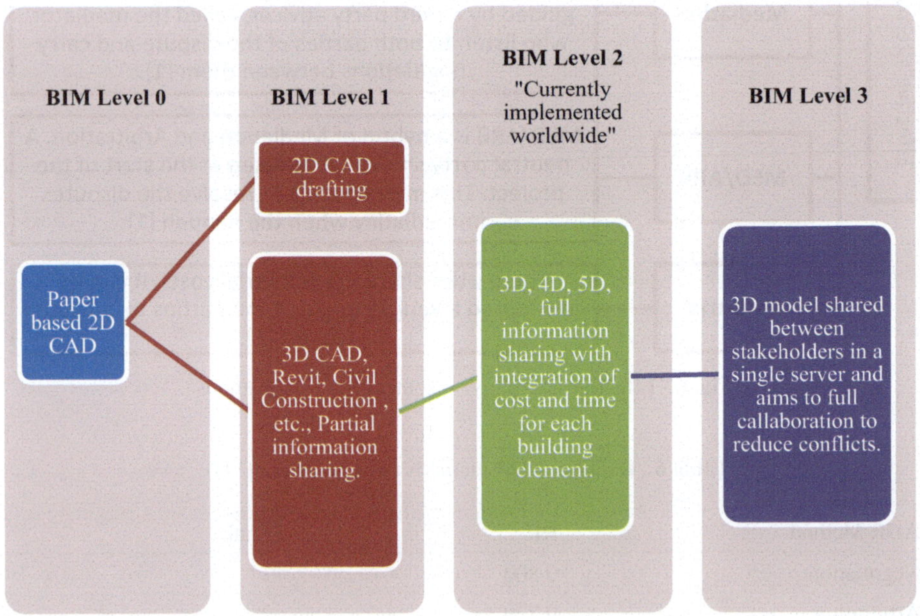

Fig. 8. The development of BIM levels to reduce conflicts between construction project stakeholders (Adapted from [12])

Dispute modeling and evaluation using fuzzy set theory and fuzzy logic

Artificial Intelligence and Fuzzy techniques have been widely used in solving problems in construction engineering research [14–17] Due to the complexity of the dispute problem, fuzzy systems may be used to solve it effectively [13]. Figure 9 introduces a flow chart of a fuzzy logic system.

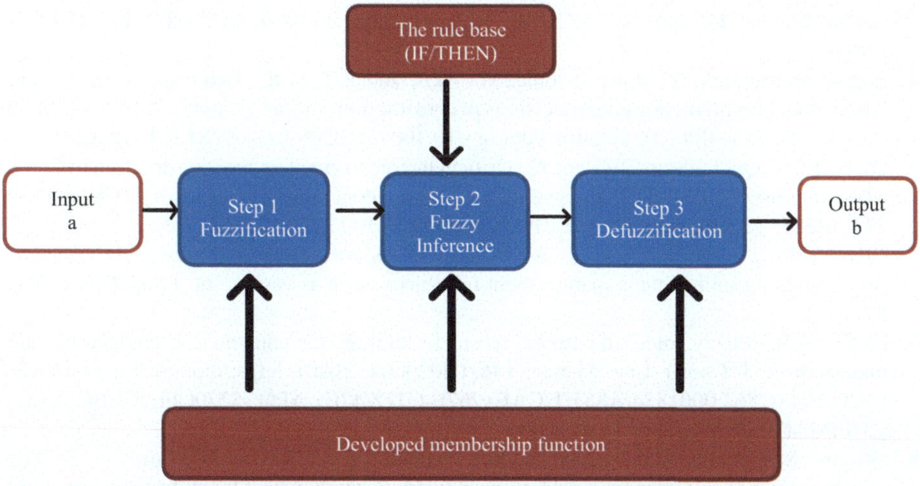

Fig. 9. A Fuzzy logic system flow chart (Adapted from [13]).

6 Conclusion

The construction industry is complicated and requires the involvement of many parties to accomplish the projects. Due to the complexity of construction projects and the conflicts between parties, disputes may arise. If the construction disputes are not managed and solved, the disputes transform directly into claims. However, the efforts and existing methods for solving conflicts are not avoidable and need extra efforts and innovative methods, tools, and integration of new technologies to be solved.

Traditional methods were widely used to solve construction disputes. Due to the complexity of the construction projects, the ADR methods were proposed as alternative dispute resolution methods.

On the other hand, Fuzzy logic systems and BIM technologies are highly recommended to deal with disputes. The full sharing of information and collaboration between project stakeholders when using BIM reduces conflicts and the probability of disputes.

The proposed dispute resolution methods proposed in published research in 10 years from 2014 to 2024 were only considered in this study. Also, construction project disputes were only included. Future studies on using different technologies to resolve conflicts in early project stages are highly recommended.

References

1. Alaloul, W.S., Hasaniyah, M.W., Tayeh, B.A.: A comprehensive review of disputes prevention and resolution in construction projects. MATEC Web Conf. **270**, 05012 (2019). https://doi.org/10.1051/MATECCONF/201927005012
2. Mathew, B.K., Kassim, R.: Fuzzy logic approach in construction delay analysis. Int. Res. J. Eng. Technol. **7** (2020)
3. Nasirzadeh, F., Carmichael, D.G., Jarban, M.J., Rostamnezhad, M.: Hybrid fuzzy-system dynamics approach for quantification of the impacts of construction claims. Eng. Constr.

Archit. Manag. **26**, 1261–1276 (2019). https://doi.org/10.1108/ECAM-08-2017-0150/FULL/PDF

4. Bagherian-Marandi, N., Ravanshadnia, M., Akbarzadeh-T, M.R.: Two-layered fuzzy logic-based model for predicting court decisions in construction contract disputes. Artif. Intell. Law (Dordr). **29**, 453–484 (2021). https://doi.org/10.1007/S10506-021-09281-9/TABLES/15
5. Qi, R., Li, S., Qu, L., Sun, L., Gong, C.: Critical factors to green mining construction in China: a two-step fuzzy DEMATEL analysis of state-owned coal mining enterprises. J. Clean. Prod. **273**, 122852 (2020). https://doi.org/10.1016/J.JCLEPRO.2020.122852
6. Chupin, M.M., Sergeevich Katasev, A., Muratovich Akhmetvaleev, A., Kataseva, D.V.: Neuro-fuzzy model in supply chain management for objects state assessing. Int. J Sup. Chain. Mgt. **8** (2019)
7. Fayek, A.R.: Fuzzy logic and fuzzy hybrid techniques for construction engineering and management. J. Constr. Eng. Manag. **146**, 04020064 (2020). https://doi.org/10.1061/(ASCE)CO.1943-7862.0001854/ASSET/CAE1267F-C372-47F5-AF6F-5270CFE9CEBC/ASSETS/IMAGES/LARGE/FIGURE4.JPG
8. Hoseini, S.A., Fallahpour, A., Wong, K.Y., Mahdiyar, A., Saberi, M., Durdyev, S.: Sustainable supplier selection in construction industry through hybrid fuzzy-based approaches. Sustainability **13**, 1413 (2021). https://doi.org/10.3390/SU13031413
9. Hipel, K.W., Fang, L.: The graph model for conflict resolution and decision support. IEEE Trans. Syst. Man Cybern. Syst. **51**, 131–141 (2021). https://doi.org/10.1109/TSMC.2020.3041462
10. Hendiani, S., Bagherpour, M.: Developing an integrated index to assess social sustainability in construction industry using fuzzy logic. J. Clean. Prod. **230**, 647–662 (2019). https://doi.org/10.1016/J.JCLEPRO.2019.05.055
11. Illankoon, I.M.C.S., Tam, V.W.Y., Le, K.N., Ranadewa, K.A.T.O.: Causes of disputes, factors affecting dispute resolution and effective alternative dispute resolution for Sri Lankan construction industry. Int. J. Constr. Manag. **22**, 218–228 (2022). https://doi.org/10.1080/15623599.2019.1616415
12. Abougamil, R.A., Thorpe, D., Heravi, A.: Investigating the source of claims with the importance of BIM application on reducing construction disputable claims in KSA. Buildings **13**, 2219 (2023). https://doi.org/10.3390/buildings13092219
13. Naji, K.K., Mansour, M.M., Gunduz, M.: Methods for modeling and evaluating construction disputes: a critical review. IEEE Access. **8**, 45641–45652 (2020). https://doi.org/10.1109/ACCESS.2020.2976109
14. El-Kholy, A.M., El-Shikh, M.Y., Abd-Elhay, S.K.: Which fuzzy ranking method is best for maximizing fuzzy net present value? Arab. J. Sci. Eng. **42**, 4079–4098 (2017). https://doi.org/10.1007/S13369-017-2573-4/METRICS
15. Elwakil, E., Zayed, T.: Data management for construction processes using fuzzy approach. Construction Research Congress 2012: Construction Challenges in a Flat World, Proceedings of the 2012 Construction Research Congress, pp. 1222–1231 (2012). https://doi.org/10.1061/9780784412329.123
16. Elhegazy, H., et al.: Artificial intelligence for developing accurate preliminary cost estimates for composite flooring systems of multi-storey buildings. J. Asian Archit. Build. Eng. **21**, 120–132 (2022). https://doi.org/10.1080/13467581.2020.1838288
17. Elhegazy, H., Badra, N., Haggag, S.A., Abdel Rashid, I.: Implementation of the neural networks for improving the projects' performance of steel structure projects. J. Ind. Intgr. Manag. **7**, 133–152 (2022). https://doi.org/10.1142/S2424862221500251/ASSET/IMAGES/LARGE/S2424862221500251FIGF11.JPEG

Radioactivity in Building Materials: An Analytical Review

Nezha Mejjad[1](✉), Mounia Messouli[2], Amine el Mahdi Safhi[3], Abdelmourhit Laissaoui[1], Ouafa El Hammoumi[2], and Ahmed Fekri[2]

[1] National Centre for Nuclear Energy, Science and Technology (CNESTEN), Rabat, Morocco
mejjadnezha@gmail.com
[2] Faculty of Sciences Ben M'Sik, Hassan II University of Casablanca, Casablanca, Morocco
[3] Department of Building, Civil and Environment Engineering, Concordia University, Montreal, Canada

Abstract. Building Materials such as stone, bricks and decorative rock among others added to concrete, may contain natural radioelements such as potassium (^{40}K) uranium (^{238}U), and thorium (^{232}Th). The radiation released from such materials depends on the type of material but is usually very low. The present study analyzes the current knowledge about radioactivity in building materials using a bibliometric tool and calculates the long-term variations of external and internal dose residents are exposed to because of ^{226}Ra, ^{232}Th, and ^{40}K in a standard room in thirteen different countries around the world using a computer code RESRAD BUILD. The bibliometric analysis showed that the research studies focus is on assessing the radiation levels and doses as well as potential risks to human health while it was noticed a lack of studies related to mitigation and remediation processes. The simulations carried out using RESRAD BUILD showed that outdoor and indoor exposure to radiation displays low variations over the average lifetime of 70 years. The inhalation doses show values more significant over the 70 years in Nigeria, followed by Australia, Finland, and the other 10 countries. The outdoor doses were more important in Finland, Nigeria and Cuba. These observed variations among countries shed light on the importance of considering each country's geographical location and geological characteristics while establishing norms and guidelines related to emitted radiation limits from building materials.

Keywords: Radioactivity · Building Materials · Bibliometric analysis · RESRAD BUILD · dose rate · radiation level

1 Introduction

According to the U.S Environmental Protection Agency (EPA, 2022), some building materials such as stone, brick, and decorative rock, among others, added to concrete may contain natural radioelements such as ^{40}K, ^{238}U, and ^{232}Th. Building materials are derived from the Earth's crust, which explains the occurrence of a small amount of natural radioactive material from the Earth. The radiation emitted by these materials depends on the type of material but is usually very low. The main naturally occurring radioactive

A. S. Mosallam et al. (Eds.): ICATH 2023, SCI, pp. 151–173, 2024.
https://doi.org/10.1007/978-3-031-76557-5_13

materials in building materials such as brick, gypsum, sandstone, and granite are radium, uranium and thorium. According to the Center for Disease Control and Prevention (CDC, 2022), these radioactive materials can decay into radon (radioactive gas), resulting in increased radiation levels. It is worthy to mention that the radiation depends mainly on the type and the amount of the used material.

It has recently been reported that building materials of both natural and anthropogenic (industrial and mining by-products) origin contributes considerably to the population external and internal exposure to natural radioactivity (Trevisi et al., 2008; Estokova et al., 2022). The decay product of natural radioelements in building materials leads to the radon release, which is a radioactive gas (Thumvijit et al.; 2020; Jasaitis and Peciuliene, 2021). According to the UNSCEAR (2000) report, 50% of the total received radiation dose by the public is originated from radon and its progeny (Radon short-living decay products) (UNSCEAR, 2000). This radioactive gas is ranked as the second most common cause of lung cancer (Kohli, 2000; WHO, 2009). Based on a report of the World Health Organization (WHO, 2009), radon gas is considered a substantial indoor air contaminant, where it is reported to be causing between 15000 and 20000 cancer cases deaths yearly in the United States (Tchorz-Trzeciakiewicz and Olszewski, 2019). In addition, in Germany, five percent of cases of lung cancer are related to public exposure to radon released from buildings. (Wichmann et al., 2006).

Thus, the knowledge of the levels of radioactivity in building materials is crucial for preventing and reducing human exposure to the emanating radiation. In this sense, this review analyzes and reviews the carried-out research associated with the building material radioactivity levels and human exposure rate using a bibliometric analysis tool. The objective then is to compare the concentrations of natural radionuclides in building materials measured in different countries and calculating the long-term variations in the effective dose that residents are exposed to because of Ra-226, Th-232, and K-40 in a standard room using a computer code called RESRAD BUILD.

2 Material and Methods

2.1 Data Collection

This analytical study investigates available data on radioactivity concentrations in building materials, including granite, marble, cements, and bricks. The study consists of analyzing the studies that measured the concentrations of radioactivity in such material, and doses that can be inhaled by the public indoors and outdoors buildings (Fig. 1). In the present study, Scopus and Dimensions databases were used to extract the data. The search query "Radioactivity AND Building AND Material" was performed on December 7th, 2023. Only the research paper in English were considered. As shown in Table 1, Scopus displayed 1570 documents published between 1954 and 2024 while dimension.ai exhibited 855 documents published between 1933 and 2024. The search was carried out within, abstract, title and keywords.

2.2 Data Analysis

In order to assess the interest in measuring and evaluating human exposure to radiation from material used for building purpose such as brick, cement, and limestone by

Table 1. The number of the found publications related to the search query.

Databases	Documents	Search query	Date Range
Scopus	1570	Radioactivity AND Building AND Material	1954–2024
Dimensions.ai	855	Radioactivity AND Building AND Material	1933–2024

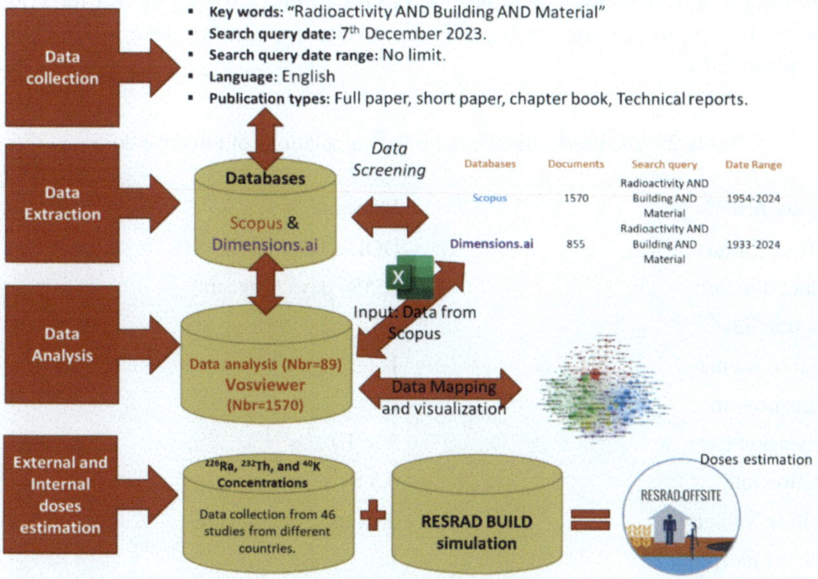

Fig. 1. Study's flowchart.

the scientific community, we opted for a bibliometric analysis of research carried out on radiation levels measurement in building materials. The analysis was carried out using VosViewer software (Fig. 1), which is used for constructing, mapping and visualizing networks of articles selected from different databases. Thus, Vosviewer provides functionality of text mining that can be used for establishing and mapping networks of important keywords or terms from a body of scientific literature.

2.3 The Use of RESRAD-BUILD Code for Simulation

In the present study, the computer code RESRAD-BUILD 4.0 was used to assess the indoor radiation exposure (Adelikhah et al., 2022). This computer code belongs to the RESRAD family codes, which has been established at *Argonne National Laboratory* in the United States to investigate potential biota and human radiation exposures from the residual radioactive materials (Pepin, 2018; Aydarous et al., 2022). This code was used in this study, to estimate the radiological dose encountered dwellers living in a contaminated building by radioactive elements. The average activity concentration of three radionuclides (^{226}Ra, ^{232}Th, and ^{40}K) measured in concrete samples from 13 countries around the world were utilized as an input parameter to determine the dose over 1, 5, 10, 20, 50 and 70 years in two main pathways, internal and external radiation exposure. The parameters of a standard model room were adopted, i.e., with the following dimensions, 4.0 × 5.0 × 2.8 m, as defined in the report of Radiation Protection 112 (EC, 1999), to carry out the simulation. The position of the exposed target person for the external dose assessment was designated in the center of the room and 1 m above the floor (Yu et al., 2000). 80 % of indoor occupancy rate was assumed (UNSCEAR, 2000). The building materials density was assumed to be around 2400 kg m^3 for the concrete. Additional key input parameters used as input data concerning dose assessments are shown in the Table 2.

Table 2. The used parameters for the calculations of effective dose.

Input parameters	Value
Dose/Risk Library	DOE STD-1196–2011
Exposure duration	25550 days (70years)
Indoor fraction	0.5
Number of rooms	1
Time integration	17
Resuspension rate	5×10^{-7} s^{-1}
Ventilation rate	0.5 h^{-1}
Deposition Velocity	0.01 m s^{-1}
Number of receptors	1
Room dimension	4 × 5 × 2.8 m
Breathing rate	18 m^3 d^{-1}
Time fraction	0.8
Shielding thickness	0 cm
Ingestion rate	0.0001 m^2 h^{-1}
Source type	Volume
Receptor location	2 × 2.5 × 1 m
Source geometry	Rectangular
Input Radionuclide unit	Bq g^{-1}
Source location	4 × 5 × 2.8 m
Air fraction	0.1

3 Results

3.1 Data Analysis: Bibliometric Analysis

The extracted documents are mainly articles and represent 78% and 90% of the database according to the Scopus and dimensions.ai databases, respectively. For both databases, *Radiation Protection Dosimetry* and *Journal of Environmental Radioactivity* are the top journals where the highest number of papers related to radioactivity in building materials were published. The extracted 1570 documents indexed in the Scopus database have a total keyword of 8770 and only those with a minimum occurrence of 20 (240 keywords) were used for constructing and generating the visualization map. Five clusters were formed as shown in Figure 2. The map displays five colored clusters linked by 20,027 links and 137,391 total link strength.

The Red cluster includes 73 terms including "radioactivity" (occurrence: 790), Radiation, radiation protection, cement, and radioactive waste. It also regroups some terms related to some specific materials as phosphogypsum, fly ash, and red mud. These keywords are related to the characterization of those materials as construction materials including metals content assessment and radiation level measurement among other processes of construction material characterization and waste assessment. The occurrence of the terms "recycling", "metals", "reuse", "durability", "disposal", "trace elements" indicate the growing interest in valorizing industrial and mining waste into construction materials, which will have substantial environmental impact by decreasing the waste volume and providing ecological raw materials (Segui et al., 2023, Safhi, 2022; Safhi et al., 2024).

The green cluster with 66 keywords covers topics related to radionuclide concentration measurement in building materials. The main keywords are natural radioactivity, dose rates, radium equivalent activity, and annual effective dose. Building materials such as granite can be a source of the exposition of humans to radiation, which explain the interest in assessing radiation levels and dose rate that can be inhaled by humans indoors and outdoors buildings so, to determine safe inhalation limits and guarantee the safe use of those building materials. The cluster also includes terms related to the technique used for the determination of radionuclides (spectrometry gamma) in construction materials such as brick, sand, and granite. This suggests that brick, granite and sand are among the most investigated samples for radiation levels and dose measurement. Indeed, those materials can contribute to indoor and outdoor radiation as they are used for building construction as well as decorative inside buildings, mainly granite.

The blue cluster contains 58 terms and covers topics related to radon inhalation and its possible effect on human health. The main keywords covered by the blue cluster are building and construction materials, radiation exposure, radioisotope, and air pollution. Building materials such as marble, granite, and brick encloses very low concentrations of naturally occurring radioactivity. These building materials can release radon gas causing the rise of its levels inside the building and pose a serious health risk. Radon in buildings can cause lung cancer. Wichmann et al. (2006), indicated that in Germany, the risk of lung cancer is found associated with residential exposure to radon, with approximately five per cent of lung cancer cases reported in Germany attributable to residential exposure to radon in the population.

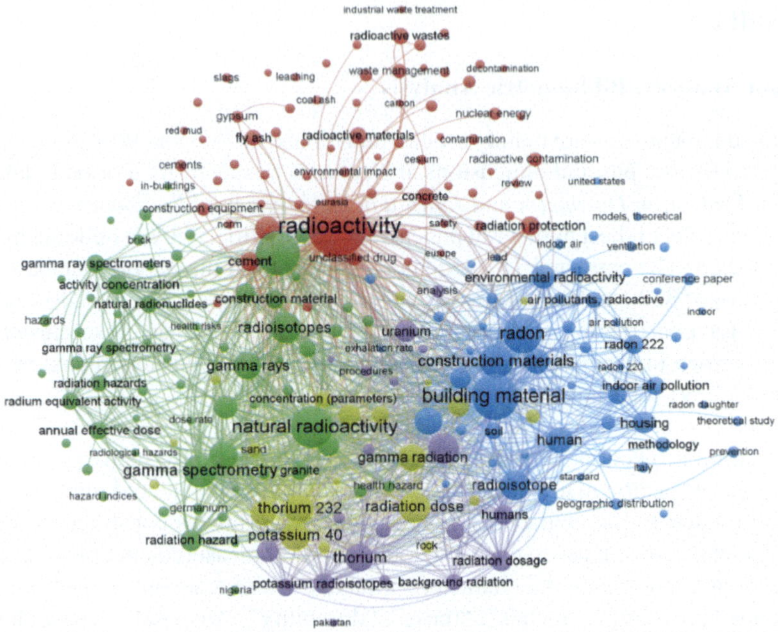

Fig. 2. Network visualization map based on keyword occurrences. *The number of publications linked to the search query or keywords governs the circle's size; the line denotes the relation between every keyword. The relation between each keyword is strong when the line is short.*

Two other minor clusters were identified. The yellow cluster (22 items) is mostly related to terms associated with controlled studies, materials testing, and dosimetry. The purple cluster (20 items) is found to be related to radiation monitoring, radiation dosage, and ionizing radiation.

The analysis of the clusters showed that the research studies focus is on the assessment of radiation levels and doses and the potential risk to human health. However, it should be noted that there is a lack of studies on mitigation and remediation processes. However, the current trend is towards the use of environmentally friendly materials produced from mining and industrial waste, which allow the conservation of natural resources (e.g., sand exploitation), to ensure sustainability and better control and management of waste. In contrast, Smetsers and Tomas (2019) pointed out that the use of such materials in response to the demand for sustainability could lead to higher indoor radiation doses.

However, the studies conducted to evaluate the risks related to radiation exposure to human health are still not well studied worldwide. The density map shows an unbalanced distribution of studies on radiation levels in building materials (Fig. 3). China ranks first with 123 documents, followed by the United States (121 documents), Italy (113 documents), India (111 documents) and Egypt (88 documents). The possible reason for this is the geological characteristic of each country, where any building material made from natural rock or mineral raw material - depending on the geological origin - contains a natural proportion of uranium and radium. The density map only shows countries with five or more documents related to radioactivity in building materials. Technological capabilities may be one of the reasons for the observed uneven progress of scientific research input between countries.

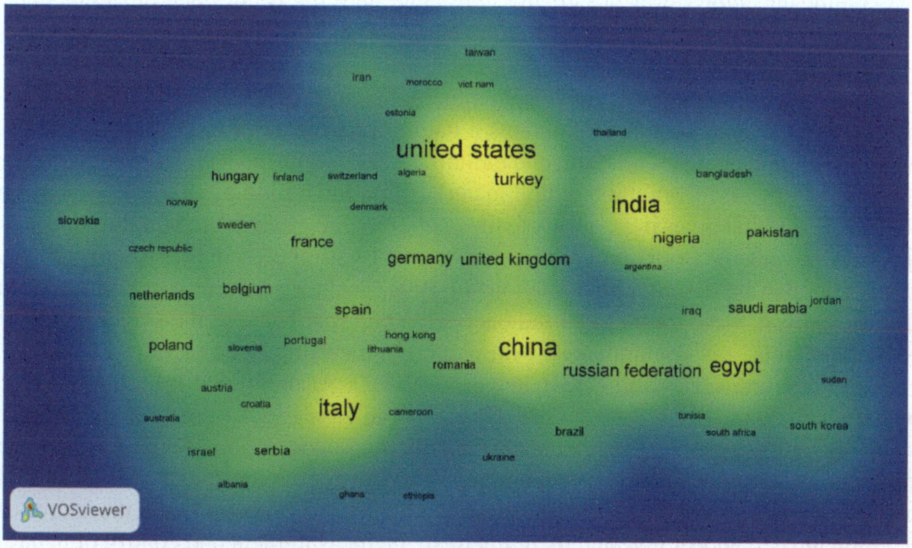

Fig. 3. Map of density established based on Bibliographic Data (bibliographic coupling analysis of the countries established based on publications weight).

3.2 Building Material as a Source of Natural Radioactivity

The radiation to which the public is exposed can originate from natural and artificial sources. Natural sources comprise cosmic radiation, terrestrial radiation and internal radiation retained in the human body because of inhalation or ingestion of radionuclides (UNSCEAR, 2000; IAEA, 2007). The magnitude and degree of this natural exposure are related to geographical location and human activities (UNSCEAR, 2000; IAEA 2007). The issue of dose rate from cosmic rays is influenced by the height above sea level, while the local geology is the main factor affecting radiation from the ground (Ramachandran, 2011).

Building materials can be a source of human exposition to radiation. Materials such as marble, granite, and bricks used as construction and decorative materials may enclose high levels of radioactivity, which varies according to the type of rock source, the geological locations and settings and the geochemical characteristics of the materials. Indeed, terrestrial materials such as granite naturally enclose significant concentrations of radioactivity, which can differ from one site to another according to geological features, including the nature, type and radionuclides content in the bedrock and mode of formation of the granite. Sedimentary rocks also contain low amounts of radionuclides, which occur naturally in the original rock; the mode of formation of sedimentary rock could affect the amount of these radionuclides. Several studies have linked the radioactivity content in rocks to the geological parameters where uranium deposits have mostly occurred in four types of rocks: granite, volcanic rocks, sandstone and metamorphic rocks (Zhang et al., 2021; Zhu et al., 2001). Indeed, Cho et al. (2015) reported that significant radon doses are greatly correlated with the occurrence of sedimentary and metamorphic rocks rich in organic matter (e.g., granitic gneiss, limestone, shale, ironstone and phosphatic rocks).

Every building material is made from mineral raw material or natural rock - depending on its geological origin – it contains a natural proportion of radium and uranium. When the natural decay of uranium and radium occurs, radon and its related products are formed and released from the building materials into the buildings (Fig. 4). Radon is a radioactive gas that gives it the ability to move up from the ground into buildings through openings in walls or floors that are in close contact with the ground (EPA.gov, 2023). Roads can also be a source of radiation. In 1994, asphalt-concrete roads in Taoyuan County, Taiwan, were found to be radioactively abnormal because the slag used in the asphalt concrete was contaminated with naturally occurring radionuclides of the thorium and uranium series and ^{40}K (Chen et al., 2002). At the global level, the average annual received radiation dose by the world's population is 2.4 mSv/y, of which 80% is from naturally occurring radioelements and the remainder is mostly from man-made sources (UNSCEAR, 2000). Since the majority of the population spends nearly 80% of their time indoors (Suliman and Alsafi, 2021), building materials represent the source of radiation inhaled and/or ingested by humans, so the measurement and control of natural radiation in buildings is of great interest for human safety.

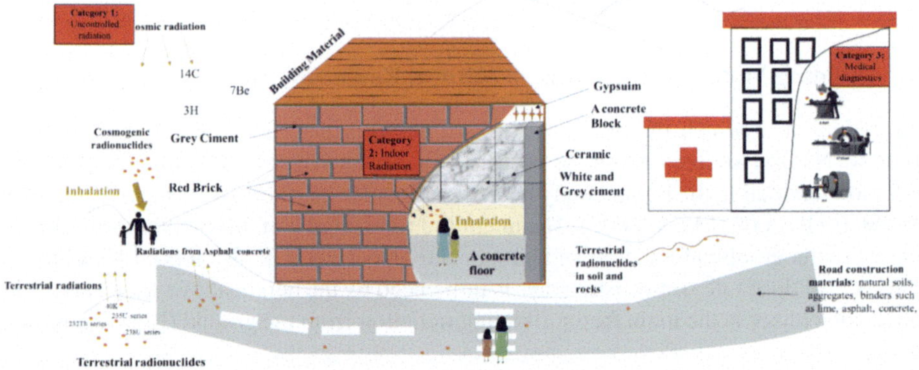

Fig. 4. Possible sources of radiation (Designed by N. MEJJAD).

The average annual dose varies according to the source. Three source categories were reported to account for around 100% of the received radiation dose by the population: uncontrolled natural sources (e.g., cosmic radiation and ^{40} K), terrestrial radionuclides in soil, rock and soil-based building materials, and medical diagnostics (Fig. 5). The other sources of ionizing radiation (e.g., nuclear industry) account only for 1% of the total received dose by the population. Thus, according to the relative importance of the source category, the received dose levels may change (Fig. 5).

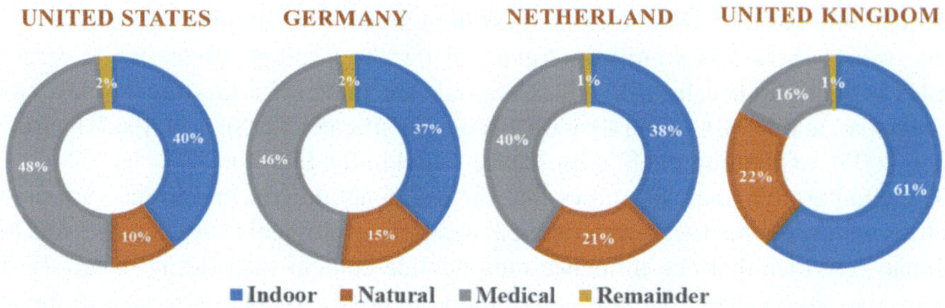

Fig. 5. Relative significance of sources that contribute to the average radiation dose received by the public in the United States of America, Germany, United Kingdom, and the Netherlands (BfS, 2017; PHE, 2016, NCRP, 2009)

3.3 Comparative Analysis of the Radionuclide's Concentrations Measured in Building Materials from Different Countries

Table 2 displays the radionuclides concentrations of ^{226}Ra, ^{40}K and ^{232}Th measured in various types of construction materials including grey cement, white cement, limestone, sand, gypsum, red clay brick, marble, ceramic and concrete collected from 46 countries around the world. Radium-226 concentrations in grey cement fluctuate between 5.85 Bq.kg^{-1} recorded in Chad and 118.7 Bq.kg^{-1} found in China. Thorium activities fluctuate between 3.34 Bq.kg^{-1} found in Qatar and 67.6 Bq.kg^{-1} registered in Palestine. The potassium values measured in cement samples were 18.5 and 952.2 Bq.kg^{-1} reported in Palestine and Bangladesh respectively. The white cement samples collected from Qatar, Egypt, Greece, and Cuba show values inferior to worldwide permissible concentrations (UNSCEAR, 1993). The concentrations of Ra-226 in sand samples collected from 16 countries are under the world average values except for samples from Romania that indicate a value of 92.8 Bq.kg^{-1} exceeding the European average concentration (35 Bq.kg^{-1}, Trevisi et al., 2008) and world average value (50 Bq.kg^{-1}, UNSCEAR, 1993). The thorium concentrations in sand material samples were lower than the standard value for the 16 countries while the concentrations of ^{40}K were reported higher in Romania, Brazil and Cameroun. Limestone and gypsum samples showed values less high than the international reference values, with the exception of Bangladesh, which showed very high values compared to other countries and standards. Same results were reported in a study on building materials from Bangladesh (Roy et al., 2015), which explained that the high levels were related to mining activities contributing to elevated concentrations of thorium and uranium.

The study highlighted that even with these high values, the building material samples were reported safe for usage in urban dwellings construction in Bangladesh and did not show any substantial source of radiation hazards (Roy et al., 2015). Except for Egypt and Albania, all samples of red clay brick collected from different countries show high concentrations of ^{226}Ra, in contrast, thorium shows values within the permissible limit values. The potassium concentrations were reported very high in almost all red clay brick samples except for Cameroun and Turkey which showed values less important than reported values by UNSCEAR. The high concentrations of naturally occurring radionuclides are mainly linked to the use of clay in the manufacture of red brick, so the clay is classified as one of the rocks richest in radioactive elements (Agnieszka, 2017). Marble values show less significant values of the three radionuclides while ceramics samples indicate high values of ^{40}K. Ceramics are made of a mixture of clay, silica, talc, feldspar, and kaolin minerals with zirconium silicates ($ZrSiO_4$) (Abu Khadra and Kamel, 2005), explaining the highest values found in the literature.

The comparative analysis shows that the radionuclide concentrations vary among countries which are due to the geographical location, geological characteristics and kind of human activities that can influence radionuclide content in building materials. The analysis also shows that ^{40}K values were found to be the major contributor to the total concentration recorded for all samples.

Table 3. Radionuclides concentrations measured in different type of building material from 46 countries.

Sample type	Country	^{226}Ra (Bq kg^{-1})	^{232}Th (Bq kg^{-1})	^{40}K (Bq kg^{-1})	Reference
Cement	Qatar	13.2 ± 0.3	3.34 ± 0.05	225.5 ± 6.1	Al-Sulaiti et al. (2011)
	Malaysia	29 ± 7	31 ± 9	205 ± 71	Abdullahi et al. (2018)
	China	118.7 ± 142	36.1 ± 17.8	444.5 ± 163.1	Lu et al. (2014)
	Brazil	14 ± 1	42 ± 4	243 ± 8	Camila et al. (2015)
	Turkey	33 ± 4	15 ± 2	220 ± 13	Özdis et al. (2016)
	Egypt	44.61 ± 2.83	10.39 ± 1.54	51.12 ± 5.96	Shoeib and Thabayneh (2014)
	Iran	34 ± 5	25 ± 5	229 ± 49	Shahrokhi et al. (2020)
	Tunisia	44.3 ± 3.1	12.47 ± 1.7	129.9 ± 11.1	Gharbi et al. (2012)
	India	38.01	34.87	188.14	Ravisankar et al.(2012)
	Serbia	37 ± 22	15 ± 29	43 ± 64	Gordana et al. (2014)
	Albania	55.0 ± 5.8	17.0 ± 3.3	179.7 ± 48.9	Xhixha et al. (2013)
	Algeria	41 ± 7	27 ± 3	422 ± 3	Amrania and Tahtatb (2001)

(continued)

Table 3. (*continued*)

Sample type	Country	^{226}Ra (Bq kg^{-1})	^{232}Th (Bq kg^{-1})	^{40}K (Bq kg^{-1})	Reference
	Romania	33.9 ± 2.2	16.1 ± 1.66	200.1 ± 10.97	Lavinia et al. (2014)
	Bangladesh	60.5 ± 2.1	64.7 ± 2.6	952.2 ± 12.6	Khandoker et al. (2015)
	Ireland	60.3	10.9	130.6	Lee et al. (2004)
	Cuba	23 ± 7	11 ± 3	467 ± 85	Brígido Flores et al. (2008)
	Cyprus	28	7	127	Michael et al. (2010)
	Senegal	112.69 ± 26.02	13.12 ± 1.88	73.35 ± 18.12	Ousmane (2020)
	Netherlands	27	19	230	Ackers et al. (1985)
	Portland (USA)	34	13	182	Ángel Sanjuán et al. (2019)
	Australia	52	48	114	Beretka et al. (1985)
	Lao People's Democratic Republic	41.12 ± 2.44	16.60 ± 2.37	141.48 ± 4.50	Xayheungsy et al. (2018)
	Hong Kong	19.2	18.9	127	Dabayneh (2007)
	Saudi Arabia	38.4	45.3	86	El-Taher (2012)
	Palestine	85.1	67.6	18.5	Dabayneh (2007)
	Italy	38	22	218	Rizzo et al. (2001)
	Austria	27	14	210	Sorantin and Steger (1984)
	Kuwait	13	9	240	Bou-Rabee et al. (1996)
	Zambia	23	32	134	Hayumbu et al. (1995)
	Brazil	61.7	58.5	564	Malanca et al. (1993)

(*continued*)

Table 3. (*continued*)

Sample type	Country	^{226}Ra (Bq kg^{-1})	^{232}Th (Bq kg^{-1})	^{40}K (Bq kg^{-1})	Reference
	Vietnam	39.86 ± 17.43	25.46 ± 4.69	243.5 ± 62.2	Le et al. (2011)
	Greece	20	13	247	Stoulos et al. (2003)
	Cameroon	27.01	15.24	276.53	Ngachin et al. (2007)
	Chad	5.85	5.73	81.31	Penabei et al. 2018
	Nigeria	46.6	24.1	75.8	Ademola (2008)
W. cement	Qatar	18.9 ± 0.5	4.9 ± 0.5	62.9 ± 22.6	Al-Sulaiti et al. (2011)
	Egypt	17.45 ± 2.33	8.44 ± 1.49	4.09 ± 4.72	Shoeib and Thabayneh (2014)
	Greece	28 ± 9	16 ± 3	32 ± 18	Papaefthymiou and Gouseti (2008)
	Cuba	45 ± 10	22 ± 12	99 ± 27	Brígido Flores et al. (2004)
Sand	Qatar	13.2 ± 0.3	3.34 ± 0.05	225.5 ± 6.1	Al-Sulaiti et al. (2011)
	Malaysia	43 ± 20	45 ± 20	451 ± 355	Abdullahi et al. (2018)
	China	32.5 ± 21.6	47.7 ± 12.1	249.6 ± 22.8	Lu et al. (2014)
	Brazil	73 ± 4	173 ± 7	550 ± 5	Camila et al. (2015)
	Cameroun	14 ± 1	31 ± 1	586 ± 13	Ngachin et al. (2007)
	Egypt	16.63 ± 1.54	13.36 ± 1.58	118.9 ± 4.79	Shoeib and Thabayneh (2014)
	Iran	32 ± 6	32 ± 6	289 ± 54	Shahrokhi et al. (2020)
	Greece	12 ± 3	2.6 ± 3.4		Papaefthymiou and Gouseti (2008)
	Tunisia	3.83 ± 1.1	4.9 ± 1.1	28.5 ± 1.1	Gharbi et al. (2012)

(*continued*)

Table 3. (*continued*)

Sample type	Country	^{226}Ra (Bq kg^{-1})	^{232}Th (Bq kg^{-1})	^{40}K (Bq kg^{-1})	Reference
	India	4.6	21.72	353.87	Ravisankar et al. (2012)
	Serbia	26 ± 41	30 ± 35	210 ± 380	Gordana et al. (2014)
	Algeria	12 ± 1	7 ± 1	74 ± 7	Amrania and Tahtatb (2001)
	Romania	92.8 ± 3.95	22.78 ± 2.7	629.4 ± 36	Lavinia Muntean et al. (2014)
	Ireland	12.4	16.9	335.2	Lee et al. (2004)
	Cuba	17 ± 4	16 ± 6	208 ± 104	Brígido Flores et al. (2004)
	Nigeria	24.3	22.3	469.6	Ademola (2008)
Limestone	Jordan	10.82 ± 0.22	0.78 ± 0.26	4 ± 0.18	Sharaf and Hamideen (2013)
	Malaysia	10 ± 1	12 ± 3	62 ± 19	Abdullahi et al. (2018)
	Turkey	28.8	8.9	76.4	Mavi and Akkurt (2010)
	Greece	14 ± 6	1.6 ± 0.5		Papaefthymiou and Gouseti (2008)
	Italy	70.5 ± 5.5	7.1 ± 0.6	46.5 ± 4	Serena Righi and Luigi Bruzzi (2006)
	Pakistan	26.97	4.08	60.65	Rahmat Gul et al. (2014)
	Serbia	84 ± 44		14 ± 10	Gordana et al. (2014)
	Algeria	16 ± 3	13 ± 2	36 ± 3	Amrania and Tahtatb (2001)

(*continued*)

Table 3. (*continued*)

Sample type	Country	^{226}Ra (Bq kg^{-1})	^{232}Th (Bq kg^{-1})	^{40}K (Bq kg^{-1})	Reference
	Romania	8.5 ± 0.9	0.6 ± 0.8	SLD	Lavinia E. Muntean et al. (2014)
Gypsum	Qatar	11.5 ± 0.2	2.5 ± 0.1	72 ± 3	Al-Sulaiti et al. (2011)
	Turkey	27.5	15.6	200.2	Mavi and Akkurt (2010)
	Egypt	8.15 ± 2.81	7.93 ± 2.90	85.1 ± 10.01	Shoeib and Thabayneh (2014)
	Iran	27 ± 4	21 ± 4	92 ± 19	Shahrokhi et al. (2020)
	Pakistan	4.59	4.47	72.58	Rahmat Gul et al. (2014)
	Serbia	20 ± 26	0.4 ± 3	23 ± 55	Gordana K. Pantelic et al. (2014)
	Bangladesh	58.4 ± 3.7	91.2 ± 3.8	1101.1 ± 19.6	Khandoker Asaduzzaman et al. (2015)
	Ireland	20	5.83	145.4	Lee et al. (2004)
Red Clay brick	China	124.7 ± 15.3	28.9 ± 1.4	3902 ± 9.8	Lu et al. (2014)
	Brazil	60 ± 3	42 ± 2	1279 ± 176	Camila et al. (2015)
	Turkey	58.9	11.7	248.8	Mavi and Akkurt (2010)
	Cameroun	49.6 ± 0.3	91 ± 2	172 ± 4	Ngachin et al. (2007)
	Tunisia	45.96 ± 9.1	51.36 ± 7.8	670.21 ± 7.8	Gharbi et al. (2012)
	Albania	33.4 ± 6.4	42.2 ± 7.6	644.1 ± 64.2	Xhixha et al. (2013)
	Egypt	23.06 ± 2.60	23.11 ± 2.99	447.84 ± 10.16	Shoeib and Thabayneh (2014)
	Algeria	65 ± 7	51 ± 5	675 ± 4	Amrania and Tahtatb (2001)

(*continued*)

Table 3. (*continued*)

Sample type	Country	^{226}Ra ($Bq\ kg^{-1}$)	^{232}Th ($Bq\ kg^{-1}$)	^{40}K ($Bq\ kg^{-1}$)	Reference
	Bangladesh	57.5 ± 2.2	75.8 ± 2.9	1080.3 ± 12.7	Khandoker et al. (2015)
	Cuba	57 ± 16	12 ± 10	857 ± 759	Brígido Flores et al. (2004)
Marble	Jordan	2.84 ± 0.38	1.24 ± 0.26	18 ± 0.84	Sharaf and Hamideen (2013)
	Cameroun	8 ± 2	0.35	19 ± 2	Ngachin et al. (2007)
	Greece	1.8 ± 1.0	2.2 ± 0.5		Papaefthymiou and Gouseti (2008)
	Serbia	17 ± 16			Gordana et al. (2014)
	Algeria	23 ± 2	18 ± 2	310 ± 3	Amrania and Tahtatb (2001)
	Nigeria	9.4	5.6	33.4	Ademola (2008)
Ceramic	Jordan	33.86 ± 0.66	28.82 ± 0.46	411 ± 26	Sharaf and Hamideen (2013)
	Iran	34 ± 5	28 ± 7	316 ± 74	Shahrokhi et al. (2020)
	Egypt	51.12 ± 2.74	40.52 ± 2.54	682.6 ± 10.13	Shoeib and Thabayneh (2014)
	Algeria	55 ± 5	41 ± 4	410 ± 4	Amrania and Tahtatb (2001)
Concrete	Jordan	9.86 ± 0.22	4.18 ± 0.36	30 ± 1.32	Sharaf and Hamideen (2013)
	Italy	15.5 ± 1.6	16.2 ± 1.8	310 ± 25	Righi and Bruzzi (2006)
	Ireland	28.9	12.05	217.3	Lee et al. (2004)
	Cuba	25 ± 10	12 ± 7	595 ± 166	Brígido Flores et al. (2004)

(*continued*)

Table 3. (*continued*)

Sample type	Country	^{226}Ra (Bq kg^{-1})	^{232}Th (Bq kg^{-1})	^{40}K (Bq kg^{-1})	Reference
	Nigeria	37.3	70.5	339.0	Ademola. (2008)
	Canada	11	17	488	Zikovsky and Kennedy (1992)
	Australia	14.8	40.7	199.8	Beretka and Mathew (1985)
	Greece	14	3.2	70	Savidou et al. (1995)
	Lithuania	32 ± 4	17 ± 3	426 ± 48	Pilkyte et al. (2005)
	Estonia	35.1 ± 2.1	11.3 ± 0.9	207 ± 5	Merle Lust and Enn Realo (2012)
	United Kingdom	18	42		Cliff et al. (1985)
	Netherlands	24 ± 8	18 ± 6	160 ± 30	De Jong et al. (2006)
	Finland	53.4	37.6	838	Raimo Mustonen (1983)
	France	10 ± 2	7 ± 1	128 ± 17	Samir Dziri (2013)
	Germany	18	12	193	Gehrcke et al. (2010)
	Hungary	11 ± 2	6 ± 1	142 ± 6	Szabó et al. (2013)

3.4 RESRAD BUILD Simulation Results

Because most residents spend nearly 80% of their time indoors, the emitted natural radioactivity from the ground, floors, and walls could have a substantial consequence on their health. The simulation and estimation of the doses to which the residents are exposed from contaminated building, or from contaminated furniture or equipment found in the building, has been carried out using the RESRAD BUILD code version 4.0. (Adelikhah et al., 2022). The mean activity concentrations of ^{40}K, ^{232}Th, and ^{226}Ra measured in concrete samples from 13 countries (Table 3) around the world were collected and used as an input parameter to the code to determine the dose over 1, 5, 10, 20, 50 and 70 years in 2 major pathways namely internal and external radiation exposure. Figure 6 exhibits the calculated internal and external doses to which a person can be exposed when he is located in the center part of a room of a dimension of 4.0 × 5.0 × 2.8 m and 1 m above the floor. The simulations were carried out over 70 years, according to the United Nations report, across the world, the life expectancy average by 2021 of a newborn is 73 (Issifou and Pewitt, 2022). The results of simulation revealed that outdoor and indoor exposure to radiation displays low variations over the average lifetime of 70 years. The doses are relatively stable in the first five years and show some slight variations between 5 and 20 years then they become stable over the last 50 years. The inhalation doses show values more significant over the 70 years in Nigeria, followed by Australia, Finland, and the other 10 countries. The outdoor doses were more important in Finland, Nigeria and Cuba.

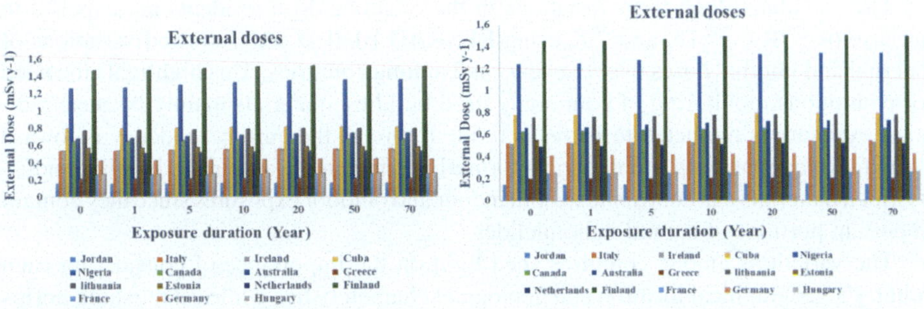

Fig. 6. Long-term variation in the effective dose residents are exposed to because of ^{226}Ra, ^{232}Th, and ^{40}K in a standard room in 13 different countries around the world.

These variations of doses over the years highlight the need for continuous monitoring of commonly used building materials. The radiation exposure is expected to increase in the near future mainly because of the current tendency toward using recycled materials and materials formerly considered waste in the building industry. Such materials can contain important portions of natural radionuclides, thus contributing to both indoor and outdoor exposures.

Besides the variations among countries shed light on the importance of considering each country's geographical location and geological characteristics while establishing norms and guidelines related to emitted radiation limits from building materials. In addition, standards limits differ by country mainly because of differences in adopted preconditions, which sometimes made the inter-comparison not always suitable.

4 Conclusions

Building materials contain a proportion of naturally occurring radionuclides. Concentrations of radionuclides differ according to the origin of the material used in building. Thus, the assessment and monitoring of their concentrations is essential to avoid any potential risk to human health. The present study reviews the current knowledge about the levels of radioactivity in diverse types of building materials from 46 countries around the world and measures internal and external human exposure to radiation in a standard room using a computer code called RESRAD BUILD.

The studies conducted to measure the risks associated with radiation exposure to human health are still not well studied worldwide. There is an unbalanced distribution of studies on radiation levels in building materials among countries, where China, USA and Italy ranked first in the field with more than 100 publications. Technological capabilities may be one of the reasons for the observed uneven progress of scientific research input between countries.

The comparative analysis among 46 countries around the world shows that the radionuclide concentrations vary among countries which are due to the geographical location, geological characteristics and kind of human activities that can influence radionuclide content in building materials. The analysis also shows that ^{40}K values were found to be the main contributor to the total concentration recorded for all samples.

The calculated long-term variations in the effective dose residents are exposed to because of ^{226}Ra, ^{232}Th, and ^{40}K using RESRAD BUILD code, showed variations of external and internal doses over the years and among countries. This highlights the need for continuous monitoring of commonly used building materials mainly because radiation exposure is predicted to increase in the future as the current tendency is toward using recycled materials and materials formerly considered waste in the building industry. Such materials can contribute to both indoor and outdoor exposures since they contain important portions of natural radionuclides.

The variations among countries shed light on the importance of considering each country's geographical location and geological characteristics while establishing norms and guidelines related to emitted radiation limits from building materials.

References

Abdullahi, S., Ismail, A.F., Samat, S.: Determination of indoor doses and excess lifetime cancer risks caused by building materials containing natural radionuclides in Malaysia. Nucl. Eng. Technol. **51**(1), 325–336 (2018)

Abu Khadra, S.A., Kamel, N.H.: Natural Radioactivity in Ceramic Materials (2005)

Ackers, J.G., Den Boer, J.F., De Jong, P., Wolschrijn, R.A.: Radioactivity and radon exhalation rates of building materials in the Netherlands. Sci. Total Environ. **45**, 151–156 (1985)

Adelikhah, M., Imani, M., Miklos hegedus, M., Kovacs, T.: Modelling of indoor external and internal exposure due to different building materials containing NORMs in the vicinity of a HNBRA in Mahallat, Iran. Heliyon 8 e08909Agnieszka Pękala 2017 IOP Conf. Ser. Mater. Sci. Eng. **245**, 022033 (2022)

Ademola, J.A.: Determination of natural radionuclides content in some building materials in Nigeria by gamma-ray spectrometry. Health Phys. **94**(1), 43–48 (2008)

Ahmed, N.K.: Measurement of natural radioactivity in building materials in Qena city, Upper Egypt. J. Environ. Radioact. **83**(1), 91–99 (2005)

Al-Jarallah, M.: Radon exhalation from granites used in Saudi Arabia. J. Environ. Radioact. **53**(1), 91–98 (2001)

Amin, S.A., Naji, M.: Natural radioactivity in different commercial ceramic samples used in Yemeni buildings. Radiat. Phys. Chem. **86**, 37–41 (2013)

Amrani, D., Tahtat, M.: Natural radioactivity in Algerian building materials. Appl. Radiat. Isot. **54**(4), 687–689 (2001)

Al-Sulaiti, H.: Determination of the natural radioactivity in Qatarian building materials using high-resolution gamma-ray spectrometry. Nucl. Instrum. Methods Phys. Res. Section A: Accelerators, Spectrometers, Detectors, and Associated Equipment **652**(1), 915–919. (2011)

Asaduzzaman, K., et al.: Assessment of natural radioactivity levels and potential radiological risks of common building materials used in Bangladeshi dwellings. PLoS ONE **10**(10), e0140667 (2015)

Asaduzzaman, K., Khandaker, M.U., Amin, Y.M., Bradley, D.A.: Natural radioactivity levels and radiological assessment of decorative building materials in Bangladesh. Indoor Built Environ. **25**(3), 541–550 (2016)

Aydarous, A., Zeghib, S., Shittu A., Al-Subaie, H.: Radiological hazard assessment and sensitivity analysis for soil samples in Taghdoua area of Ranyah, Saudi Arabia. J. Radiat. Res. Appl. Sci., 119–128 (2022)

Beretka, J., Mathew, P.J.: Natural radioactivity of Australian building materials, industrial wastes and by-products. Health Phys. **48**(1), 87–95 (1985)

BfS: Umweltradioaktivität und Strahlenbelastung im Jahr 2015: Unterrichtung durch die Bundesregierung Bundesamt für Strahlenschutz (BfS), Berlin (2017)

Bou-Rabee, F., Bem, H.: Natural radioactivity in building materials utilized in the State of Kuwait. J. Radioanal. Nucl. Chem. **1996**(213), 143–149 (1996)

Cazula, C.D., Campos, M.P., Mazzilli, B.P.: Gamma exposure due to building materials in a residential building at Peruíbe, São Paulo, Brazil. J. Radioanal. Nucl. Chem. **306**, 637–640 (2015)

CDC: Radiation from Building Materials (2022). https://www.cdc.gov/nceh/radiation/building.html. Accessed 08 Dec 2022

Cho, B.-W., Choo, C.O., Kim, M.S., Hwang, J., Yun, U., Lee S.: Spatial relationships between radon and topographical, geological, and geochemical factors and their relevance in all of South Korea Environ. Earth Sci. **74**, 5155–5168 (2015)

Chen, W., Huang, J., Chin, C., Chen, Y.: Management of roads with abnormal radiation (No. IAEA-TECDOC--1271)

Cliff, K.D., Green, B.M.R., Miles, J.C.H.: The levels of radioactive materials in some common UK building materials. Sci. Total. Environ. **45**, 181–186 (1985)

Dabayneh, K.: Natural radioactivity in different commercial ceramic samples used in Palestinian buildings as construction materials (2007)

De Jong, P., Van Dijk, W., van der Graaf, E.R., de Groot, T.J.: National survey on the natural radioactivity and 222Rn exhalation rate of building materials in the Netherlands. Health Phys. **91**(3), 200–210 (2006)

Dziri, S., Nachab, A., Nourreddine, A., Sellam, A., Gelus, D.: Experimental and simulated effective dose for some building materials in France (2013)

El-Taher, A.: Assessment of natural radioactivity levels and radiation hazards for building materials used in Qassim area, Saudi Arabia. Rom. J. Phys. **57**(3-4), 726–735 (2012)

European Commission: Radiological protection principles concerning the natural radioactivity of building materials. Radiation Protection 112EPA, 2022 (1999)

Estokova, A., Singovszka, E., Vertal, M.: Investigation of building materials' radioactivity in a historical building—a case study. Materials **15**, 6876 (2022)

Eštoková, A., Palaščáková, L.: Assessment of natural radioactivity levels of cements and cement composites in the Slovak Republic. Int. J. Environ. Res. Public Health, 2013, 10 (2023)

European Commission (EC): radiation protection 112 (1999) Radiological protection principles concerning the natural radioactivity of building materials directorate-general environment, nuclear safety and civil protection

EPA: Natural Radioactivity in Building Materials (2022) https://www.epa.gov/radtown/natural-radioactivity-building-materials. Accessed 08 Dec 2022

EPA Environmental Protection Agency: A Citizens Guide to Radon (2016). https://www.epa.gov/sites/production/files/2016-12/documents/2016_a_citizens_guide_to_radon.pdf. Accessed 09 Dec 2022

EPA: Radionuclide Basics: Radon (2002) https://www.epa.gov/radiation/radionuclide-basics-radon. Accessed 08 Dec 2023

Flores, O.B., Estrada, A.M., Zerquera, J.T.: Natural radioactivity in some building materials in Cuba and their contribution to the indoor gamma dose rate. Radiat. Prot. Dosimetry **113**(2), 218–222 (2005)

Frutos-Puerto, S., Pinilla-Gil, E., Andrade, E., Reis, M., Madruga, M.J., Rodríguez, C.M.: Radon and thoron exhalation rate, emanation factor and radioactivity risks of building materials of the Iberian Peninsula. PeerJ **8**, e10331 (2020)

Für Strahlenschutz, B.: Umweltradioaktivität und Strahlenbelastung (2005)

Gbenu, S.T., Oladejo, O.F., Olukotun, S.F., Makinde, O.W., Fasasi, M.K., Balogun, F.A.: Assessment of radioactivity and radiological hazards in commercial ceramic tiles used in Ife-Central, local government area of Osun State, Nigeria. Egypt. J. Basic Appl. Sci. **3**(4), 377–382 (2016)

Gharbi, F., Oueslati, M., Abdelli, W., Samaali, M., Ben Tekaya, M.: Exposure to radiation from the natural radioactivity in Tunisian building materials. Radiat. Prot. Dosimetry **152**(4), 418–422 (2012)

Gehrcke, K., Hoffmann, B., Schkade, U., Schmidt, V., Wichterey, K.: Radiation exposure from the use of NORM in building materials in Germany. In: Naturally Occurring Radioactive Material (NORM VI). Proceedings of an International Symposium (2011)

Gjeçi, E., Karaja, T., Xhixha, G., Shyti, M.: Characterization of natural radioactivity in building materials. Proceedings Book, 144 (2015)

Gul, R., Ali, S., Hussain, M.: Estimation of radioactivity level and associated radiological hazards of limestone and gypsum used as raw building materials in Rawalpindi/Islamabad region of Pakistan. Radiat. Prot. Dosimetry **158**(3), 340–349 (2014)

Hayumbu, P., Zaman, M., Lubaba, N., Munsanje, S., Muleya, D. Natural radioactivity in Zambian building materials collected from Lusaka. J. Radioanal. Nucl. Chem. Art. **199**(3), 229–238 (1995)

Iqbal, M., Tufail, M., Mirza, S.M.: Measurement of natural radioactivity in marble found in Pakistan using a NaI (Tl) gamma-ray spectrometer. J. Environ. Radioact. **51**(2), 255–265 (2000)

International Atomic Energy Agency Naturally occurring radioactive material (NORM V); Proceedings of the International Symposium on Naturally Occurring Radioactive Material; Seville, Spain, 19–22 March 2007. 549p.

Issifou, I., Pewitt, J.: On the importance of monitoring inequality in life expectancy. United Nations Department of Economic and Social Affairs. Policy Brief No145 (2022). https://www.un.org/development/desa/dpad/wpcontent/uploads/sites/45/publication/PB_145.pdf. Accessed 08 Dec 2023

Jasaitis, D., Pečiulienė, M.: Natural radioactivity and radon exhalation from building materials in underground parking lots. Appl. Sci. **11**, 7475 (2021)

Kassi, B., Boukhair, A., Azkour, K., Fahad, M., Benjelloun, M., Nourreddine, A.M.: Assessment of exposure due to technologically enhanced natural radioactivity in various samples of Moroccan building materials. World J. Nucl. Sci. Technol. **8**(04), 176 (2018)

Kohli, S.: Childhood leukaemia in areas with different radon levels: a spatial and temporal analysis using GIS. J. Epidemiol. Community Health **54**(2000), 822–826 (2000). https://doi.org/10.1136/jech.54.11.822

Lee, E.M., Menezes, G., Finch, E.C.: Natural radioactivity in building materials in the Republic of Ireland. Health Phys. **86**(4), 378–383 (2004)

Le Nhu Sieu, N.T.B., Truong, Y., Nguyen Trong Ngo, N.T.L., Phuc, N.D.T., Thanh, N.K., Khoa, T.D.: Natural Radioactivity in Commonly Building Materials Used in Vietnam. 11255 WM 2011 Conference, February 27-March 3 2011

Lewicka, S., Piotrowska, B., Łukaszek-Chmielewska, A., Drzymała, T.: Assessment of natural radioactivity in cements used as building materials in Poland. Int. J. Environ. Res. Public Health **19**(18), 11695 (2022)

Lu, X., Chao, S., Yang, F.: Determination of natural radioactivity and associated radiation hazard in building materials used in Weinan, China. Radiat. Phys. Chem. **99**, 62–67 (2014)

Lust, M., Realo, E.: Assessment of natural radiation exposure from building materials in Estonia. Proc. Estonian Acad. Sci. **61**(2), 107 (2012)

Majid, A.A., Ismail, A.F., Yasir, M.S., Yahaya, R., Bahari, I.: Radiological dose assessment of naturally occurring radioactive materials in concrete building materials. J. Radioanal. Nucl. Chem. **297**, 277–284 (2013)

Malanca A., Pessina V., Dallara G.: Radionuclide-content of building materials and gamma-ray dose rates in dwellings of Rio-Grande Do-Norte Brazil. Radiat. Prot. Dosim. **48**, 199–203 (1993)

Mavi, B., Akkurt, I.: Natural radioactivity and radiation hazards in some building materials used in Isparta, Turkey. Radiat. Phys. Chem. **79**(9), 933–937 (2010)

Michael, F., Parpottas, Y., Tsertos, H.: Gamma radiation measurements and dose rates in commonly used building materials in Cyprus. Radiat. Prot. Dosimetry **142**(2–4), 282–291 (2010)

Muntean, L.E., Cosma, C., Moldovan, D.V.: Measurement of natural radioactivity and radiation hazards for some natural and artificial building materials available in Romania. J. Radioanal. Nucl. Chem. **299**, 523–532 (2014)

Mustonen, R.: Natural radioactivity in and radon exhalation from Finnish building materials. Health Phys. **46**(6), 1195–1203 (1984)

NCRP: Ionizing Radiation Exposure of the Population of the United States. NCRP Report No. 160, (2009)

Ndour, O., et al.: Assessment of natural radioactivity and its radiological hazards in several types of cement used in Senegal. SN Appl. Sci. **2**, 1–8 (2020)

Ngachin, M., Garavaglia, M., Giovani, C., Njock, M.K., Nourreddine, A.: Assessment of natural radioactivity and associated radiation hazards in some Cameroonian building materials. Radiat. Meas. **42**(1), 61–67 (2007)

Ngachin, M., et al.: 226Ra, 232Th and 40K contents and radon exhalation rate from materials used for construction and decoration in Cameroon. J. Radiol. Prot. **28**(3), 369 (2008)

Onjefu, S.A., Johannes, N.N., Abah, J., Onjefu, L.A., Mwiya, S.: Natural radioactivity levels and evaluation of radiological hazards in Usakos marble, Erongo region, Namibia. Int. J. Radiat. Res. **20**(2), 403–409 (2022)

Özdiş, B.E., Çam, N.F., Canbaz Öztürk, B.: Assessment of natural radioactivity in cements used as building materials in Turkey. J. Radioanal. Nucl. Chem. **311**, 307–316 (2016)

Penabei, S., et al.: MK Assessment of natural radioactivity levels and the associated radiological hazards in some building materials from Mayo-Kebbi region, Chad. Radioprotection **53**, 265–278 (2018)

Pantelic, G K., Todorovic, D.J., Nikolic, J.D., Rajacic, M.M., Jankovic, M. M., Sarap, N.B.: Measurement of radioactivity in building materials in Serbia. J. Radioanal. Nucl. Chem. **303**, 2517–2522 (2015)

Papaefthymiou, H., Gouseti, O.: Natural radioactivity and associated radiation hazards in building materials used in Peloponnese, Greece. Radiat. Meas. **43**(8), 1453–1457 (2008)

Pekala, A.: 2017 IOP Conf. Ser. Mater. Sci. Eng. **245**, 022033 (2022)

Pepin, S.: Using RESRAD-BUILD to assess the external dose from the natural radioactivity of building materials. Constr. Build. Mater. **168**, 1003–1007 (2018)

PHE: Ionising Radiation Exposure of the UK Population: 2010 Review. Public Health England, Chilton (2016)

Pilkytė, L., Butkus, D., Morkūnas, G.: Assessment of external dose indoors in Lithuania. Radiat. Prot. Dosimetry **121**(2), 140–147 (2005)

Ramachandran, T.V.: Background radiation, people and the environment Iran. J. Radiat. Res. **9**, 63–76 (2011)

Ravisankar, R., et al.: Measurement of natural radioactivity in building materials of Namakkal, Tamil Nadu, India using gamma-ray spectrometry. Appl. Radiat. Isot. **70**(4), 699–704 (2012)

Righi, S., Bruzzi, L.: Natural radioactivity and radon exhalation in building materials used in Italian dwellings. J. Environ. Radioact. **88**(2), 158–170 (2006)

Rizzo, S., Brai, M., Basile, S., Bellia, S., Hauser, S.: Gamma activity and geochemical features of building materials: estimation of gamma dose rate and indoor radon levels in Sicily. Appl. Radiat. Isot. **55**(2), 259–265 (2001)

Roy, S., Alam, M.S., Begum, M., Alam, B.: Radioactivity in building materials used in and around Dhaka city. Radiat. Prot. Dosimetry **114**(4), 527–532 (2015)

Safhi, A.E.M. (ed.): Valorization of Dredged Sediments as Sustainable Construction Resources, 1st edn. CRC Press (2022). https://doi.org/10.1201/9781003315551

Safhi, A.E.M., Mejjad, N., El Fadill, H., Bortali, M.: Dredged materials in Morocco: current practices, policies, and roadmap for sustainable management. Case Stud. Constr. Mater. **20**, e03045 (2024). https://doi.org/10.1016/j.cscm.2024.e03045

Sanjuán, A., Suarez-Navarro, M., Argiz, J.A., Mora, C.P.: Assessment of radiation hazards of white and grey Portland cements. J. Radioanal. Nucl. Chem. **322**, 1169–1177 (2019)

Savidou, A., Raptis, C., Kritidis, P.: Natural radioactivity and radon exhalation from building materials used in Attica Region, Greece. Radiat. Prot. Dosimetry **59**(4), 309–312 (1995)

Segui, P., Safhi, A.e.M., Amrani, M., Benzaazoua, M.: Mining wastes as road construction material: a review. Minerals **13**, 90 (2023). https://doi.org/10.3390/min13010090

Shahrokhia A., Adelikhaha, M., Chalupnikb, S., Kocsisa, E., Toth-Bodrogia, E., Kovács, T.: Radioactivity of building materials in Mahallat, Iran – an area exposed to a high level of natural background radiation – attenuation of external radiation doses. Materiales de Construcción **70**(340), e233 (2020). ISSN-L 0465-2746

Sharaf, J.M., Hamideen, M.S.: Measurement of natural radioactivity in Jordanian building materials and their contribution to the public indoor gamma dose rate. Appl. Radiat. Isot. **80**, 61–66 (2013)

Shoeib, M.Y., Thabayneh, K.M.: Assessment of natural radiation exposure and radon exhalation rate in various samples of Egyptian building materials. J. Radiat. Res. Appl. Sci. **7**(2), 174–181 (2014)

Smetsers, R.C., Tomas, J.M.: A practical approach to limit the radiation dose from building materials applied in dwellings, in compliance with the Euratom Basic Safety Standards. J. Environ. Radioact. **196**, 40–49 (2019). https://doi.org/10.1016/j.jenvrad.2018.10.007

Sorantin, H., Steger, F.: Natural radioactivity of building materials in Austria. Radiat. Prot. Dosimetry. **7**(1-4), 59–61 (1984)

Stoulos, S., Manolopoulou, M., Papastefanou, C.: Assessment of natural radiation exposure and radon exhalation from building materials in Greece. J. Environ. Radioact. **69**(3), 225–240 (2003)

Suliman, I.I., Alsafi, K.: Radiological risk to human and non-human biota due to radioactivity in coastal sand and marine sediments, Gulf of Oman. Life **11**, 549 (2021). https://doi.org/10.3390/life11060549

Szabó, Z., Völgyesi, P., Nagy, H.É., Szabó, C., Kis, Z., Csorba, O.: Radioactivity of natural and artificial building materials–a comparative study. J. Environ. Radioact. **118**, 64–74 (2013)

Tchorz-Trzeciakiewicz, D., Olszewski, S.: Radiation in different types of building, human health. Sci. Total. Environ. **667**, 511–521 (2019). https://doi.org/10.1016/j.scitotenv

Thumvijit, T., et al.: Identifying indoor radon sources in Pa miang, chiang mai. Thailand. Sci. Rep. **10**, 1–14 (2020). https://doi.org/10.1038/s41598-020-74721-6

Trevisi, R., D'Alessandro, M., Nuccetelli, C., Risica, S.: Radioactivity in building materials: a first overview of the European scenario (2008)

Tufan, M.Ç., Dişci, T.: Natural radioactivity measurements in building materials used in Samsun, Turkey. Radiat. Prot. Dosimetry **156**(1), 87–92 (2013)

United Nations Scientific Committee on the Effects of Atomic Radiation UNSCEAR: Report to the General Assembly, with Scientific Annexes, United Nations, New York (1993). United Nations sales publication E.94.IX.2 (1993). https://www.unscear.org/docs/publications/1993/UNSCEAR_1993_Report.pdf. Accessed 09 Dec 2023

United Nations Scientific Committee on the Effects of Atomic Radiation (UNSCEAR) Sources, Effects and Risks of Ionizing Radiation. United Nations; New York, NY, USA: 2000. Report to the General Assembly with Annex B: Exposures from Natural Sources of Radiation

Viruthagiri, G., Rajamannan, B., Suresh Jawahar, K.: Radioactivity and associated radiation hazards in ceramic raw materials and end products. Radiat. Prot. Dosimetry **157**(3), 383–391 (2013)

WHO: WHO Handbook on Indoor Radon: a Public Health Perspective (2009). https://www.who.int/ionizing_radiation/env/9789241547673/en/

Wichmann, H.E., Menzler, S., Schaffrath-Rosario, A., Kreienbrock, L.: Lung cancer risk in Germany attributable to radon in homes. Epidemiology **17**(6), S68–S69 (2006)

Xayheungsy, S., Khiem, L.H., Nam, L.D.: Assessment of the natural radioactivity and radiological hazards in Lao cement samples. Radiat. Prot. Dosimetry **181**(3), 208–213 (2018)

Xhixha, G., et al.: The worldwide NORM production and a fully automated gamma-ray spectrometer for their characterization. J. Radioanal. Nucl. Chem. **295**, 445–457 (2023)

Yu, C., et al.: Development of probabilistic RESRAD 6.0 and RESRAD-BUILD 3.0 computer codes (2000)

Zhang, L. Chen, Z., Wang, F., White, N.C., Zhou T.: Release of uranium from uraninite in granites through alteration: implications for the source of granite-related uranium ores Econ. Geol. **116**, 1115–1139 (2021). https://doi.org/10.5382/econgeo.4822

Zhu, H.C., Charlet, J.M., Poffijn A.: Radon risk mapping in southern Belgium: an application of geostatistical and GIS techniques. Sci. Total Environ. **272**, 203–210 (2001). https://doi.org/10.1016/S0048-9697(01)00693-3

Zikovsky, L., Kennedy, G.: Radioactivity of building materials available in Canada. Health Phys. **63**(4), 449–452 (1992)

A Comprehensive Analysis of Container vs Prefabricated Emergency Shelters: Case Form Syria Post-disaster Reconstruction

Lina A. Khaddour[1(✉)], Rania Obead[1], Siegfried Yeboah[2], Temidayo Osunsanmi[1], and Ismail Elhassnaoui[1]

[1] School of Computing, Engineering and The Built Environment, Edinburgh Napier University, Edinburgh, UK
L.Khaddour@napier.ac.uk

[2] School of the Built Environment and Architecture, London South Bank University, London, UK

Abstract. The escalation of both natural and man-made disasters globally has led to profound challenges for nations grappling with the aftermath, particularly in the provision of post-disaster emergency shelters. These shelters hold critical significance in disaster risk mitigation and resilience efforts, serving as essential havens for displaced individuals amidst the environmental and socio-economic repercussions of calamities. Despite their significance, post-disaster shelters have faced criticism due to their substantial cost requirements, prolonged construction timelines, inadequate thermal comfort standards, and limited community involvement.

This study conducts a comparative analysis between two types of temporary housing, namely prefabricated and container shelters, utilized in refugee camps in Syria following the conflict. Employing a comparative Life Cycle Assessment (LCA) approach, the study evaluates the performance of these emergency shelters, both sharing the same base area. Energy/carbon analysis and Life Cycle Costing (LCC) calculations were conducted for these shelters. Utilizing the Integrated Environmental Solutions Virtual Environment (IESVE) software, the study assesses the energy performance of both shelter types. The analysis of embodied carbon revealed a marginal increase of 3.89% in the container shelter design compared to prefabricated structures, with values of 2133KgCO2 and 2050 KgCO2, respectively. Additionally, operational emissions for container shelters were observed to be 28% lower than for prefabricated designs, recording 336690KgCO2 and 470505KgCO2, respectively. Moreover, the total cost for prefabricated shelters was determined to be 33.61% higher than that for container shelters. Overall, the comparative LCA energy/emission and costing approach revealed that shipping containers have a reduced environmental impact and lower cost implication compared to prefabricated emergency shelters of similar design.

The recommendation and the criteria developed in this research can serve as a decision-making tool aiding the selection of suitable post-disaster emergency shelters across diverse affected regions globally, incorporating considerations of climatic zones and design parameters.

A. S. Mosallam et al. (Eds.): ICATH 2023, SCI, pp. 174–186, 2024.
https://doi.org/10.1007/978-3-031-76557-5_14

Keywords: Post-disaster · Reconstruction · Energy · LCA · Simulation · Sustainability

1 Introduction

In the last 40 years, natural and man-made disasters have increased globally. Over this period, about 21,000 incidents have been recorded with about 1.7 million associated casualties and even more displaced people [1]. In the last 20 years, more than 500 natural disasters per year were recorded globally, four times more than pre-1980 numbers [2]. Many of these disaster incidents led to major loss of lives as well as properties, making mitigation of natural disaster consequences crucial [3]. The expansion in natural and man-made disasters in recent years has led to an increased focus on emergency shelters. Post-disaster reconstruction, mainly construction of emergency and temporary housing, is challenging and complicated for many reasons. The urgent need to shelter affected victims and the fact that emergency shelters are designed to accommodate disaster survivors for a limited time usually from 6 month to 5 years, although in many cases this estimated time is exceeded due to the length of time required to build permanent housing [1]. With the present impacts of natural and conflict disasters associated refugees, the need for emergency shelters is expected to increase. As emergency shelters are usually built after a disaster to quickly house displaced survivors, the construction is usually very fast, and unsustainable methods are frequently used. For this reason, emergency shelter development tends to have significant failures such as their high cost and ignoring of cultural and environmental factors [4]. For these reasons, the need to improve the impacts of emergency shelters on survivors has become a necessity.

Post-disaster emergency shelters assume a pivotal role in the provision of relief following catastrophic events, particularly in densely populated urban cities that have undergone extensive devastation. These metropolises often feature towering residential structures, escalating the probability of both human casualties and structural damage, thus amplifying the complexities involved in post-disaster reconstruction. An illustrative case is Damascus, the capital of Syria, which confronted numerous challenges during its reconstruction phase following the conflict spanning 2011 to nearly 2020. These challenges encompassed economic downturns, an unstable and costly energy supply, and a substantial decline in overall economic activities, notably within the construction sector [4].

The conflict in Syria constituted a disaster of significant magnitude, resulting in a high toll of human lives, widespread destruction, and extensive socio-economic and environmental ramifications. Foremost among the affected sectors was housing, followed closely by energy. The Syrian war incurred losses amounting to $4,958 million and $1,445 million in the housing and energy sectors, respectively [4]. Notably, the aftermath of the conflict led to an astonishing 11 million displaced individuals, with 8 million internally displaced, and a staggering 55% of city buildings reduced to rubble [5]. The energy sector bore an immense brunt, witnessing a drastic reduction in electricity supply to a mere 15% of its 2010 capacity [5]. This diminished capacity covered less than half of the residential sector's energy requirements, consequently escalating expenses for alternative energy sources like natural gas [5].

Therefore, post-disaster emergency strategies must adopt a phased approach encompassing immediate, medium-term, and long-term measures for post-disaster reconstruction [6]. Moreover, beyond the extensive impact on all economic sectors, especially the energy domain, any post-disaster emergency plan should emphasize the imperative of "building back better". In the context of Syria, this underscores the need to address the residential sector's implications for energy consumption and carbon emissions, accounting for 49% and 40%, respectively, of the country's annual energy consumption and carbon emissions [6].

To ensure competent disaster recovery, Life Cycle Analysis (LCA) could offer a comparative tool for measuring the environmental impacts and providing decision-making criteria to compare between alternative emergency shelters [8]. Using LCA in the built environment provides a comprehensive view of the environmental impact of the construction process and enhances decision-making sustainability. LCA evaluates the environmental impact of different alternatives considering all construction stages from the design to the end of life [9]. LCA measures the whole process emissions from the material extraction, transportation, construction, operation, and end of life. The calculations for the LCA are usually done in the building's life stages and it includes material extraction and manufacturing (A1–A3), construction stages (A4–A5), operation stages (B1–B7), end-of-life stage (C1–C4) and behind life (D) [8]. Due to the significant immigration following disasters, the reconstruction and use of emergency shelters have been growing rapidly and are most likely to increase in the future. Therefore, improving post-disaster buildings' energy consumption, and carbon emissions besides reducing their overall cost is becoming important and highly linked with the overall global efforts in the construction sector to reduce energy consumption, carbon emissions, and cost.

An overview of the existing literature concerning post-disaster reconstruction reveals a limited focus on life cycle and life cost analysis. Numerous studies concentrate on the design and location measures of post-disaster shelters, yet the dearth of research offering guidance to decision-makers in determining optimal emergency shelters remains evident [1, 13]. Notably, previous international research employing life cycle assessment (LCA) for emergency shelters utilizes disparate data sets across various countries, thereby conferring limited value upon their findings [9]. The prevailing research primarily emphasizes environmental assessment or energy concerns, neglecting to holistically encompass sustainability [3]. Furthermore, a comprehensive comparative analysis assessing diverse emergency shelters based on criteria such as building materials, design, energy usage, and cost is conspicuously absent [10].

This research endeavours to bridge these gaps in the literature by devising a comparative evaluation framework employing LCA approach for emergency shelters. The evaluation encompasses both prefabricated and container-based typical post-disaster emergency shelters. Architectural specifications, operational data, and functional aspects of these shelters were procured from working drawings, utility bills, and reports associated with emergency shelters. The outcomes of this study are envisioned to serve as a decision-making tool aiding decision-makers in evaluating various emergency shelter techniques in areas affected by disasters. Consequently, this study delineates the following research objectives:

1. To develop a comprehensive comparative framework integrating Life Cycle Assessment (LCA) encompassing Energy/Carbon LCA and Life Cycle Costing (LCC) for post-disaster emergency shelters.
2. To assess prefabricated versus container post-disaster emergency shelters through IESVE simulation, Energy/Carbon LCA, and LCC methodologies.

2 Methodology

This research constitutes a case study examining the utilization of prefabricated versus container emergency shelters in the post-conflict scenario of Syria. The study is structured into four distinct phases. Initially, the investigation involves simulation to compute the operational carbon/energy metrics for two categories of emergency shelters (containers versus prefabricated) within the specific climate zone of Damascus. Subsequently, the second phase entails determining the embodied carbon/energy and end-of-life carbon/energy metrics. These LCA (Life Cycle Assessment) phases are then amalgamated. The fourth and final phase involves the computation of Life Cycle Costing (LCC) for both shelter types, utilizing input data gathered from Damascus during the timeframe spanning June 2023 to September 2023.

This study employs the LCA methodology to compare the environmental impact of two distinct forms of temporary shelters deployed in post-disaster Damascus: prefabricated versus container shelters. The LCA calculations encompass various parameters and inputs, commencing with the determination of operational energy/emissions, computed through the utilization of IESVE software. Subsequently, the embodied energy/emissions are derived using the Inventory of Carbon and Energy as a database. Finally, the end-of-life energy/emissions are estimated by considering the recovery potential of a portion of these energy/emissions.

2.1 The Case Study Design and Construction Inputs

The chosen locale for this research case study is Damascus, the principal city of Syria situated in the southern region of the country. Damascus experiences an elongated summer season spanning from early June to late September, characterized by high temperatures peaking at approximately 40 °C. Conversely. The winter period, in Damascus, spans from mid-November to mid-April, featuring temperatures that can plummet to around −2 °C.

In the context of this analysis and evaluation of prefabricated and container emergency shelters, Fig. 1 illustrates the standardized layout design adopted for emergency shelters. This design emphasizes compactness, facilitating ease of construction, transportation, and lifting. It comprises two bedrooms, a bathroom, and a living area, catering to the accommodation needs of a family of four.

In the realm of LCA within the construction sector, various parameters and assumptions are often necessary due to the intricate nature of buildings, characterized by numerous interrelated systems and subsystems. Many of these assumptions are intricately connected to the future utilization and performance of buildings [9, 15]. While integral to the LCA process, these simplifications and assumptions may also be perceived as limitations within the study [13].

Fig. 1. The pilot project-compliant building

Within the scope of this research, several simplifications and assumptions were employed, drawing from construction practices and prior LCA case studies literature. These encompass:

- Considering a 15-year life span, aligned with the actual duration of post-disaster emergency shelters constructed along the borders between Syria and neighbouring countries (Lebanon, Jordan, and Turkey).
- Gathering project design specifications from construction standards prevalent in Syria.
- Assuming no fluctuation in energy prices and consumption over the subsequent 15 years, despite the probable impact of economic instability and inflation in post-disaster scenarios.
- Presuming uniform energy requirements for heating and cooling throughout the shelter's life span, disregarding potential energy poverty situations post-disaster where electricity availability may be limited.
- Anticipating recovery and reuse of certain shelter components during the end-of-life stage, contingent upon material specifications.
- Estimating the shelter occupancy at four individuals.
- Encompassing all life costs of the shelter over its duration, such as construction, use, maintenance, and end-of-life expenses, yet excluding certain costs like land, furnishings, and insurance from the study.
- Acknowledging the unpredictability of shelter energy consumption over 15 years, thus introducing uncertainty into the study results.
- Estimating maintenance and repair costs, acknowledging their significant variability among buildings and dependence on multifaceted factors.
- Employing IEVES for operational energy calculations, which also factor into the shelter's operational energy cost.
- Relying on market costs and the shelter's lifespan for LCA and energy cost computations.

- Recognizing that energy and emissions calculations are contingent upon design, construction materials, climate, and users, all of which are integrated into the study inputs specific to Syria.

2.2 Simulation Methodology

The energy/carbon cases were simulated using IESVE to evaluate the prefabricated vs container shelter types. IESVE is a simulation tool that can be used for both new build and refurbishment projects, the software calculates the buildings' energy consumption, carbon emissions, indoor air quality, and thermal performance. The software also can help the designers evaluate the performance of different materials and help in selecting the ones associated with suitable thermal performance [14]. The IESVE simulation program was used to calculate the operation energy and emissions, as shown in Table 1. The input date was based on the selected simulation location from Damascus, Syria. The internal thermal gains calculated depends on the following: 4 people, 4 fluorescent lights, and 1 cooker. 1 heating and cooling equipment was added for the occupant's comfort.

Table 1. The operation energy and operation emissions results from IESVE

The Container	The prefabricated
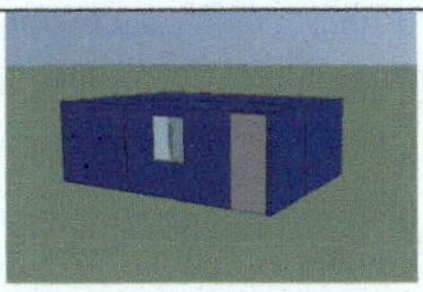	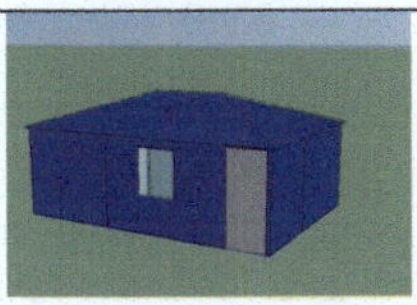
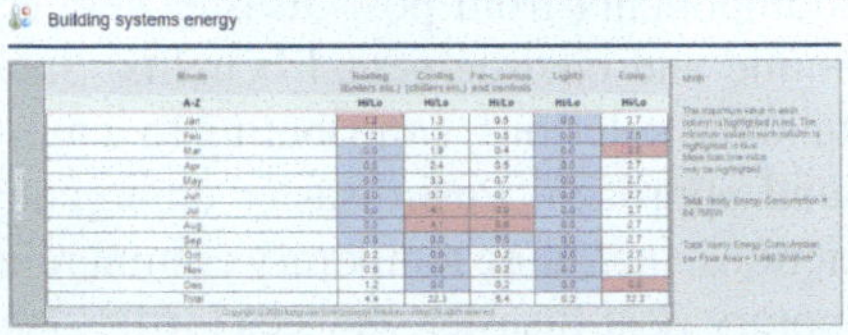	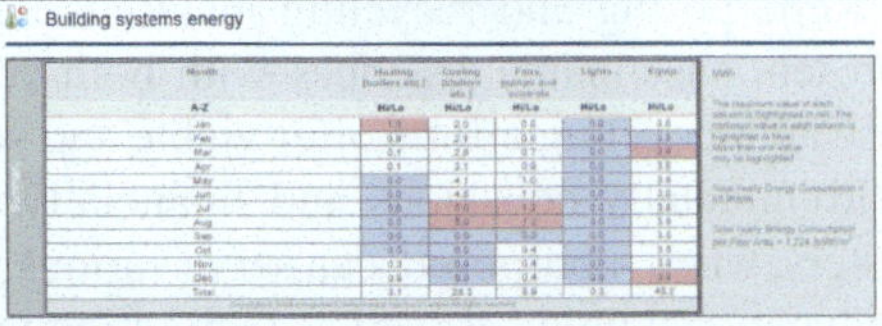
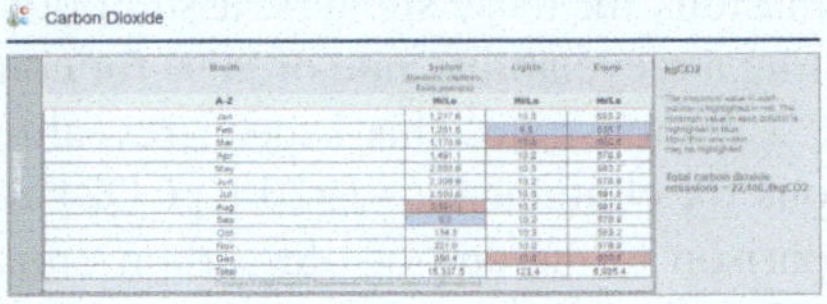	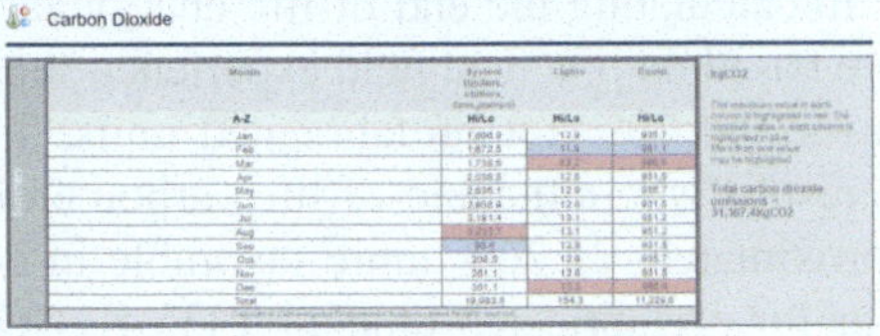

3 Embodied and End-of-Life Carbon/Energy Calculation

For calculating Embodied Energy (EE) and Embodied Carbon (EC) data from the Inventory of Carbon and Energy (ICE) Version 1.6a, was used. For the substructure, a raft foundation of 50 mm reinforced concrete was suggested as a foundation and ground slab at the same time. This is due to its resilience and suitability for a wide range of soil including low bearing capacity one. Also, this type of foundation allows for the distribution of the load equally, has better earthquake and flood resistance, and needs less workforce in excavation tasks [15]. To calculate the total EE and EC for each type of emergency shelter the total quantity for each element was initially calculated and the volume in m^3 was converted into kg, as EE and EC in ICE are all in kg. From ICE data, the EE and EC for each item were multiplied by their total quantities. The same procedure was applied to all other building elements. Finally, the total EE and EC for each type of emergency shelter were accumulated for the EE and EC for all components as can be seen in Table 2. The embodied carbon analysis demonstrated that container design has insignificant increase of 3.89% compared to prefabricated results with 2133KgCO2 and 2050 KgCO2 respectively.

Table 2. EE and EC calculation for the container & prefabricated

Prefabricated EC ($KgCO_2$)	Prefabricated EE (MJ)	Container EC ($KgCO_2$)	Container EE (MJ)	
74	583	74	583	Substructure
1634	28392	1717	31366	Superstructure
342	6340	342	6340	Door and windows
2050	35315	2133	38289	Total

Repurposing and recycling materials and components from post-disaster shelters represent sustainable practices with both economic and environmental benefits. Such reuse and recycling efforts offer viable and sustainable solutions, contributing to a more environmentally conscious and financially sound approach. Post-disaster emergency shelters, when repurposed, can serve various functions, including temporary commercial or educational units, as well as facilitating temporary post-disaster services [16].

In calculating the end-of-life energy and emissions for these shelters, assumptions were based on practical field experience suggesting a recyclability rate of 90% for containers and 70% for prefabricated units. Table 3 demonstrates this aspect, revealing the results where the end-of-life carbon emissions for containers (−1920 KgCO2) are approximately 25.26% more favorable in comparison to the end-of-life carbon emissions for prefabricated units (−1435 KgCO2). This signifies a notable environmental advantage for container shelters concerning their end-of-life carbon emissions.

Table 3. End of Life EE and EC for the container & prefabricated

	EE (MJ)	End of life EE (MJ)	EC($KgCO_2$)	End of life EC ($KgCO_2$)
Container	38289	−34460	2133	−1920
Prefabricated	35315	−24721	2050	−1435

4 Life Cycle Costing

The comprehensive LCC assessment considered the total costs attributed to both shelter types. Khaddour and Yeboah's Life Cycle Costing (LCC) method [5] was employed in this study to compute the expenses associated with prefabricated and container shelters. Construction costs, encompassing material and labour expenses, were contingent upon estimations derived from prevailing Syrian construction cost projections. Similarly, services and maintenance costs were estimated based on current market rates. The operational cost accounted for the energy consumption in operation, calculated using IEVES software, factoring in energy prices and the average exchange rate in Syria from June 2023 to September 2023. However, the research did not account for energy poverty prevalent in camp areas, leading to limited electricity availability for a few hours daily.

Furthermore, end-of-life costs were quantified in USD, relying on an assumed material recovery rate of 90% for containers and 70% for prefabricated shelters. Table 4 delineates the outcomes of the LCC analysis for container versus prefabricated emergency shelters. The findings revealed that the total cost for prefabricated shelters was approximately 33.61% higher than that for container shelters.

Table 4. LCC for the container & prefabricated

Cost in USD	Container	Prefabricated
Construction cost	64067	67800
Services cost	16600	16600
Maintenance costs	3844	4068
Operation costs	2912	3825
End-of-life cost	−57660	−47460
Total	29763	44833

5 Simulation Results and Discussion

The IESVE simulation program calculated the operation energy/emissions while embodied energy and emissions, as well as the end-of-life energy/emissions, were calculated. The operation energy and operation emissions were then multiplied by 15 years, the expected life of each shelter type.

According to the findings presented in Fig. 2, the operational energy was identified as having a more substantial impact than the embodied energy for both types of emergency shelters, aligning with previous conclusions from LCA studies [17, 18, 19]. The operational phase relates to the energy needed to sustain thermal comfort within the shelters. Notably, the notable increases in both the operational energy (OE) for prefabricated shelters (4638600MJ) and the operational carbon (OC) (470505KgCO2) suggest superior thermal performance in comparison to containers with OE (3493800MJ) and OC (336690KgCO2) respectively.

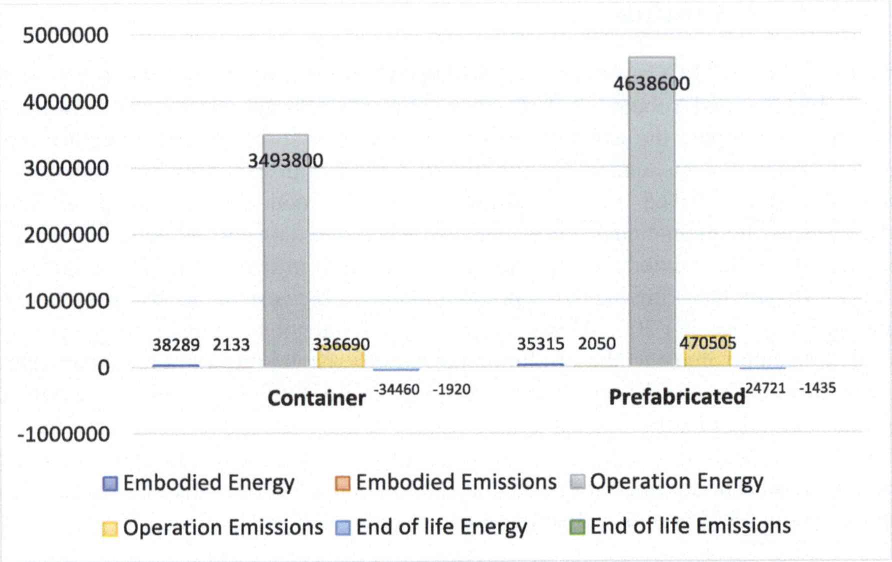

Fig. 2. Prefabricated vs container emergency shelters energy and emissions

Furthermore, the analysis of embodied energy reveals considerable disparities. Specifically, container shelters demonstrated an embodied energy of EE = 38289MJ and EC = 2133KgCO2, while prefabricated shelters exhibited EE = 35315MJ and EC = 2050 KgCO2. Figure 2 underscores a marked enhancement in end-of-life energy/carbon for container units (−1920KgCO2) compared to the (−1435KgCO2) carbon emissions from the end-of-life cycle of prefabricated units. This substantial improvement is attributed to a higher percentage of recovery materials from the container shelters compared to the prefabricated shelters.

It is evident from the results in Fig. 3 that the overall life cycle energy/carbon demonstrates better performance by the container units compared to the prefabricated ones.

Based on the findings depicted in Fig. 4, the Life Cycle Costing (LCC) analysis indicates the cost-effectiveness of container shelters, presenting a total cost of $29763, in contrast to $44833 for prefabricated units. In essence, the total cost for prefabricated shelters was identified to be 33.61% higher than that of container shelters. The primary reasons contributing to the elevated LCC of prefabricated shelters are the exorbitant building material costs and remarkably low household energy tariffs in Syria. However, this cost dynamic does not resonate in the context of emergency shelters on a broader scale, as developers and governmental planning entities tend to grapple with soaring building expenses irrespective of considerations for thermal comfort or potential cost savings.

Moreover, the holistic evaluation encompassing Life Cycle Assessment (LCA) energy/emissions and costing for shipping containers reveals a lower environmental impact and cost implications compared to similarly designed prefabricated emergency shelters. This comparative analysis of LCA and LCC underscores the advantageous

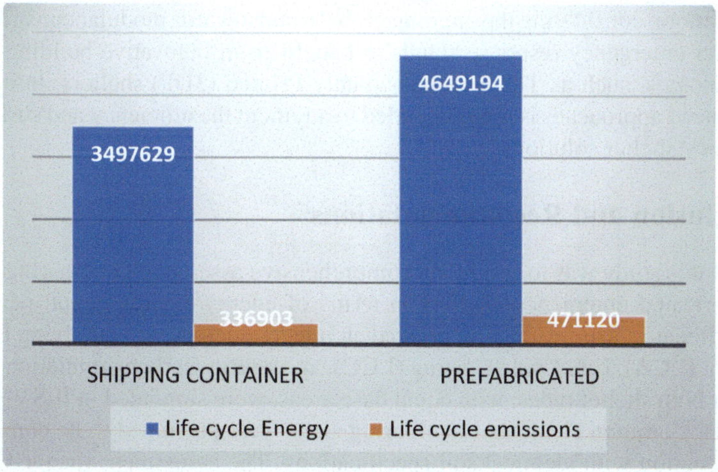

Fig. 3. Whole life cycle energy and emissions for container vs prefabricated unit

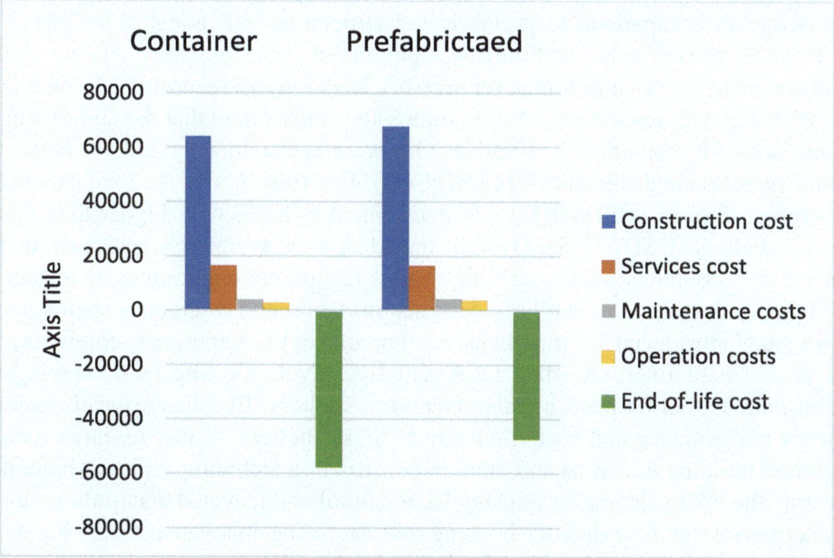

Fig. 4. LCC for the container & prefabricated

position of container shelters over prefabricated ones. Nonetheless, despite these advantages, container emergency shelters are not universally perceived as optimal sustainable solutions due to their inadequate thermal performance. In addressing the challenges linked to post-war shelters in Syria, leveraging local materials and enhancing insulation emerge as effective strategies for performance enhancement, ultimately leading to cost amelioration.

Furthermore, considering the ongoing global trend towards modular construction, the post-disaster emergency response stands to benefit from innovative building materials and technologies, such as Three-Dimensionally Printed (3DP) shelters. Integration of these advanced approaches is recommended to augment the efficiency and sustainability of emergency shelter solutions.

6 Conclusion and Recommendations

The aim of this study was to conduct a comprehensive assessment comparing container and prefabricated emergency shelters in terms of energy/environmental performance and cost efficiency. This evaluation utilized an integrated approach involving Life Cycle Assessment (LCA), Life Cycle Costing (LCC), and energy/carbon simulation. Identical designs for both shelter types, with equal base areas, were simulated in IESVE software for analysis. Common elements like flooring slabs were presumed to be constructed in the same location with identical soil specifications. The assessment spanned the entire lifespan of both shelter types, encompassing embodied energy/emissions, operational energy/emissions, and demolition embodied energy/emissions over a 15-year period.

The analysis of embodied carbon revealed a marginal increase of 3.89% for container design in comparison to prefabricated structures, indicating 2133KgCO2 and 2050 KgCO2, respectively. Additionally, operational emissions for container shelters were observed to be 28% lower than for prefabricated designs, recording 336690KgCO2 and 470505KgCO2, respectively. Furthermore, the study found that the end-of-life carbon emissions for containers (−1920KgCO2) were approximately 25.26% better than those for prefabricated shelters (−1435KgCO2). In terms of cost, the total expense for prefabricated shelters (USD44833) was determined to be 33.61% higher than that for container shelters (USD29763). Overall, the LCA energy/emission and cost analysis suggested that container shelter exhibit a more favourable environmental impact and cost efficiency compared to similarly designed prefabricated emergency shelter.

This study introduced a comprehensive comparative LCA approach, combining Life Cycle Assessment Energy/Carbon LCA and Life Cycle Costing LCC, serving as a decision-making tool for post-disaster emergency shelters. To enhance the environmental/energy performance and cost efficiency of these shelters, further research focusing on different building materials and modern construction technologies is recommended. Evaluating the life cycle impact of using local natural and recycled materials as sustainable alternatives for post-disaster housing and exploring innovative technologies like 3DP in emergency shelter construction should be considered.

Drawing lessons from prior emergency responses presents an opportunity to offer improved shelters for post-disaster camp residents. Constructing customized emergency shelter designs tailored to specific regions can potentially conserve energy/emissions, reduce costs, save time after emergencies, and contribute to enhanced resilience and quality of life.

The criteria developed in this study can serve as a decision-making tool for selecting appropriate post-disaster emergency shelters in diverse affected regions worldwide, accounting for climatic zones and design inputs. Despite their advantages, container shelters are perceived as suboptimal sustainable solutions due to their inadequate thermal performance. The research concluded that enhancing shelter material selection, construction methods, and considering location are crucial parameters to improve energy/emission performance and reduce the costs for the emergency response.

References

1. Atmaca, A., Atmaca, N.: Comparative life cycle energy and cost analysis of post-disaster temporary housings. Appl. Energy **171**, 429–443 (2016). https://doi.org/10.1016/j.apenergy.2016.03.058
2. Ismail, F.Z., Halog, A., Smith, C.: How sustainable is disaster resilience? an overview of sustainable construction approach in post-disaster housing reconstruction. Int. J. Disaster Resil. Built. Environ. **8**(5), 555–572 (2017). https://doi.org/10.1108/IJDRBE-07-2016-0028
3. KamacI-Karahan, E., Kemeç, S.: Residents' satisfaction in post-disaster permanent housing: Beneficiaries vs. non-beneficiaries. Int. J. Disaster Risk Reduction **73**, April 2022. https://doi.org/10.1016/j.ijdrr.2022.102901
4. Hosseini, S.M.A., Farahzadi, L., Pons, O.: Assessing the sustainability index of different post-disaster temporary housing unit configuration types. J. Build. Eng. **42**, October 2021. https://doi.org/10.1016/j.jobe.2021.102806
5. Khaddour, L.A., Yeboah, S.K.: Multiple-criteria optimization of residential buildings envelope toward nZEBs: simplified approach for Damascus post-war (2022)
6. Khaddour, L.A., Deng, W.: Multi-criteria sustainability risk management for post-war residential re-construction: the case of Damascus. J. Housing Built Environ. (2023). https://doi.org/10.1007/s10901-023-10024-2
7. Khaddour, L.A.: Comparative analysis of residential building envelopes newly implementing the building insulation code in Damascus. Int. J. Environ. Sci. Technol. (2023). https://doi.org/10.1007/s13762-023-05053-x
8. Su, X., Huang, Y., Chen, C., Xu, Z., Tian, S., Peng, L.: A dynamic life cycle assessment model for long-term carbon emissions prediction of buildings: a passive building as case study. Sustain Cities Soc. **96**, September 2023. https://doi.org/10.1016/j.scs.2023.104636
9. Barbhuiya, S., Das, B.B.: Life Cycle Assessment of construction materials: methodologies, applications and future directions for sustainable decision-making. Case Studies in Construction Materials, vol. 19, December 2023. https://doi.org/10.1016/j.cscm.2023.e02326
10. Carvalho, J.P., Alecrim, I., Bragança, L., Mateus, R.: Integrating BIM-based LCA and building sustainability assessment. Sustainability **12**(18), September 2020. https://doi.org/10.3390/SU12187468
11. Khaddour, L.A., Yeboah, S.K., Dodoo, J.K.: Book Chapter : Ecological and Carbon Footprints of Cities Encyclopedia of Sustainable Technologies, 2nd Edition 2023 Elsevier (2023).https://doi.org/10.1016/B978-0-323-90386-8.00044-9
12. Pomponi, F., Moghayedi, A., Alshawawreh, L., D'Amico, B., Windapo, A.: Sustainability of post-disaster and post-conflict sheltering in Africa: What matters? Sustain Prod. Consum. **20**, 140–150 (2019). https://doi.org/10.1016/j.spc.2019.06.007
13. Khasreen, M.M., Banfill, P.F.G., Menzies, G.F.: Life-cycle assessment and the environmental impact of buildings: a review. Sustainability **1**(3), 674–701 (2009). https://doi.org/10.3390/su1030674
14. Integrated Environmental Solutions. Accessed 14 Nov 2023. https://www.iesve.com/
15. Shankar Magar, J., Kudtarkar, A., Magar, J., Pachpohe, J., Nagargoje, P.: Study and analysis of types of foundation and design construction. Int. Res. J. Eng. Technol. (2020). https://doi.org/10.5281/zenodo.3995061
16. Félix, D., Branco, J.M., Feio, A.: Temporary housing after disasters: a state of the art survey. Habitat Int. **40**, 136–141 (2013). https://doi.org/10.1016/j.habitatint.2013.03.006

17. Asdrubali, F., Baldassarri, C., Fthenakis, V.: Life cycle analysis in the construction sector: guiding the optimization of conventional Italian buildings. Energy Build. **64**, 73–89 (2013). https://doi.org/10.1016/j.enbuild.2013.04.018
18. Bastos, J., Batterman, S.A., Freire, F.: Life-cycle energy and greenhouse gas analysis of three building types in a residential area in Lisbon. Energy Build **69**, 344–353 (2014). https://doi.org/10.1016/j.enbuild.2013.11.010
19. Ryberg, M.W., Ohms, P.K., Møller, E., Lading, T.: Comparative life cycle assessment of four buildings in Greenland. Build Environ. **204**, October 2021. https://doi.org/10.1016/j.buildenv.2021.108130

Testing and Reviewing Technologies in Cities' Buildings and Infrastructure

The Impact of Cortical Bone Dimensional Deterioration on the Propagation Characteristics of Guided Ultrasonic Waves

Azdine Drissi[1](✉), Mhammed El Allami[2], and Abdelali Tajmouati[1]

[1] Labo MIET, Department of Physics Applied FST, University Hassan 1St, Settat, Morocco
a.drissi@uhp.ac.ma

[2] Labo MIET, Department of Physics Applied FST, CRMEF, Settat, Morocco

Abstract. This study investigates the possible application of ultrasonic guided waves in the assessment of mechanical characteristics of human bones, with the objective of establishing a non-invasive and economically viable approach for the detection of disorders such as osteoporosis. This study aims to examine the influence of variations in cortical bone thickness on the properties of guided waves using numerical simulation and analytic techniques. This work centres on the analysis of guided wave propagation in tubular waveguides that emulate the skeletal system. The investigation employs homogeneous single-layer structures consisting of bone-like tubes devoid of marrow and tissue. The research findings indicate that geometric parameters have a major impact on the velocity of guided waves in the bone system. The study further examines the fundamental assumptions inherent in the models employed and finds appropriate guided wave modes for the practical evaluation of bone health.

Keywords: Guided wave · Cortical Bone · Dispersion Curve · Longitudinal Wave · Phase Velocity

1 Introduction

The evaluation of human bone is critical in determining fracture risk and the extent of osteoporosis, as both conditions involve a decline in bone quality. Ultrasound has emerged as a cost-effective, portable, and non-invasive technique for characterizing bones, making it an attractive alternative to traditional methods that involve harmful radiation. Bone degradation can present as mechanical changes such as decreased strength, modulus, and density, increased porosity, or dimensional alterations such as reduced thickness.

This investigation focuses on the impact of dimensional changes in cortical bone on guided ultrasonic wave propagation. Previous research has employed guided ultrasonic waves to estimate cortical bone thickness and investigate the effects of thickness on wave propagation. This research has investigated the association between ultrasonically measured thickness and real bone thickness through the utilisation of 2D simulation

A. S. Mosallam et al. (Eds.): ICATH 2023, SCI, pp. 189–200, 2024.
https://doi.org/10.1007/978-3-031-76557-5_15

models. Furthermore, investigations have been conducted on the axial transmission phenomenon at both high and low frequencies. Additionally, three-dimensional models have been utilised to assess the impact of cortical bone's anisotropy, thickness, and shape on the velocity of ultrasound. Nevertheless, the aforementioned research has predominantly relied on plate theory as a means to estimate cortical thickness, a method that may not precisely capture the intricacies of irregular cross-sectional shapes.

The primary objective of this work is to utilise cylindrical guided waves as a means to more accurately simulate the irregular cross-sectional shapes that are frequently observed in real bones. Nevertheless, this methodology fails to take into account the intricate composition of bone, which encompasses marrow and several types of tissue. The main aim of this study is to ascertain cortical thickness by the analysis of phase velocity dispersion curves of cylindrical guided waves, with the intention of selecting the most suitable modes for this specific objective. The methodologies and simulation models employed are elucidated in Sect. 2, whilst the "best section" methodology is utilised to construct phase velocity dispersion curves for different cortical thicknesses. Section 3 offers a comprehensive analysis of the most advantageous modes and their associated ramifications. The report finishes by providing a thorough analysis of the obtained results.

2 Methods

2.1 Cortical Bone Model (Single Layer)

This study concerns the modeling of cortical bone as a hollow cylindrical structure with a 4 mm thickness and an internal radius of 8 mm, neglecting the central core filled with marrow and soft tissue. The dimensions were based on the average dimensions found in the skeletal system. The cortical bone is considered uniform in the axial direction and elastic, homogeneous, and isotropic in material. Only the 2D cross-section was modeled due to this assumption. The cortical degradation was modeled as a decrease in cortical thickness, while the material properties of the base were held constant. Three levels of cortical thickness, namely 2, 3, and 4 mm, were examined. The 4 mm thickness level is considered a healthy bone and is compared to two levels of osteoporotic degradation. Table 1 displays the properties of the cortical bone, as presented in [5].

The long bone possesses a limited cortical thickness, denoted by r_{int} *and* r_{ext} which represent the inner and outside radii of the bone, respectively.

The suggested model will undergo examination at several degrees of dimensional deterioration, involving the reduction of cortical bone thickness. The dispersion properties will be assessed and compared for each level.

2.2 Theoretical Study

The issue pertaining to guided waves in tubes has garnered significant attention within the field of nondestructive testing [6]. However, there is a dearth of comprehensive research conducted specifically in the context of bone application.

The development of the characteristic equations starts with the well-known Navier equation of motion, Eq. (1):

$$\mu\Delta^2 u + (\lambda + \mu)\nabla(u.\nabla) = \rho\frac{\partial^2 u}{\partial t^2} \tag{1}$$

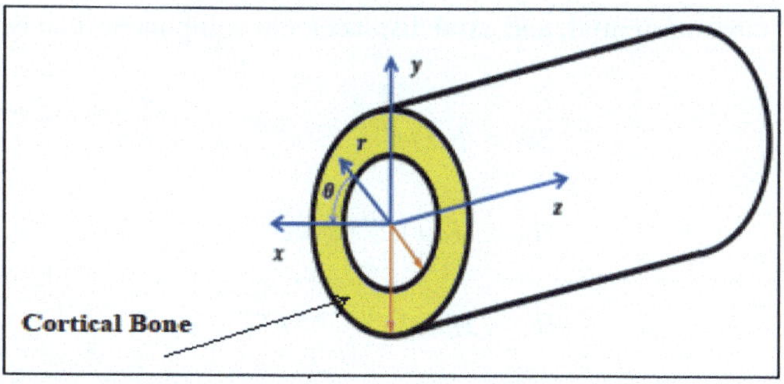

Fig. 1. The cortical bone is represented as a hollow cylinder within a cylindrical coordinate system.

Table 1. Material properties for healthy cortical bone.

Material	Young's modulus (Gpa)	Poisson's Ratio	Density (Kg /m3)	Inner Radius(mm)	Outer Radius(mm)
Cortical Bone	14.95	0.31	2300	8	12

In the given equation, "u" represents the displacement field,, λ and μ denote the Lamé constants associated with the material, and ρ is the density. To derive a solution for Eq. (1), we make the assumption that the displacement field [6] may be expressed as a linear combination of the gradient of a scalar potential field, Φ, and the curvature of a vector potential field, H with the additional stress such that:

$$\nabla.H = 0 \tag{2}$$

$$u = \nabla\Phi + \nabla \wedge H \tag{3}$$

Substituting (3) into (1) gives scalar and vector wave equations:

$$\nabla^2\Phi = \frac{1}{c^2}\frac{\partial^2\Phi}{\partial t^2}; \nabla^2 H = \frac{1}{c^2}\frac{\partial^2 H}{\partial t^2} \tag{4}$$

With $.c_T = \sqrt{\mu/\rho}; c_L = \sqrt{(\lambda + 2\mu)/\rho}$

The initial publication of the precise solutions to the axially propagating guided waves in hollow traction-free tubes was conducted by Gazis [7]. Subsequently, further investigations were carried out by Dr. Graff [8] and Rose [9]. Unlike isotropic plates, where two-dimensional modeling was adequate, the modeling of guided waves in tubes necessitates consideration of three dimensions. For the purposes of convenience, it is advisable to address the tube problem within the framework of cylindrical coordinates, namely r, θ, and z (see Fig. 1).

The traction-free boundary conditions are:

$$\sigma_{rr} = \sigma_{r\theta} = \sigma_{rz} = 0 for r = r_{int} and r = r_{ext} \tag{5}$$

The radial, circumferential, and axial displacement components can be expressed as follows:

$$\Phi = \Phi(r)e^{i(kz-\omega t)}e^{im\theta} \tag{6}$$

$$H_r = H_r(r)e^{i(kz-\omega t)}e^{im\theta} \tag{7}$$

$$H_\theta = H_\theta(r)e^{i(kz-\omega t)}e^{im\theta} \tag{8}$$

$$H_z = H_z(r)e^{i(kz-\omega t)}e^{im\theta} \tag{9}$$

where m is the circumferential order of the wave modes, k is the wave number and ω is the angular frequency while que $\Phi(r)$ et $H_j(r)$ ($j = r, \theta, z$) denote the displacement amplitudes, which are composed of Bessel and modified Bessel functions.

When considering the axial transmission, particularly the propagation of waves down the elongated axis of the tube, the guided wave modes may be classified into three separate categories [10].

The three types of modes considered in this study are longitudinal modes L(0, m) (axisymmetric modes), torsional modes T(0, m) (axisymmetric modes), and flexural modes F(n, m) (non-axisymmetric modes).

The dispersion equation governing the behaviour of a hollow tube that is free from external traction may be mathematically represented as:

$$|G(m, \omega, k)| = 0 \tag{10}$$

where [G] is a eight-by-eight global matrix "Appendix.".

For $G_{ij} with i = 4, 5, 6$ *and* $1 \leq j \leq 8$ are identical to the previous expressions with $r = r_{ext}$.

The dispersion relations of a free tube can be obtained by solving the roots of Eq. (10) by numerical methods. The concept of a numerical solution algorithm will be explained in Sect. 3.

2.3 Implementation of Numerical Solution.

By solving Eq. (10), one may derive the dispersion curves of tubular structures, which represent the correlation between the phase velocities and frequencies of the directed waves. Due to the unavailability of precise analytical solutions, numerical methods are frequently employed in order to get dispersion curves.

The dispersion equation (Eq. 10) governing the propagation of longitudinal and torsional waves (with m = 0) in a hollow cylinder may be mathematically represented as follows:

$$f(k, F) = 0 \tag{11}$$

In addition, a valuable approach involves converting the wavenumber roots into phase and group velocities, and subsequently representing these variables as a function of frequency.

$$V_{ph} = \frac{\omega}{k}, V_{gr} = \frac{\partial\omega}{\partial k} \tag{12}$$

What gives the curves of dispersion of phase and group velocities. So the dispersion equation of guided waves in the hollow cylinder can be expressed as follows:

$$f\left(V_{ph(gr)}, f\right) = 0 \tag{13}$$

The provided equation is exclusively reliant on the variable $V_{ph(gr)}$. In order to ascertain the roots of this equation, it is possible to utilise both the Muller technique and the bisection approach. When comparing the Muller technique to the bisection method, it can be observed that the bisection approach exhibits more resilience while being less efficient. In the present scenario, we employ the latter methodology as seen in Fig. 2. It is important to acknowledge that some dispersion curves of phase velocity, which are in close vicinity to longitudinal or flexural directed waves, may demonstrate closeness to certain spots. As a result, the effectiveness of root-finding may be considerably hindered. This requires the utilization of a significantly reduced step size, denoted as ΔV_{ph}, in situations when the roots are in close proximity to one another.

For a given frequency f, the roots of the frequency equation generally consist of complex wavenumbers $k = k_r + jk_i$. Wavenumbers with a value of $k_i = 0$ correspond to the propagating or proper modes of the waveguide, while those with non-zero values represent evanescent modes that attenuate along the axial distance z. The symmetry of the coefficients in the frequency equation gives rise to pairs of purely real or imaginary solutions ($\pm k_r$ and $\pm\, jk_i$), as well as quartets of complex solutions ($\pm k_i \pm jk_i$). The validity of these solutions in a waveguide problem is determined by the principle of the conservation of energy. For instance, waves propagating towards the positive z axis of an infinite waveguide must have an imaginary part of the wavenumber $\mathrm{Im}\{k\} \geq 0$.

The roots of the frequency equation often comprise complex wavenumbers, denoted as $k = k_r + jk_i$, where f represents the provided frequency. Wavenumbers characterised by $k_i = 0$ correspond to the propagating or proper modes of the waveguide, whereas wavenumbers with non-zero values indicate the presence of evanescent modes that experience attenuation along the axial distance z. The presence of symmetrical coefficients in the frequency equation results in the emergence of pairs of solutions that are either fully real or purely imaginary ($\pm k_r$ and $\pm\, jk_i$), as well as quartets of solutions that are complex in nature ($\pm k_i \pm jk_i$). The determination of the correctness of these solutions in a waveguide problem is contingent upon the principle of energy conservation. For example, when waves propagate towards the positive z axis of an infinite waveguide, it is necessary for the imaginary component of the wavenumber, denoted as $\mathrm{Im}\{k\} \geq 0$.

In our research, our focus is solely directed towards Wavenumbers characterized by.

$k_i = 0$ are associated with the propagating modes of the waveguide due to our primary focus on waves that propagate without experiencing attenuation [11].

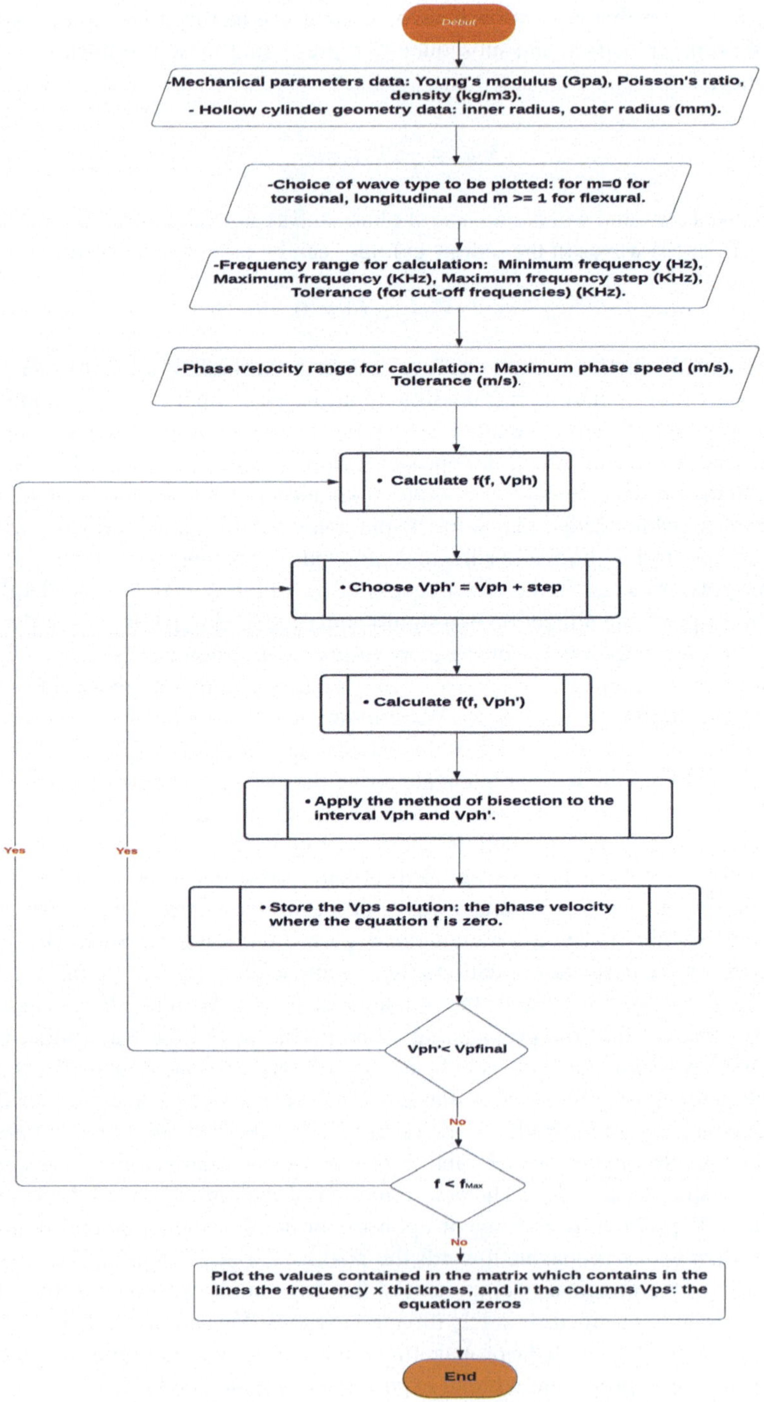

Fig. 2. The algorithm for computing dispersion curves.

3 Result and Discussion

3.1 SDispersion Curve

We will now plot the dispersion curves for cortical bone this simplified model is isotropic and homogeneous with a single solid layer. The phase velocity and groupe velocity dispersion curves for Longitudinal and flexural guided wave obtained for the healty cortical bone are shown in Fig. 3(a, b) and compared with SAFE method using **GUIguw** software.

It is important to highlight that our focus in this discussion is solely on the actual solutions to the dispersion relations, specifically those that relate to the propagation of longitudinal and flexural modes. We do not consider the non-propagating modes, which are characterized by pure imaginary wave numbers, or the attenuated modes, which possess complex wave numbers. These particular cases are not included in our analysis [12].

The dispersion curve of the phase velocity is in accordance with the SAFE outcomes obtained from GUIGW software.

In Fig. 4, the dispersion curve is depicted, illustrating the phase velocity achieved for the three different cortical thicknesses under investigation.

Table 2 provides a numerical comparison of the phase velocities. Based on the data provided in Fig. 4 and Table 2, it is evident that the L(0,3) and F(1,4) modes exhibit the most pronounced disparities in phase velocity when comparing the three distinct cortical thickness scenarios. This assertion is further substantiated by the quantitative comparison presented in Table 2. When employing a cortical thickness of 4 mm as a baseline, a decrease in thickness of 1 mm (resulting in a cortical thickness of 3 mm) leads to an average disparity in phase velocity of roughly 2803 m/s (~14.1%) for the

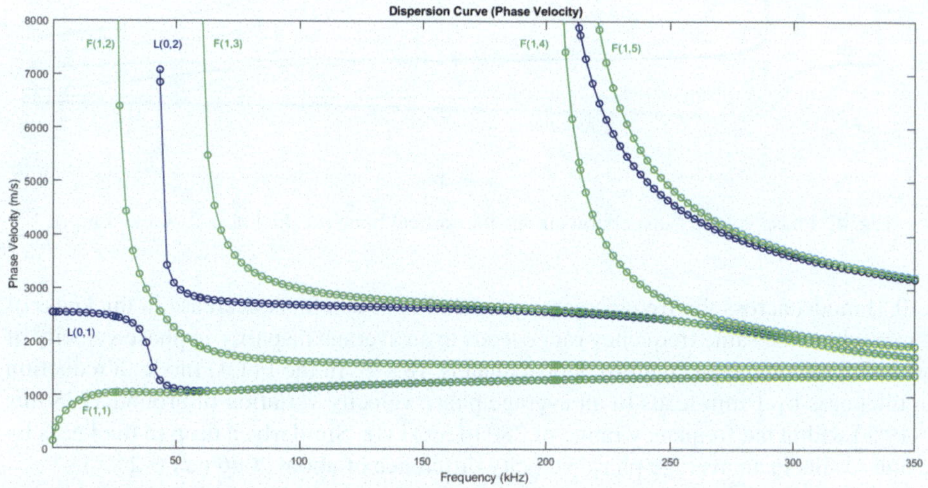

Fig. 3. **a** Dispersion curves of phase velocity of healthy cortical bone by the Bisection method (--) and SAFE Method (o). **b** Dispersion curves of Group velocity of healthy cortical bone by the Bisection method (--) and SAFE Method (o)

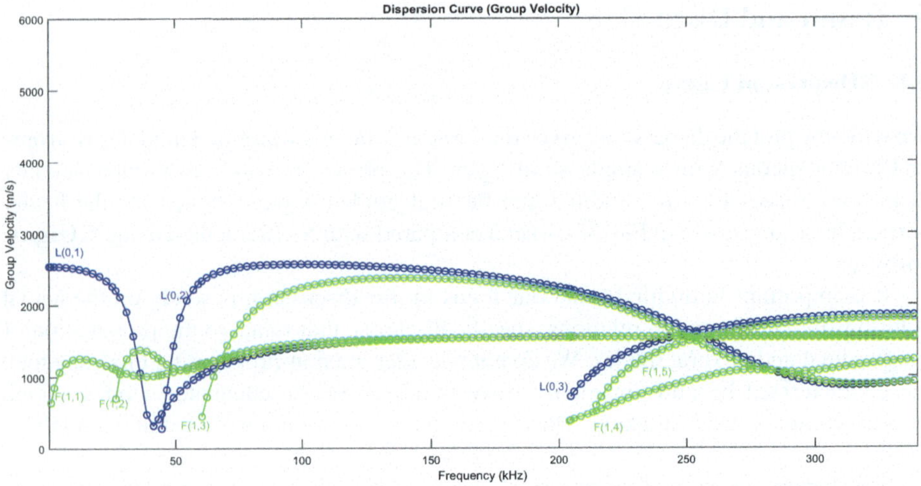

Fig. 3. (*continued*)

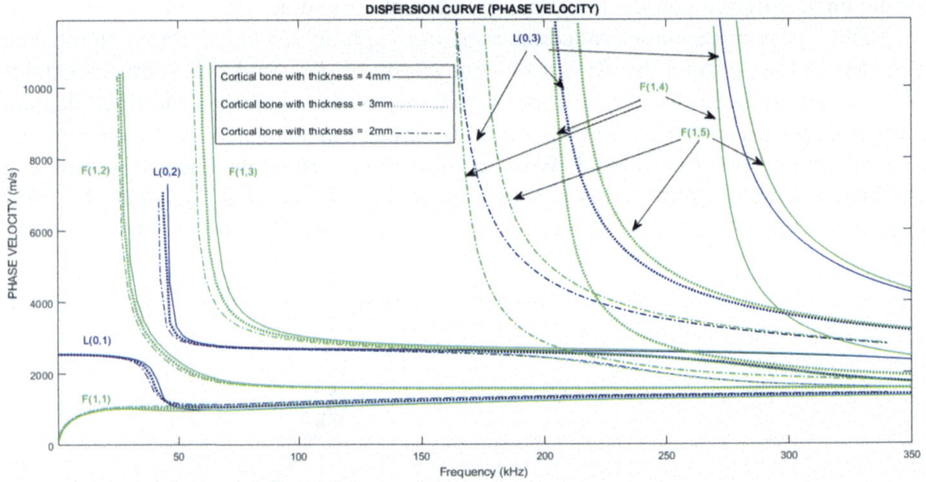

Fig. 4. Phase velocity dispersion curves for cortical bone (of thickness 2, 3 and 4 mm).

L(0,3) mode, across the frequency range of 280 to 330 kHz. A decrease in thickness of 2 mm within the same frequency range leads to an average disparity in phase velocity of around 3390 m/s, equivalent to approximately 18.4%. In the F(1,4) mode, a reduction in thickness by 1 mm leads to an average phase velocity variation of around 1778 m/s (~19%) within the frequency range of 280 to 330 kHz. Similarly, a drop in thickness by 2 mm results in an average phase velocity difference of about 2066 m/s (~24%).

Table 2. Comparison of phase velocities for the identified suitable modes, L(0,3) and F(1,4).

Mode	Freq (Khz)	Vel. Diff. Between	
		Thk = 2 mm and Thk = 4 mm	Thk = 3 mm and Thk = 4 mm
L(0,3)	280	5303	4597
	290	3819	3190
	300	2972	2395
	310	2437	1912
	320	2083	1586
	33O	1817	1350
F(1,4)	280	3624	3240
	290	2281	1953
	300	1660	1381
	310	1304	1059
	320	1069	859
	330	904	717

4 Conclusion

The aim of this research is to evaluate the feasibility of utilising focused ultrasonic waves for the purpose of distinguishing between different stages of bone degradation, namely osteoporosis. The investigation primarily concentrates on dimensional deterioration, notably the reduction in cortical thickness, while neglecting the existence of soft tissue and marrow. Nevertheless, this particular element will be duly considered in future research that employ a multilayer model. The bisection method was employed to leverage the phase velocity dispersion properties of guided waves. Through this analysis, it was ascertained that the points L(0,3) and F(1,4) are suitable for assessing alterations in cortical thickness within the frequency range of 280 to 330 kHz. It is important to note that this evaluation does not consider the influence of soft tissue and viscous marrow. The investigation of how the propagation of guided waves is influenced by the presence of soft tissue and viscous marrow is an intriguing avenue to explore. This technique allows for a more realistic representation of bone structures, where cortical bone can display nonlinearity and anisotropy, along with an uneven cross-section. Moreover, the propagation of guided waves can be influenced by additional factors, such as changes in tissue thickness or variations in viscoelastic characteristics. Consequently, this facilitates a more comprehensive assessment of the feasibility of utilising guided waves as a modality for detecting bone deterioration.

This study demonstrates the potential of guided waves in assessing bone thickness and detecting problems such as osteoporosis.

Appendix

$$
\begin{aligned}
G_{11} &= -\frac{\lambda}{2\mu} k_L^2 Z_m(pr_{int}) + Z_m''(pr_{int}) \\
G_{12} &= ikZ_{m+1}'(qr_{int}) \\
G_{13} &= -ikZ_{m-1}'(qr_{int}) \\
G_{14} &= \frac{im}{r_{int}}(-\frac{1}{r} Z_m(pr_{int}) + Z_m'(pr_{int})) \\
G_{15} &= -\frac{\lambda}{2\mu} k_L^2 W_m(pr_{int}) + W_m''(pr_{int}) \\
G_{16} &= ikW_{m+1}'(qr_{int}) \\
G_{17} &= -ikW_{m-1}'(qr_{int}) \\
G_{18} &= \frac{im}{r_{int}}(-\frac{1}{r_{int}} W_m(pr_{int}) + W_m'(pr_{int})) \\
G_{21} &= 2ikZ_m'(pr_{int}) \\
G_{22} &= \left(k_T^2 - 2k^2 - \frac{m(m+1)}{{r_{int}}^2}\right) Z_{m+1}(qr_{int}) - \frac{m}{r_{int}} Z_{m+1}'(qr_{int}) \\
G_{23} &= \left(2k^2 - k_T^2 + \frac{m(m-1)}{{r_{int}}^2}\right) Z_{m-1}(qr_{int}) - \frac{m}{r_{int}} Z_{m-1}'(qr_{int}) \\
G_{24} &= -\frac{km}{r_{int}} Z_m(qr_{int}) \\
G_{25} &= 2ikW_m'(pr_{int}) \\
G_{26} &= \left(k_T^2 - 2k^2 - \frac{m(m+1)}{{r_{int}}^2}\right) W_{m+1}(qr_{int}) - \frac{m}{r_{int}} W_{m+1}'(qr_{int}) \\
G_{27} &= \left(2k^2 - k_T^2 + \frac{m(m-1)}{{r_{int}}^2}\right) W_{m-1}(qr_{int}) - \frac{m}{r_{int}} W_{m-1}'(qr_{int}) \\
G_{28} &= -\frac{km}{r_{int}} W_m(qr_{int}) \\
G_{31} &= \frac{2im}{r_{int}}(-\frac{1}{r_{int}} Z_m(pr_{int}) + Z_m'(pr_{int})) \\
G_{32} &= k(-\frac{1+m}{r_{int}} Z_{m+1}(qr_{int}) + Z_{m+1}'(qr_{int}))
\end{aligned}
\tag{21}
$$

$$G_{33} = k(\frac{m-1}{r_{int}} Z_{m-1}(qr_{int}) + Z'_{m-1}(qr_{int}))$$

$$G_{34} = q^2 Z_m(qr_{int})$$

$$G_{35} = \frac{2im}{r_{int}}(-\frac{1}{r_{int}} W_m(pr_{int}) + W'_m(pr_{int}))$$

$$G_{36} = k(-\frac{1+m}{r_{int}} W_{m+1}(qr_{int}) + W'_{m+1}(qr_{int}))$$

$$G_{37} = k(\frac{m-1}{r} W_{m-1}(qr_{int}) + W'_{m-1}(qr_{int}))$$

$$G_{38} = q^2 W_m(qr_{int})$$

For $G_{ij} with i = 4, 5, 6 and 1 \leq j \leq 8$ are identical to the previous expressions with $r = r_{ext}$.

The matrices k^1, k^2 have the following form::

$$\text{If } C \geq C_T k^1 = \begin{bmatrix} G_{71}\, G_{72}\, G_{73}\, G_{74} \\ G_{81}\, G_{82}\, G_{83}\, G_{84} \end{bmatrix} = \begin{bmatrix} 0 -q\ -q\ k \\ 0\ 0\quad 0\, 0 \end{bmatrix}; \tag{22}$$

$$k^2 = \begin{bmatrix} G_{75}\, G_{76}\, G_{77}\, G_{78} \\ G_{85}\, G_{86}\, G_{87}\, G_{88} \end{bmatrix} = \begin{bmatrix} 0\ 0\ 0\, 0 \\ 0 -q\ q\ k \end{bmatrix}$$

$$\text{If } C < C_T k^1 = \begin{bmatrix} G_{71}\, G_{72}\, G_{73}\, G_{74} \\ G_{81}\, G_{82}\, G_{83}\, G_{84} \end{bmatrix} = \begin{bmatrix} 0 -q\ -q\ k \\ 0\ 0\quad 0\, 0 \end{bmatrix}; \tag{23}$$

$$k^2 = \begin{bmatrix} G_{75}\, G_{76}\, G_{77}\, G_{78} \\ G_{85}\, G_{86}\, G_{87}\, G_{88} \end{bmatrix} = \begin{bmatrix} 0\,0\,0\,0 \\ 0\ q\ q\ k \end{bmatrix}$$

where $Z_j and W_j$ represent incoming and outgoing Bessel functions, k is the axial wavenumber, μ one of the Lamé constants, $p = \left| k\sqrt{\left(\frac{c^2}{c^2{}_L}\right)} - 1 \right| and q = \left| k\sqrt{\left(\frac{c^2}{c^2{}_T}\right)} - 1 \right|$, $k_T = \frac{\omega}{C_T}, k_L = \omega / C_L$.

Here $Z_j and W_j$ are the Bessel functions and modified Bessel functions of first and second order j, respectively. The choice of the Bessel function is proposed in Table 3.

Table 3. Criteria for choosing the appropriate Bessel functions.

Intervalle	*Fonctions de Bessel* $(Z_j(pr), W_j(pr), Z_j(qr), W_j(qr))$
$C_L k < \omega$	$J_j(pr), Y_j(pr), J_j(qr), Y_j(qr)$
$C_T k < \omega < C_L k$	$J_j(pr), K_j(pr), J_j(qr), Y_j(qr)$
$C_T k > \omega$	$J_j(pr), K_j(pr), J_j(qr), K_j(qr)$

References

1. Moilanen, P., et al.: Ultrasonically determined thickness of long cortical bones: two-dimensional simulations of in vitro experiments. J. Acous. Soc. Am. **122**(3), 1818–1826 (2007)
2. Lee, K.I., Yoon, S.W.: Feasibility of bone assessment with leaky Lamb waves in bone phantoms and a bovine tibia. J. Acous. Soc. Am. **115**(6), 3210–3217 (2004)
3. Bossy, E., Talmant, M., Laugier, P.: Three-dimensional simulations of ultrasonic axial transmission velocity measurement on cortical bone models. J. Acous. Soc. Am. **115**(5), 2314–2324 (2004)
4. Ta, D.A., Huang, K., Wang, W.Q., Wang, Y.Y., Le, L.H.: Identification and analysis of multimode guided waves in tibia cortical bone. Ultrasonics **44**, e279–e284 (2006)
5. Thakare, D.R., Rajagopal, P.: Ultrasonic guided waves for assessing the mechanical properties of human long bone. e-Journal of Nondestructive Testing, 20(6), Proc. Indian. Soc. NDT, National Seminar & Exhibition on Non-Destructive Evaluation, NDE (2015)
6. Achenbach, J.D.: Wave Propagation in Elastic Solids. North-Holland Publishing Company, Amsterdam (1973)
7. Gazis, D.C.: Three-dimensional investigation of the propagation of waves in hollow circular cylinders. I. Analytical foundation. J. Acoust. Soc. Am. **31**(5), 568–573 (1959)
8. Graff, K.F.: Wave Motion in Elastic Solids. Dover, New York (1991)
9. Rose, J.L.: Ultrasonic Guided Waves in Solid Media. Cambridge University Press, New York (2014)
10. Silk, M.G., Bainton, K.F.: The propagation in metal tubing of ultrasonic wave modes equivalent to Lamb waves. Ultrasonics **17**, 11–19 (1979)
11. Pochhammer, L.: Uber die fortpflanzungsgeschwindigkeiten kleiner schwingungen in einem unbegrenzten isotropen kreiscylinder. J. fur Math. **81**, 324–336 (1876). (Crelle). (in German)
12. Le Clézio, E.: Diffraction des ondes de lamb par des fissures verticales. Thèse de doctorat, Université de bordeaux I (2001)

Seismic Behaviors of Reinforced Concrete Column with Different Axial Compression Ratio Strengthened by Bidirectional Basalt Fiber-Reinforced Polymer (BFRP) Laminates

Gao Peng[1], Chen Zhonghui[1], and Ayman S. Mosallam[2](✉)

[1] School of Civil Engineering, Hefei University of Technology, Hefei 230009, China
penggao@hfut.edu.cn, 2021110561@mail.hfut.edu.cn
[2] Department of Civil and Environmental Engineering, University of California, Irvine, USA
mosallam@uci.edu

Abstract. When existing reinforced concrete (RC) frame structures need seismic retrofitting, composite strengthening abilities including flexural, shear, and compressive load-bearing capacities are often required to be enhanced. The experimental research was conducted to investigate the seismic performance of RC columns confined with bidirectional fibers reinforced polymer (FRP) composite laminates. Three parameters were considered in this study: bidirectional fiber laminates, axial load ratio, and FRP anchors. Eight large-scale column specimens were evaluated experimentally and were subjected to low-frequency full-reversal cyclic lateral loadings up to failure. Experimental results indicated that the typical failure mode of columns confined with bidirectional fiber laminates differed significantly from that of unidirectional fiber laminates. Specimens strengthened with bidirectional fiber laminates exhibited a transition from bending-shear failure to bending failure, which resulted in fuller hysteresis loops, improved pinching behavior, and significantly enhanced energy dissipation and deformation capacity. After the specimen was confined, as the axial compression ratio increased within a certain range, the increase rate of bearing capacity decreased, and the increase of yield bearing capacity was more obvious than that of peak bearing capacity. The increase of axial compression ratio had little effect on the increase rate of displacement ductility, while the energy dissipation capacity increased obviously. The use of FRP anchors can further improve the bearing capacity and energy dissipation capacity of the specimen.

Keywords: Bidirectional fiber laminates · fiber anchor · axial compression ratio · seismic performance

1 Introduction

In seismic structural design, column fracture plays a crucial role in causing structural damage and collapse [1, 2]. Seismic performance of reinforced concrete (RC) structures depends on the deformability of columns [3, 4]. For the last decades, Fiber Reinforced

A. S. Mosallam et al. (Eds.): ICATH 2023, SCI, pp. 201–212, 2024.
https://doi.org/10.1007/978-3-031-76557-5_16

Polymer (FRP) composites have been evaluated in retrofitting and repairing reinforced concrete members [5]. For example, when this technique is used for RC columns' confinement, it can greatly improve bearing capacity and ductility of columns when subjected to axial loads [6, 7].

Published research [8–13] demonstrated that the hysteresis curve of columns retrofitted with unidirectional FRP is fuller than that of unconfined specimens and energy dissipation of confined specimens increased significantly. Shamim[a,b] [14, 15] conducted low-frequency full-reversal cyclic lateral loadings on RC columns combined with different axial load ratios, which indicated that the bearing capacity of columns increased with higher axial load ratios. Wang [16] retrofitted five square RC columns with CFRP (carbon fiber reinforced polymer) composites and revealed a significant improvement in energy dissipation for retrofitted concrete columns under higher axial load ratio rather than at a lower axial load ratio. Several researchers Promis [17] and Gu [18] concluded that strengthened RC columns with CFRP significantly increased bearing capacity at higher axial load ratios.

Based on the literature review conducted in this study, it is evident that column's confinement with unidirectional FRP composites increases both column's bearing capacities and ductility, however, at higher axial load levels, ductility may be affected.

In this study, columns confined with both bidirectional and multidirectional composite laminates is discussed. The experimental evaluation process involved subjecting several column specimens to low-frequency and full-reversal cyclic lateral loadings to assess seismic performance of RC columns confined with bidirectional FRP laminates considering the effect of both axial load level and the use of FRP composite anchors.

2 Description of Quasi-static Tests

2.1 Description of Column Specimens

A total of 8 reinforced concrete column specimens of 1300mm height were made, which had rectangle cross shape of 300 mm × 300 mm. The specimens were designed such that potential shear failure was avoided. The internal longitudinal steel reinforcement of each specimen comprised of four 22.0 mm-diameter steel rebars placed at the corners of cross-section, while the stirrup is in the form of 8.0 mm-diameter equally spaced at 200.0 mm. A smaller stirrup spacing of 60.0 mm was used at a distance of 300.0 mm from column top.

The axial load ratio, defined as, n, was calculated as $n = N/(f_c{}' A)$, where N represented the axial load applied at the top of the specimen, A is the column cross-sectional area, and $f_c{}'$ is the concrete axial compressive strength determined.

2.2 Design Parameters

In all tests, the values of axial load ratio, n, were taken as 0.00, 0.30, and 0.45. It should be noted that column specimens with an axial load ratio of 0.45, it approached the axial load ratio limit specified in the Code for Seismic Design of Buildings' (GB50011-2016) [19] for ordinary RC frame columns.

As mentioned earlier, three variables were tested in the experimental program included axial load ratio, bidirectional fiber laminates, and FRP anchors (See Table 1). C0, C0.3, and C0.45 represented unconfined contrast columns with axial load ratios of 0, 0.3, and 0.45, respectively. B0, B0.3, and B0.45 referred to bidirectional fiber laminates confined columns with axial load ratios of 0, 0.3, and 0.45, respectively. B0A and B0.45A correspond to bidirectional fiber laminates confined columns implemented with FRP anchors with axial load ratios of 0 and 0.45.

2.3 Material Properties

All specimens were cast by the same batch of commercial concrete, which were tested on the same day to determine the 28-day concrete compressive strength. According to China's Standard for Test Methods of Mechanical Properties of Ordinary Concrete (GB/T 50081-2016) [20], the dimensions of the cubes are 150.0 mm × 150.0 mm × 150.0 mm and the prisms are 150.0 mm × 150.0 mm × 300.0 mm. The average compressive strength of the concrete cube is 32.8 MPa, the average compressive strength of the prismatic body is 21.9 MPa, and the standard elastic modulus is 3.18 × 104 MPa.

Table 1. Column Specimens Details

Specimen ID	Axial Load Ratio	Confinement Configuration
C0	0.00	0
C0.3	0.30	0
C0.45	0.45	0
B0	0.00	2-ply bidirectional FRP laminate
B0.3	0.30	2-ply bidirectional FRP laminate
B0.45	0.45	2 layers of bidirectional FRP laminate
B0A	0	2 layers of bidirectional FRP laminate + FRP anchors
B0.45A	0.45	2 layers of bidirectional FRP laminate + FRP anchors

Notes: ***C*** *refers to contrast columns,* ***B*** *represents the use of bidirectional BFRP (Basalt Fiber Reinforced Polymer) full wrapping strengthening,* ***A*** *indicates the use of FRP anchors, and the numbers denote the axial load ratio*

Rebar mechanical properties were obtained by tensile coupon tests following the Chinese Standard for the Mechanical Properties of Steel (GB/T 228.1–2010) [21]. The yield strength of the longitudinal rebars and stirrups was 478.0 MPa and 358.0 MPa, respectively. The ultimate strength was 608 MPa and 470 MPa, respectively. The elastic modulus was 2.18×10^5 MPa and 2.18×10^5 MPa, respectively.

The on-axis mechanical properties of the bidirectional BFRP laminates and the two-part epoxy resin used in this study were determined experimentally (see Table 2) by the manufacturer according to the standard [22]. After conducting uniaxial tensile tests on the bidirectional laminates, the tensile strength in both orthogonal directions was almost the same.

Table 2. Mechanical Properties of BFRP Laminates and Two-Part Epoxy Resin

Composite Material	Longitudinal Compressive Strength, MPa	Longitudinal Elastic Modulus, MPa	Rupture Strain, %	Longitudinal Tensile Strength, MPa	Thickness, mm
BFRP	—	82,000	2.00	1330.00	0.33
Resin	82.6	3200	1.90	48.5	—

2.4 Loading Scheme

The test was carried out in Anhui Provincial Key Laboratory of Civil Engineering Structure and Materials, and the loading device was shown in Fig. 1. A horizontal low-cycle reciprocating load was applied using 100-ton MTS hydraulic servo actuator. The self-reaction force apparatus was used to apply high axial loads to the specimens. The specimen was placed on top of a steel base, with a rectangular groove in the middle to accommodate the reaction beam. A calibrated hydraulic jack was placed above the column, and a steel rod connected the reaction beam above the jack with the reaction beam in the steel rectangular groove. An axial load was applied using the jack, and the actual reading on the hydraulic pressure gauge was recorded once it stabilized. The base of the specimen was secured to the ground using two reaction beams and two vertical hydraulic jacks. The experiment at different axial load rations n of 0.00, 0.30, and 0.45 corresponded to actual axial loads of 0, 800.0 kN, and 1,200.0 kN, respectively.

3 Experimental Observations

All three contrast columns exhibited a flexural-shear failure mode. For column C0, during the loading process, the X-shaped diagonal tension cracks appeared on the backside of the concrete, as shown in Fig. 2a. In the case of specimen C0.3, the concrete at the column's ends crushed, and diagonal cracks appeared, as depicted in Fig. 2b. For specimen C0.45 with a higher axial load, the crack pattern at the top changed to vertical and short diagonal compression cracks, as illustrated in Fig. 2c. The increase in the axial load ratio led to a tendency for the concrete at the column's end to be more significantly crushed.

All the specimens confined with bidirectional fiber laminates (B0, B0.3, B0.45, B0A, and B0.45A) exhibited a flexural failure mode.

Near peak load of specimen B0 (refer to Fig. 3a), the final failure mode involved vertical cracks appearing on the compressed side of the column ends, and these cracks propagated diagonally upwards with each loading cycle. In the plastic hinge zone, the fiber laminates cracks widened until one side eventually delaminated. After the removal of the composite laminates, it was observed that larger concrete fragments fell and the longitudinal rebars were exposed.

In the case of B0.3, as depicted in Fig. 3b, several short transverse or diagonal cracks appeared in the plastic hinge zone in addition to the primary tearing crack as B0. After removing the FRP laminates, found that the concrete fragments were smaller compared to C0.

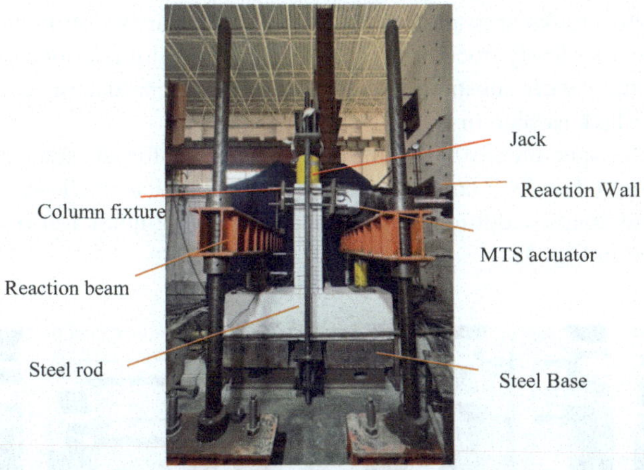

Fig. 1. Test setup

(a) C0 *(b)* C0.3 *(c)* C0.45

Fig. 2. Ultimate failure modes of unstrengthened specimens

As for specimen B0.45, the column exhibited numerous transverse short cracks in the FRP laminates, as shown in Fig. 3c. Upon removing the FRP laminates, it was evident that the concrete at the base was more thoroughly crushed.

Comparing specimens B0, B0.3, and B0.45, with the increase in the axial load ratio, the FRP laminates cracks in B0.45 gradually transitioned from diagonal primary cracks to vertical ones, and shorter transverse cracks appeared at the base. After removing the FRP laminates, it was observed that concrete expansion at the base was more pronounced, and the concrete was more thoroughly crushed.

Some obvious cracks appeared of specimen B0A near the ultimate load, and the dowel of anchor completely fractured before reaching the ultimate state. After removing FRP laminates, there were substantial diagonal cracks observed at the column base, and the concrete crushed, as seen in Fig. 3d.

In contrast, for specimen B0.45A, the dowel broke at ultimate state, and no obvious cracks were observed on FRP laminates. Only a single vertical crack was observed after removing the laminates. Additionally, fewer concrete fragments fell from the column base, as depicted in Fig. 3e.

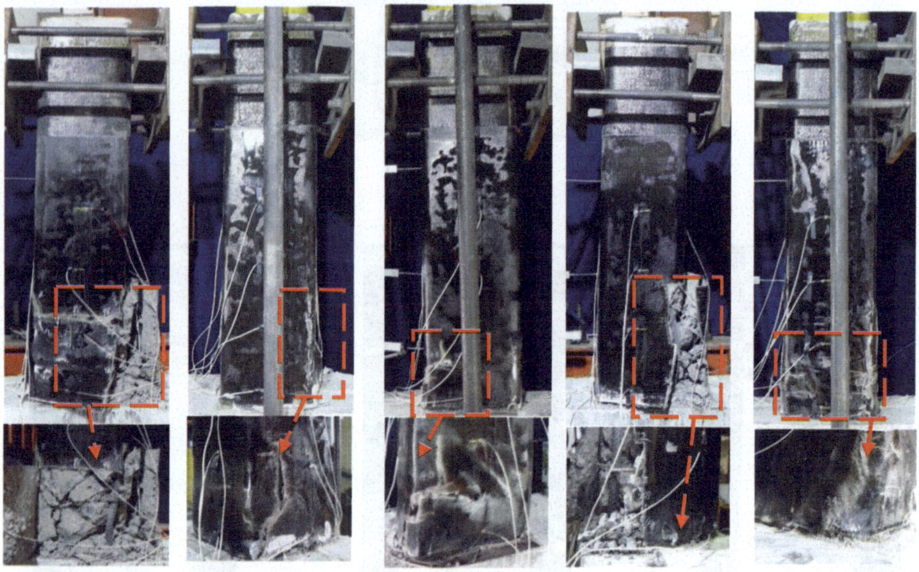

(a) B0 *(b)* B0.3 *(c)* B0.45 *(d)* B0A *(e)* B0.45A

Fig. 3. Ultimate failure mode of confined specimens

4 Experimental Results and Analysis

4.1 Hysteresis Curves

For specimens B0, B0.3, and B0.45, the hysteresis loops of the specimens' transitioned from a plump shape to a bow-shaped one as the axial load ratio increased within a certain range, which showed a pinching phenomenon (Fig. 4). The unloading stiffness of the specimens increased and the ultimate displacement reduced.

As shown in Fig. 4e and 4f, for the specimens C0.45 and B0.45, first of all, the shape of the hysteresis curve of B0.45 with confinement was fuller, and the curve pinching phenomenon was obviously improved. Secondly, the bearing capacity decreased more slowly after the peak load. Finally, the ultimate displacement of the specimen with confinement was greatly improved, which indicated that the bidirectional FRP laminates can effectively improve the deformation capacity after yielding.

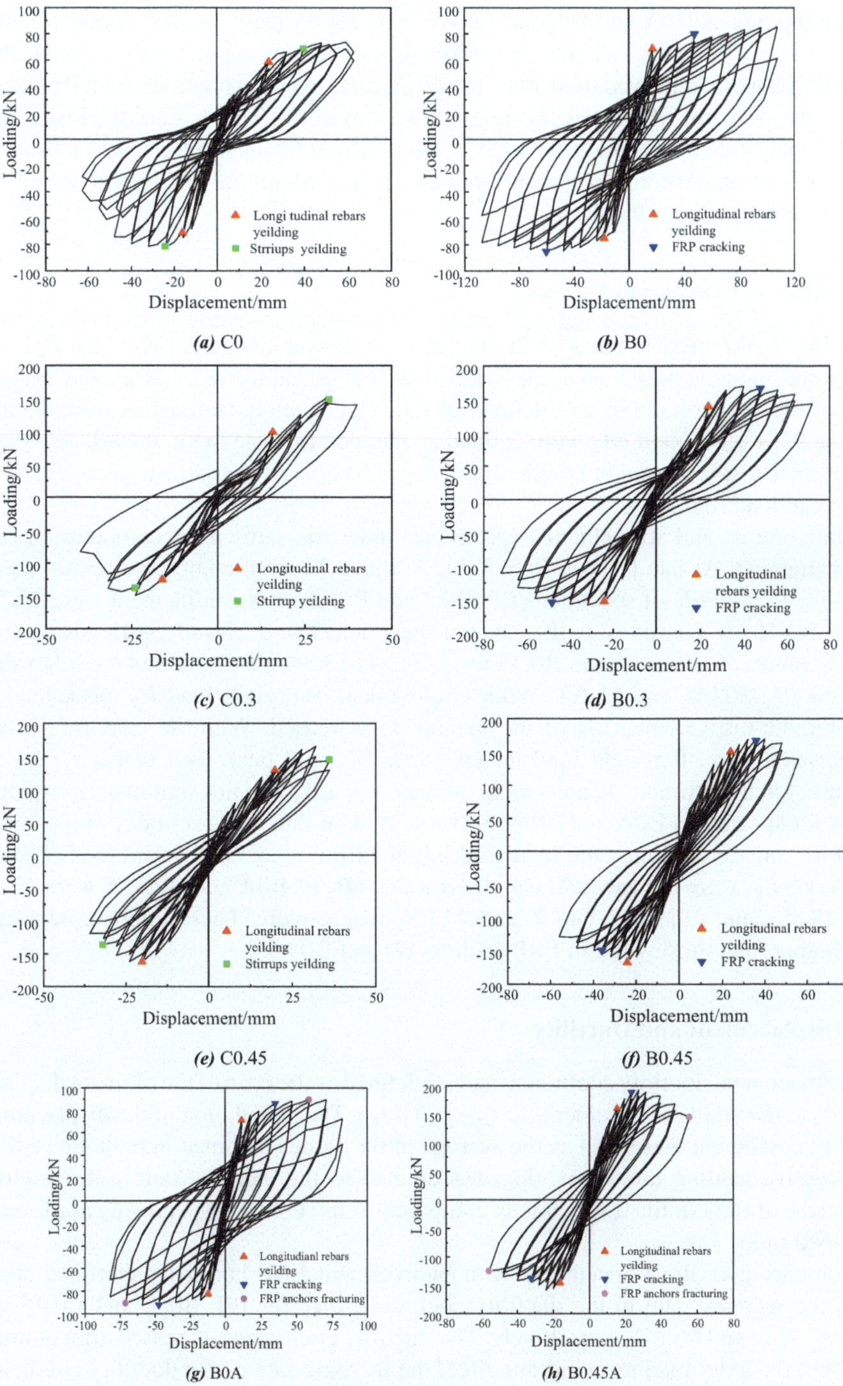

Fig. 4. Lateral load-displacement hysteretic loop curves

For specimens B0A and B0, and B0.45A and B0.45 (Fig. 4g, 4b, 4h and 4f), after installation of FRP anchors, both yield and peak load-bearing capacity of specimens B0A and B0.45A were improved. However, the displacement corresponding to FRP cracking in B0A was 30 mm, which was less than the 40 mm in B0. This is since the installation of FRP anchor increases the utilization of longitudinal fibers, which makes it easier to create stress concentration at the intersection of circumferential and longitudinal fiber, and then cracks appear in advance.

4.2 Skeleton (Envelope) Curves

As in Fig. 5, the highest point of the skeleton curve was the peak point ($\varDelta_c$, P_c). The limit point when the peak load dropped below 85% was defined as ($\varDelta_u$, P_u), and the yield point of the skeleton curve was defined as ($\varDelta_y$, P_y). The characteristics were detailed in Table 3. δP_y, δP_c and $\delta \varDelta_u$ were defined as the increase rate in yield load, peak load, and ultimate displacement of confined specimens compared to contrast specimens at the same axial load ratio.

The contrast and strengthened specimens under the same axial compression ratio were compared. As can be seen from Table 3, when the axial compression ratio was 0, 0.3 and 0.45, the δP_y of specimens B0, B0.3 and B0.45 after confinement were 26.7%, 14.1% and 10.1% compared to the contrast specimens C0, C0.3 and C0.45. Meanwhile, the δP_c values of B0, B0.3 and B0.45 were 16.3%, 13.5% and 2.7% and the $\delta \varDelta_u$ values were 68.8%, 80.0% and 55.6%. After confinement, the yield load P_y, peak load P_c and ultimate displacement $\varDelta_u$ of the specimens increased. With the increase of axial compression ratio, the yield load increase rate δP_y and peak load increase rate δP_c decrease, but the ultimate displacement increase rate $\delta \varDelta_u$ was not significantly affected.

As for specimens B0A and B0.45A, it was evident that the specimens implemented with FRP anchor exhibited the increased initial stiffness, and both yield load (P_y) and peak load (P_c) were enhanced. It can be seen that δP_y of B0A and B0.45A were 72.5% and 145.8%, and δP_c were 64.4% and 24.6%, respectively. These improvement rates were higher than those without FRP anchors B0 and B0.45.

4.3 Displacement and Ductility

The displacement ductility coefficient μ_Δ was defined as the ratio of the ultimate displacement $\varDelta_u$ to the yield displacement $\varDelta_y$, ($\mu_\Delta = \varDelta_u/\varDelta_y$). The calculation of the displacement ductility coefficient was taken as the average of the values obtained in both the positive and negative loading directions. $\delta\mu_\Delta$ was defined as the increase rate in the ductility coefficient of the confined specimens compared to the contrast specimens at the same axial load ratio.

Comparing contrast specimens with bidirectional FRP laminates confined specimens, the increase rate in the ductility coefficient $\delta\mu_\Delta$ for B0, B0.3, and B0.45 was 68.88%, 62.2, and 66.6%, respectively. The ductility coefficient increased after confinement, but the axial load ratio did not affect the increase rate of the ductility coefficient (Table. 1).

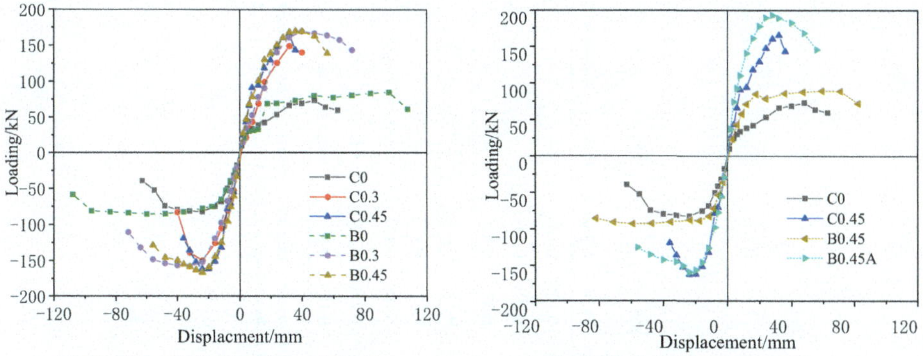

(a) C0, C0.3, C0.45, B0, B0.3 and B0.45 *(b)* C0,C0.45, B0.45, and B0.45A

Fig. 5. Skeleton curves of load-displacement

4.4 Energy Dissipation

Quantitative analysis of the energy dissipation capacity of the specimens was performed using cumulative energy E_{sum} and equivalent viscous damping coefficient ξ_e, as shown in Table 3. $\delta\xi_e$ was defined as the increase ratio in the equivalent viscous damping coefficient of the confined specimens compared to the contrast specimens at the same axial load ratio.

Comparing the three sets of specimens B0, B0.3, and B0.45, it can be observed that the increase ratio in the equivalent viscous damping coefficient $\delta\xi_e$ were 25.3%, 26.4%, and 46.9%, respectively. This indicated that the equivalent viscous damping coefficient increased with an increase in the axial load ratio after confinement.

Comparing B0A and B0.45A with C0 and C0.45 revealed that after the installation of FRP anchors, the $\delta\xi_e$ for B0A and B.45A are 68.3% and 109.2%, respectively, which were higher than that for B0 and B0.45 of 23.0% and 28.3%. This implied that the installation of FRP anchors can enhance the energy dissipation capacity of the specimens, and as the axial load ratio increased, the increase ratio in energy dissipation capacity became more pronounced.

Table 3. Summary of experimental results and increase rate of the index.

Specimen	Δ_y, (mm)	P_y, (kN)	Δ_c (mm)	Pc (kN)	Δ_u (mm)	Pu (kN)	δPy (%)	δPc (%)	δΔu (%)	μ_Δ (mm)			$\delta\mu_\Delta$ (%)	ESum (kN·mm)		ξ_e		$\delta\xi_e$	
										+	-	Avg		Peak	Ultimate	Peak	Ultimate	Peak	Ultimate
C0	18.0	47.5	48.0	73.2	64.0	59.7	–	–	–	60	4.6	4.1	–	9,517	16,695	0.126	0.188	–	–
C0.3	23.3	125.5	32.0	149.3	40.0	140.4	–	–	–	40	2.2	2	–	6,951	12,345	0.104	0.154	–	–
C0.45	21.5	137.2	32.0	165.5	36.0	143.3	–	–	–	34	1.9	1.8	–	3,882	8,726	0.078	0.115	–	–
B0	16.3	60.2	60.0	85.1	108.0	61.4	26.7	16.3	68.8	108	7	6.4	66.6	17,242	53,087	0.155	0.236	23.0%	25.3%
B0.3	24.1	143.2	40.0	169.5	72.0	143.5	14.1	13.5	80.0	72	3.4	3.2	62.2	14,563	42,291	0.133	0.195	27.2%	26.4%
B0.45	22.6	151.1	36.0	170.0	56.0	131.5	10.1	2.7	66.6	52	3.2	3	66.6	9,067	19,256	0.100	0.172	28.3%	49.6%
B0A	12.5	78.1	60.0	92.5	84.0	72.5	64.4	26.4	31.3	84	7.6	7.3	77.4	20,746	37,594	0.212	0.201	68.3%	6.8%
B0.45A	18.9	165.5	28.0	193.1	56.0	145.8	20.6	16.7	55.6	56	3.2	3.1	71.1	11,883	23,034	0.163	0.193	109.2%	68.1%

Note: *δ_{Py}, δ_{Pc}, $\delta_{\Delta u}$, $\delta_{\mu\Delta}$ and $\delta\xi_e$ are the increase rate of yield load, peak load, ultimate displacement, displacement ductility coefficient and equivalent viscous damping coefficient of the specimen under confinement compared with contrast members. For example, the yield load increase rate δP_y of the bidirectional fiber laminate specimen B0.3 is the difference between P_y of B0.3 and P_y of C0.3 divided by P_y of C0.3*

5 Conclusions

The study focused on investigating the influence of axial load ratios and FRP anchors on the seismic performance of RC columns with bidirectional FRP laminates. Eight specimens were divided into three groups and subjected to low-cycle cyclic loading tests. Through analysis of the test results and theoretical research, the following conclusions were as follows:

The failure modes of bidirectional FRP laminates differed significantly from traditional unidirectional fiber sheets. The confined specimens changed from flexural-shear failure to flexural failure. The hysteresis curves became more plump, pinching effects were improved and energy dissipation and deformation capacity were enhanced.

As the axial load ratio increased for confined specimens B0, B0.3 and B0.45, the increase rate of bearing capacity decreased with higher axial load ratios. The increase rate in yield load capacity was more noticeable than the peak load capacity. The yield load capacity increased by 16.3%, 13.5%, and 2.7%, and the peak load capacity increased by 26.7%, 14.1%, and 10.1%. The axial load ratio had a less pronounced effect on the increase in displacement ductility, but energy dissipation capacity increased with higher axial load ratios. The displacement ductility coefficient increased by 66.6%, 62.2%, and 66.6%, and energy dissipation capacity increased by 25.3%, 26.4%, and 49.6% with increasing axial load ratios after confinement for B0, B0.3 and B0.45.

The installation of anchor bolts enhanced the energy dissipation capacity of the specimens, and the increase rate of energy dissipation capacity became more significant with higher axial load ratios. After the installation of anchors, the $\delta\ \xi_e$ for B0A and B.45A were 68.3% and 109.2%, respectively, which were higher than the values for B0 and B0.45 at 23.0% and 28.3%.

References

1. Qin, Z., Na, W., Jinxin, G.: State of the art of seismic performance including shear effects and failure modes of reinforced concrete columns. J. Build. Struct. **38**(8), 1–13 (2017)
2. Hosseinpour, F., Abdelnaby, A.E.: Effect of different aspects of multiple earthquakes on the nonlinear behavior of RC structures. Soil Dyn. Earthq. Eng. **92**(2), 702–725 (2017)
3. Jian, Y., Sla, B., Xz, C., et al.: Experimental research and finite element analysis on the seismic behavior of CFRP-strengthened severely seismic-damaged RC columns. Structures **34**, 3968–3981 (2021)
4. Cui, Z., Alice, A., Behrouz, S.: Structural performance of fiber-reinforced concrete columns. Mater. Struct. **191**, 460–468 (2019)
5. Mosallam, A.S., Bayraktar A, Elmikawi M, Pul S, Adanur S. Polymer composites in construction: an overview. SOJ Mat. Sci. Eng. (2014). https://symbiosisonlinepublishing.com/materialsscience-engineering/materialsscience-engineering07.php
6. Haroun, M.A., Mosallam, A.S., Feng, M.Q., Elsanadedy, H.M.: Experimental investigation of seismic repair and retrofit of bridge columns by composite jackets. J. Reinf. Plast. Compos. **22**(14), 1243–1268 (2003)
7. Youssef, M.N., Feng, M.Q., Mosallam, A.S.: Stress–strain model for concrete confined by FRP composites. Compos. B Eng. **38**(5–6), 614–628 (2007)
8. Ouyang, L.J., Gao, W.Y., Zhen, B., Lu, Z.D.: Seismic retrofit of square reinforced concrete columns using basalt and carbon fiber-reinforced polymer sheets: a comparative study. Compos. Struct. **162**, 294–307 (2017)

9. Promis, G., Ferrier, E.: Performance indices to assess the efficiency of external FRP retrofitting of reinforced concrete short columns for seismic strengthening. Constr. Build. Mater. **26**, 32–40 (2012)
10. Jun, J., Zenga, B., Yongchang, G., et al.: Stress-strain behavior of concrete in circular concrete columns partially wrapped with FRP strips. Compos. Struct. **200**, 810–828 (2018)
11. Wang, D., Huang, L., Yu, T., Wang, Z.: Seismic performance of CFRP-retrofitted large-scale square RC columns with high axial compression ratios. J. Compos. Constr. **21**(5), 04017031 (2017)
12. Paultre, P., Boucher-Trudeau, M.: Behavior of circular reinforced-concrete columns confined with carbon fiber-reinforced polymers under cyclic flexure and constant axial load. Compos. Struct. **20**(3), 04015065 (2016)
13. Gu, D.S., Wu, Y.F., Wu, G., Wu, Z.S.: Plastic hinge analysis of FRP confined circular concrete columns. Constr. Build. Mater. **27**(1), 223–233 (2012)
14. [a] Shamim, A.S., Yau, G.: Seismic behavior of concrete columns confined with steel and fiber-reinforced polymers. ACI Struct. J. **99**(01), 71–80 (2002)
15. Shamim, A.S.: Performance of concrete structures retrofitted with fiber-reinforced polymers. Eng. Struct. **24**, 869–879 (2002)
16. Daiyu, W., Le, H., Tao, Y.: Seismic performance of CFRP-retrofitted large-scale square RC columns with high axial compression ratios. Compos. Struct. **21**(5), 04017031 (2017)
17. Promis, G., Ferrier, E.: Performance indices to assess the efficiency of external FRP retrofitting of reinforced concrete short columns for seismic strengthening. Constr. Build. Mater. **26**(3), 32–40 (2012)
18. Dongsheng, G., Gang, W., Zhishen, W.: Confinement effectiveness of FRP in retrofitting circular concrete columns under simulated seismic load. Compos. Struct. **14**(5), 531–540 (2010)
19. Ministry of Construction of the People's Republic of China. GB 50011–2016, Code of seismic design of buildings. China Architecture and Building Press, Beijing (2016). (in Chinese)
20. Ministry of Construction of the People's Republic of China. GB/T 50081–2016, Standard for test method of mechanical properties on ordinary concrete. China Architecture and Building Press, Beijing (2016). (in Chinese)
21. General Administration of Quality Supervision, Inspection and Quarantine of the People's Republic of China. GB/T 228.1–2010, Metallic Materials-Tensile testing-Part 1: Method of test at room temperature. Standards Press of China, Beijing (2010). (in Chinese)
22. Ministry of Housing and Urban-Rural Development of the People's Republic of China. GB 50728-2011, Technical code for safety appraisal of engineering structural strengthening materials. China Architecture and Building Press, Beijing (2011). (in Chinese)

Vertical Thermal Stratification in Large Buildings: Analyzing the Influence of Building Height, Internal Heat Sources, and Climatic Conditions

Nisrine Laghmich[1](✉), Zaid Romani[2], and Abdeslam Draoui[3]

[1] Laboratory of Intelligent Systems and Applications, LSIA Team, Moroccan School of Engineering Sciences (EMSI), Tangier, Morocco
n.laghmich@emsi.ma

[2] Laboratory of Built Environment and Landscape (LaBEL), National School of Architecture, Tetouan, Morocco

[3] FSTT - Materials, Systems and Energy Engineering Laboratory (MaSEEL), - ETTE (UAE/U10FST) - Abdelmalek Essaâdi University, BP 416 Tangier, Morocco

Abstract. The thermal assessment of large building requires a deep analysis of the indoor temperature according to the building height, internal gains and climatic conditions. Thermal stratification of the air in indoor environments is a phenomenon that can impact the comfort of the occupants. This study is dedicated to the thermo-aeraulic modeling of a large-volume building, aiming to proactively predict its response to three critical conditions: the incorporation of variations of building height, internal gains and external climatic conditions. Such large volume buildings are typically characterized by expansive internal spaces and intermittent, high-density occupancy, making interior thermal stratification a prevalent source of discomfort. Utilizing natural ventilation, a temperature contrast of around 7 °C was noted between 8 a.m. and 12 p.m. Subsequently, this contrast expanded to approximately 9 °C by 4 p.m. and eventually peaked at approximately 10 °C at midnight. In this context, our research introduces a zonal thermo-aeraulic modeling approach that integrates dynamic thermal simulation tools, TRNSYS and CONTAM, to address this phenomenon.

Keywords: Thermal stratification · thermo-aeraulic coupling · large volume building · mesh study · building energy simulation

1 Introduction

Large volume buildings, such as warehouses, shopping malls, and industrial facilities, represent a significant portion of the built environment and play a pivotal role in energy consumption [1]. Achieving optimal thermal comfort and energy efficiency in these building is not only essential for occupants' well-being but also for reducing the carbon footprint associated with their operation [2]. An important aspect to consider in achieving these objectives is the deliberate manipulation of thermal stratification, a phenomenon that occurs when temperature gradients form inside a building [3].

A. S. Mosallam et al. (Eds.): ICATH 2023, SCI, pp. 213–226, 2024.
https://doi.org/10.1007/978-3-031-76557-5_17

It should be noted that global warming and climate change are among the defining issues of our time and have received significant attention in recent decades [4]. Human activities have been recognized as the main causes of global warming and other environmental problems. According to the International Energy Agency (IEA), the building sector alone accounts for about 35% of final energy consumption and nearly 38% of direct and indirect CO_2 emissions [5]. The energy efficiency in buildings has become a major concern for many countries and organizations and a considerable part of the energy in buildings is used to create a comfortable indoor environment for the occupants [6].

Thermal stratification refers to the vertical layering of air masses with varying temperatures within a space. This phenomenon occurs naturally due to factors such as solar gains, building height [7], internal gains [8], climate conditions, etc. Understanding and harnessing thermal stratification is vital for optimizing energy usage, as it can lead to significant energy savings through improved heating and cooling system performance, enhanced indoor air quality, and increased occupant comfort [9].

This research paper focuses on the many aspects of thermal stratification in large buildings. Our primary objective is to explore innovative strategies that can exploit the benefits of thermal stratification, ultimately leading to more sustainable and energy-efficient building designs. We will delve into the fundamental principles governing thermal stratification, the factors influencing its formation, and the implications for energy consumption.

By analyzing the interplay between building characteristics, climate conditions, and thermal stratification dynamics, we seek to offer insights that can inform architects, engineers, and building operators in their pursuit of energy-efficient designs and operations for large volume buildings. To do that, three types of numerical models exist in the literature: nodal models [10], zonal models [11] and the computational fluid dynamics (CFD) approach [12]. The choice of an appropriate model is essential and is generally determined by four factors, namely, the desired degree of accuracy, the phenomena to be treated, the computation time, and the level of expertise of the users [13].

The research presented in this study offers valuable insights into the intricate thermal behavior of large buildings. It introduces a modeling approach that holds the potential to assist architects, engineers, and building designers in optimizing indoor climate control and improving the comfort of occupants, especially in spaces characterized by expansive interiors and fluctuating occupancy patterns. The research's primary objective is to evaluate thermal stratification, achieved through the utilization of a zonal thermo-aeraulic modeling approach integrated with dynamic thermal simulation tools such as TRNSYS and aeraulic tool CONTAM.

2 Methodology

2.1 Description of the Thermo-aeraulic Model

The National Institute of Standards and Technology (NIST) has developed the "CONTAM" aeraulic modeling tool to better integrate the variation of air flows between the connecting interfaces of different sub-volumes, the interaction between related phenomena and the integration of ventilation systems. This tool can be used to determine air flows and contaminant concentrations, and can be coupled with other dynamic thermal simulation tools such as TRNSYS. In general, the TRNSYS tool provides an interface for simulations and coupling, either by using its own subroutines called Types, or by adding new Types written in Fortran, Matlab or another compatible program [14].

The choice of these numerical simulation tools, TRNSYS and CONTAM, has been based on two criteria. The first was the facility with which the CONTAM aeraulic modeling tool could be connected to another thermal simulation environment such as TRNSYS. The second criterion is the ability to implement and solve the complete zonal thermo-aeraulic model.

TRNSYS software offers an extensive library of various components, including Type56, a multi-zone building model for simulating and calculating the various heat transfer modes in a building. In addition, Type98 has been created to perform thermo-aeraulic coupling to ensure communication and data sharing between TRNSYS and CONTAM [15]. In fact, Type98 acts as a digital server for CONTAM, enabling execution control and data transfer via TCP/IP (Transmission Control Protocol/Internet Protocol) sockets. In the latter, four categories of request can be executed: (i) sending information about the project in progress, (ii) receiving modifications to the project, and (iii) sending simulation results. Another necessity of the coupling between TRNSYS and CONTAM is the provision of a graphical user interface, called CONTAMW, which is used to build and display the image of the airflow network components [16].

Another interface named "TRNSYS 3D" has been created to facilitate the integration of the building's three- dimensional (3D) design. The elements contained in this file will subsequently be read by the TRNBUILD data manager [15].

The ping-pong and onion methods are the two existing methods for coupling the thermal and aeraulic models [17]. The first "ping-pong" method enables the numerical equations of the thermal and aeraulic models to be solved sequentially at each time step. In the second, so-called "onion" method, each model is solved in a regular method until a global convergence criterion is satisfied at each time step.

It should be noted that in the model developed, the coupling between Type98 and Type56 can only be achieved using the ping-pong method [18]. Further details of the thermo-aerodynamic coupling methodology between Type56 and Type96 are shown in Fig. 1 below.

2.2 Description of the Vertical Subdivision of an Indoor Air Volume

To investigate the vertical stratification of indoor temperature, a thermo-aeraulic coupling methodology was developed employing the software tools TRNSYS [19] and CONTAM [20]. During the implementation phase of this coupling approach, a one-dimensional vertical division of the building's air volumes was undertaken. As illustrated in Fig. 2 below, the air volumes were categorized into three distinct types: those situated in the middle, those in direct proximity to the ceiling, and those in direct proximity to the floor. Taking, for instance, the sub- volume denoted as [i, j, k], it maintains contact with two neighboring sub-volumes, one located above [i, j, k + 1], and the other situated below [i, j, k-1].

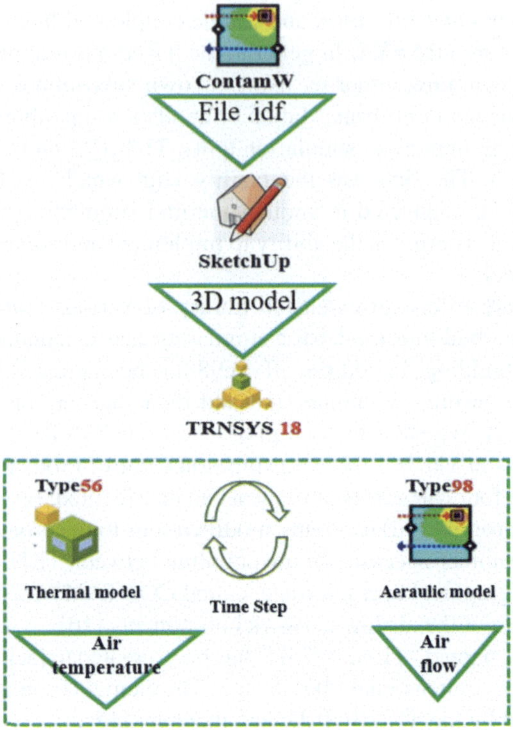

Fig. 1. Thermo-aeraulic coupling between TRNSYS and CONTAM

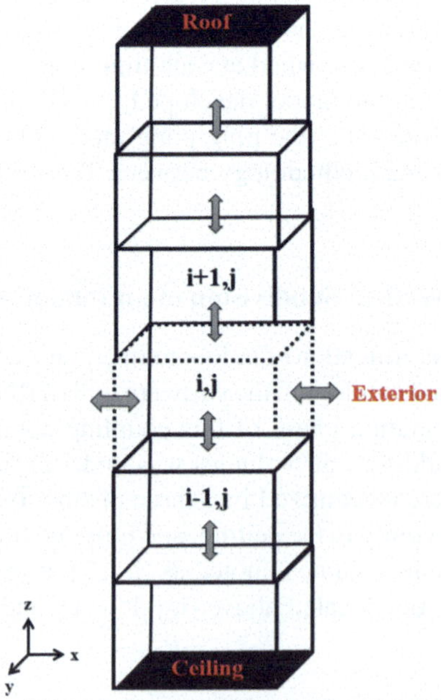

Fig. 2. Vertical decomposition of air volume

2.3 Description of the Equations of the Heat-Air Balance of the Vertical Subdivision

The process of characterizing physical phenomena needs the development of mathematical equations that establish the fundamental framework for numerical simulation tools.

In the case of each sub-volume type, the description of the physical phenomena is formulated as functions dependent on divers parameters: air flow rate, surface temperature, and the temperature of the air contained within the respective sub-volume. They can be expressed as:

•The central sub-volumes

$$Q_{cplg,i} = \dot{m}_{i+1}C_p(T_{i+1} - T_i) + \dot{m}_{i-1}C_p(T_{i-1} - T_i) \quad (1)$$

•The ceiling sub-volume

$$Q_{cplg,i} = U_{ceiling}(T_{ext} - T_i) + \dot{m}_{i-1}C_p(T_{i-1} - T_i) \quad (2)$$

•The floor sub-volume

$$Q_{cplg,i} = U_{floor}(T_{floor} - T_i) + \dot{m}_{i+1}C_p(T_{i+1} - T_i) \quad (3)$$

With **Cp** the heat capacity of the air ($kJ \cdot kg^{-1} \cdot K^{-1}$), $\dot{m}$ the mass flow rate ($Kg \cdot h^{-1}$) et **Ti** the air temperature of the representative sub-volume. **Tceiling** the floor temperature, Text the ambient outdoor temperature. ***Qcplg***, the thermal energy rate and ***Uceiling*** and ***Ufloor*** are thermal transmittance successively for ceiling and floor.

3 Result and Discussion

3.1 Presentation of the Studied Building

The focus of our scientific investigation pertains to a specific building characterized by its structural composition comprising four facades, one floor, and a roof (Fig. 3). The building area is 528 m^2 and a ceiling height of 7.5 m. Each facade is equipped with windows, collectively accounting for a glazed area of 20 m^2. This glazed surface area constitutes approximately 23% of the total external surface area of the facades. This building is considered as receiving the public and it is occupied every day from 8 am to 1 pm and from 2 pm to 7 pm.

The building under examination is situated within the Mediterranean climatic zone, specifically at the geographical coordinates of 35° 46′ N latitude and 5° 48′ W longitude, corresponding to the city of Tangier in the northwestern region of Morocco. To acquire meteorological data representative of a typical year for this location, the Meteonorm (TMY) tool has been used.

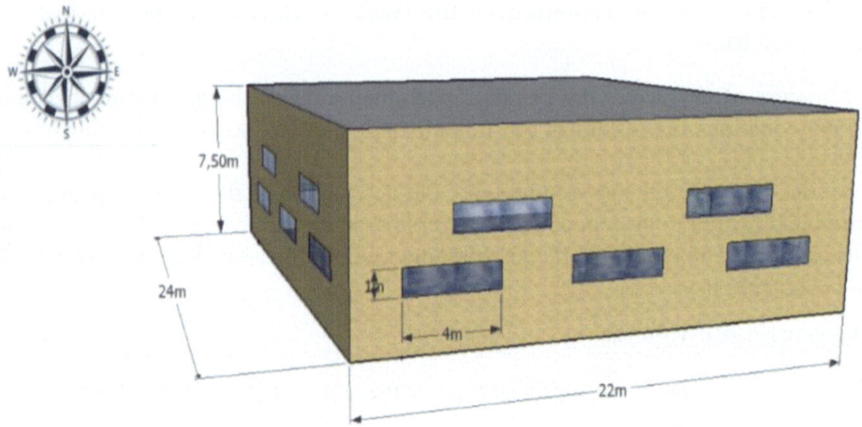

Fig. 3. Geometric configuration of the studied building [21]

3.2 Choice of the Optimal Mesh Size

The mesh sensitivity study is an indispensable step in the simulations we wish to perform. To do this, simulations with a progressively refined mesh (Fig. 4) in the upper and lower parts of the building volume (occupancy zone and artificial lighting zone), namely, 2 sub-volumes, 5 sub-volumes, 8 sub-volumes and 13 sub-volumes, were performed and analyzed to check if there are differences between the results obtained. The choice of sub-volume mesh size along the z-axis was made on the basis of the assessment of thermal stratification as the ground and roof are approached. Therefore, a refined mesh (case of 13 sub-volumes) is used in the areas near the floor and ceiling.

After calculating air temperatures for each sub-volume in the four mesh configurations, a comparative analysis of indoor air temperature results was conducted, focusing on a one-hour interval during both winter and summer. Furthermore, the average annual temperature was computed to assess the potential impact of mesh size selection on the precision of numerical outcomes.

As shown in Fig. 5, variations in the number of sub-volumes within the mesh, ranging from 2 to 13, resulted in an increase of approximately 0.5 °C in both hourly temperature and the mean annual temperature. This observation affirms the fact that finer mesh refinement contributes to a reduction in the uncertainty of the obtained numerical results.

In the subsequent phase of this research, we have employed a 13-subvolume mesh to investigate the vertical stratification within the expansive building volume. This choice was driven by the observation that the computational time necessary for simulations with this mesh is little different from that of the 8 sub-volumes mesh. Additionally, the difference in average annual temperatures ($\Delta T = T13 - T8$) between these two mesh configurations amounts to approximately 0.15 °C.

3.3 Analysis of Vertical Temperature Distribution

Based on the mentioned findings, it is imperative to include a representation of the vertical distribution of indoor air temperature within the studied building to fully comprehend

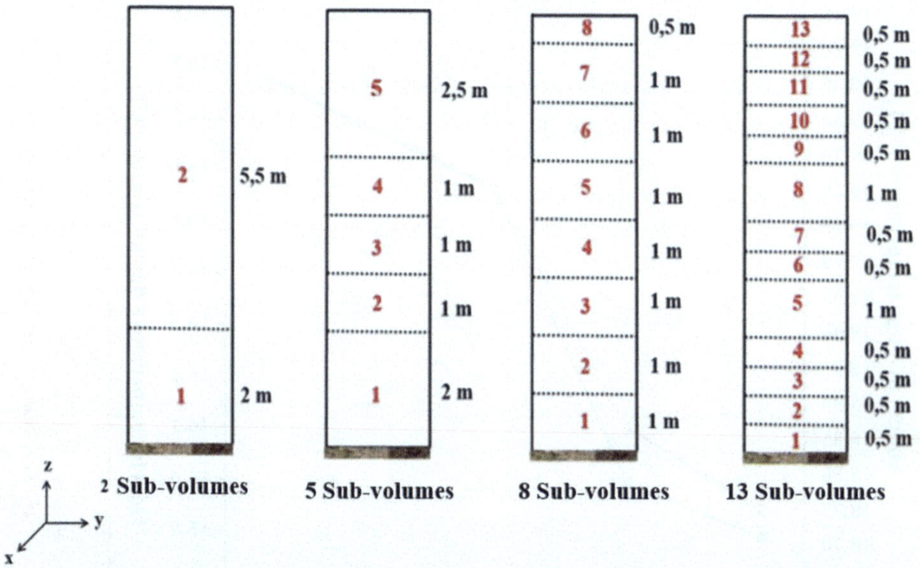

Fig. 4. Different vertical subdivisions of the studied building (height = 7,5 m)

and interpret the outcomes obtained through vertical subdivision. Figure 6 provides insight into the evolution of vertical thermal stratification in a scenario where windows are open on a typical summer day (July 21).

Numerous conclusions can be drawn from this analysis, which can be summarized as follows:

- A distinct variability in air temperature along the vertical axis was observed during the three time points (08:00, 12:00, and 00:00) represented, leading to the emergence of vertical thermal stratification. This phenomenon can be attributed, in part, to the influence of thermal drafts and, on the other hand, to the positioning of windows at specific heights, namely between 2 m and 3 m and between 4 m and 5 m. These windows correspond to glazed surface areas of 50% and 38%, respectively, for sub-volumes numbered 5 and 8 (as illustrated in Fig. 6) within the mesh comprising 13 sub-volumes.
- The temperature difference of the air in the sub-volumes in contact with the floor and ceiling varies throughout the day. Indeed, a difference of approximately 7 °C was observed between 8 a.m. and 12 p.m. This difference increased to about 9 °C by 4 p.m. and then reached approximately 10 °C at midnight. Several factors are responsible for these variations. The airflow in the sub-volumes in contact with the ceiling is lower than that in the central sub-volumes, leading to a stagnation of heat that can only be dissipated through strategically positioned windows. Moreover, in the type of building under study, namely, a large volume structure, approximately 528 m^2 of the roof is exposed to solar radiation. Consequently, a portion of the solar radiation received by the external surface of this roof is transmitted inside the building. This underscores the necessity of incorporating passive solutions such as "Cool Roof" or green roofing

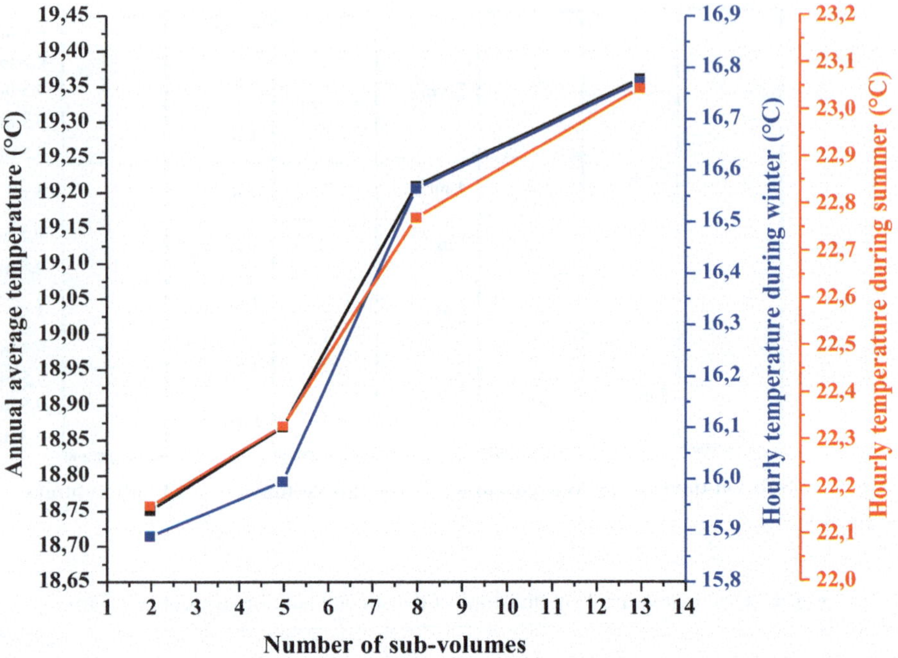

Fig. 5. Mesh sensitivity and representation of the evolution of the hourly temperature during winter and summer and the mean annual temperature for each case of the mesh (2, 5, 8 and 13 sub-volumes)

to reduce heat transfer into the building and, consequently, the temperature difference between the sub-volumes near the floor and ceiling. According to the results obtained from our simulations, around 4 p.m., we observe an increase in air temperature in the sub-volumes near the windows of the studied building. This can be explained by the fact that the two-dimensional zonal thermo-aeraulic modeling approach does not account for the horizontal distribution of temperature and airflow. This leads to a temperature stagnation throughout the day in the sub- volumes near the openings;

Based on the outcomes obtained from our simulations, an increase in air temperature in the sub-volumes adjacent to the windows of the investigated building is noticeable around 4 pm. This can be explained by the fact that the horizontal temperature distribution was not taken into account in the vertical one-dimensional thermo-aeraulic modeling approach. This causes a stagnation of the temperature during the day in the sub-volumes near the openings.

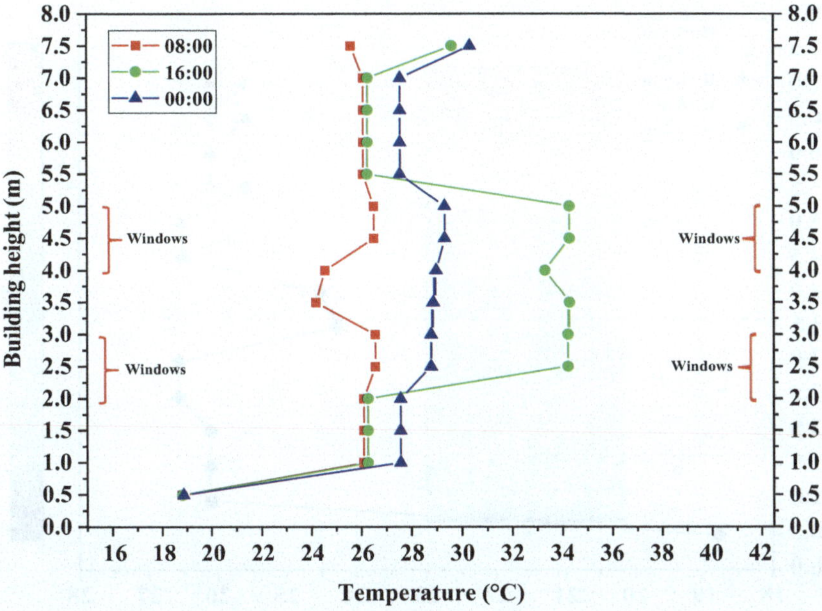

Fig. 6. Indoor thermal stratification of the building studied in the case of natural ventilation through windows

3.4 Impact of the Building Height on the Thermal Stratification

In Fig. 7 below, we present our investigation into the influence of the height of the large volume building under study on thermal stratification during a summer period extending from 21/07 to 24/07. Notably, it's essential to mention that the window positions remain the same and are maintained in a closed state. It is important to note that the reduction in building height has an inconsequential impact on the thermal stratification behavior.

3.5 Impact of Internal Heat Gains on the Thermal Stratification

The results of the study investigating the impact of internal heat gains on vertical thermal stratification, in the case of daily natural ventilation, during both a summer period (from 21/07 to 24/07) and a winter period (from 21/02 to 24/02), are presented in Fig. 8 below.

These internal heat gains are attributable to the presence of occupants with an internal heat source of 130 W per person, assuming a seated position and engaging in light activity. This occupancy scenario was applied in this subsection, considering a total of 112 individuals. This heat source was distributed across the first four vertical sub-volumes, assuming a standard average height of 1.8 m for each occupant [22]. The occupancy scenario corresponds to a presence from 08:00 to 13:00 and from 14:00 to 19:00.

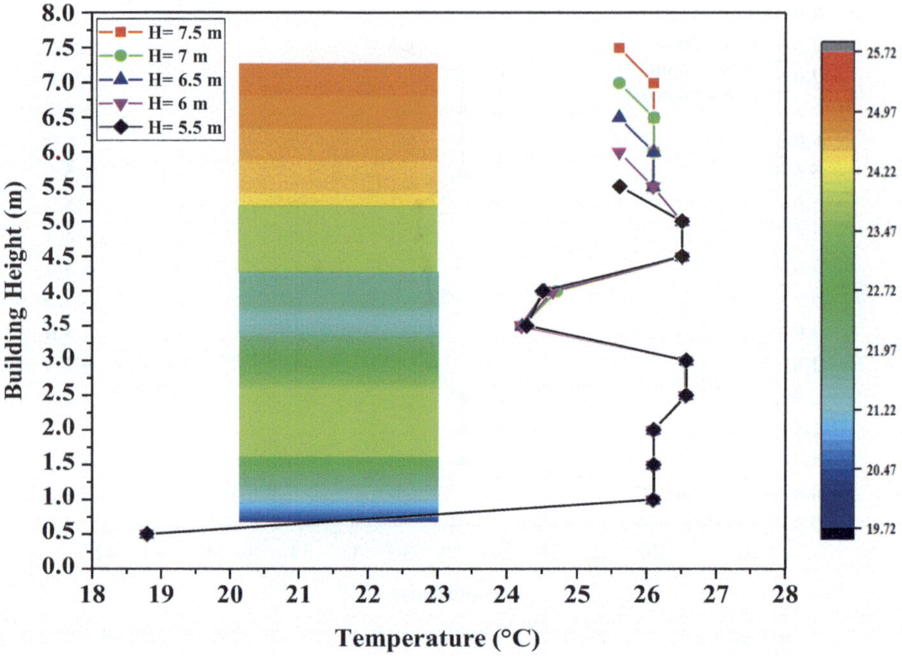

Fig. 7. Impact of the building height on the thermal stratification

According to Fig. 8 below, a slight increase in temperature occurred in the sub-volume in contact with the floor, with an approximate temperature difference of 8 °C observed between the first sub-volume and the other three occupied sub-volumes. However, no variation was observed in the other sub-volumes or in the central and upper zones, as the excess heat generated by the occupants was vented through the windows, which remained open throughout the day.

In accordance with ASHRAE Standard 55, the vertical thermal stratification in the occupied sub-volumes will lead to local thermal discomfort, as the temperature difference between the air at the level of occupants' feet, i.e., at floor level, and that at head level exceeds 3 °C. Hence, the integration of air conditioning systems is necessary to ensure more acceptable comfort conditions for occupants inside large-volume buildings.

3.6 Impact of Climate Conditions on the Thermal Stratification

The impact of external climatic conditions on the indoor air temperature and vertical thermal stratification of the examined building has been studied and analyzed. The following five climates has been investigated: Equatorial, Continental, Oceanic, Desert, and Mediterranean. These climates respectively represent the climatic data of the cities of Jakarta (Indonesia), New York (USA), Paris (France), Riyadh (Saudi Arabia), and Tangier (Morocco) [23].

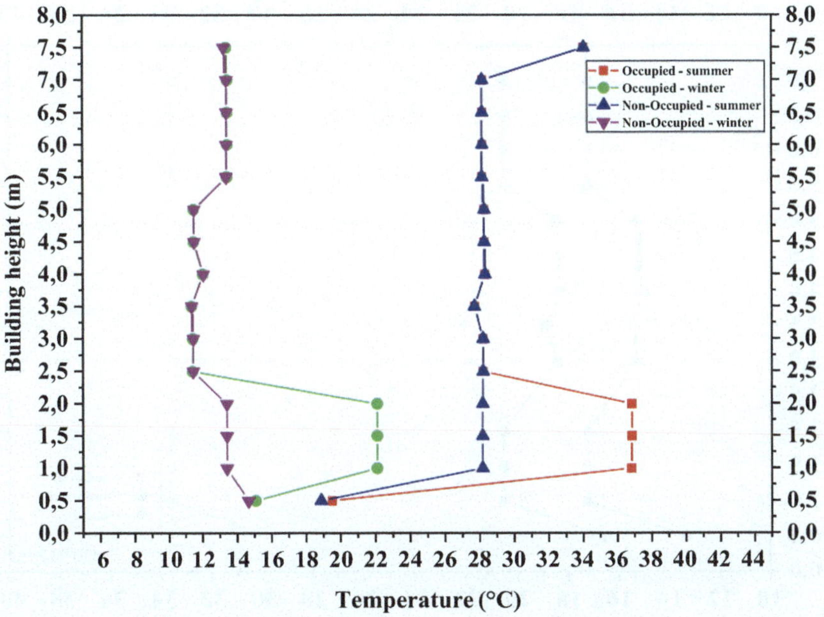

Fig. 8. Impact of internal heat gains on thermal stratification in a large-volume building with daily natural ventilation during both a summer period (from 21/07 to 24/07) and a winter period (from 21/02 to 24/02)

As demonstrated in Fig. 9 below, climatic conditions exert a significant impact on the indoor air temperature of the large-volume building under study. This is attributed, on one hand, to the consideration of thermal and airflow exchanges between the interior and exterior environments. On the other hand, it is due to the strong correlation between air flow rate and the temperature difference between the interior and exterior of the building. The temperature differential between the sub-volumes near the floor and those near the ceiling is approximately 4.5 °C for the Oceanic climate, 4.5 °C for the Continental climate, 6 °C for the Mediterranean climate, 5.5 °C for the Equatorial climate, and 5.5 °C for the Desert climate. Notably, we can observe that the behavior of vertical thermal stratification remains consistent across all five climates, except for the Oceanic climate, where a slight variation is noted in the last sub-volume near the ceiling. These results underscore the importance of predicting vertical thermal stratification, which is highly dependent on the external climate.

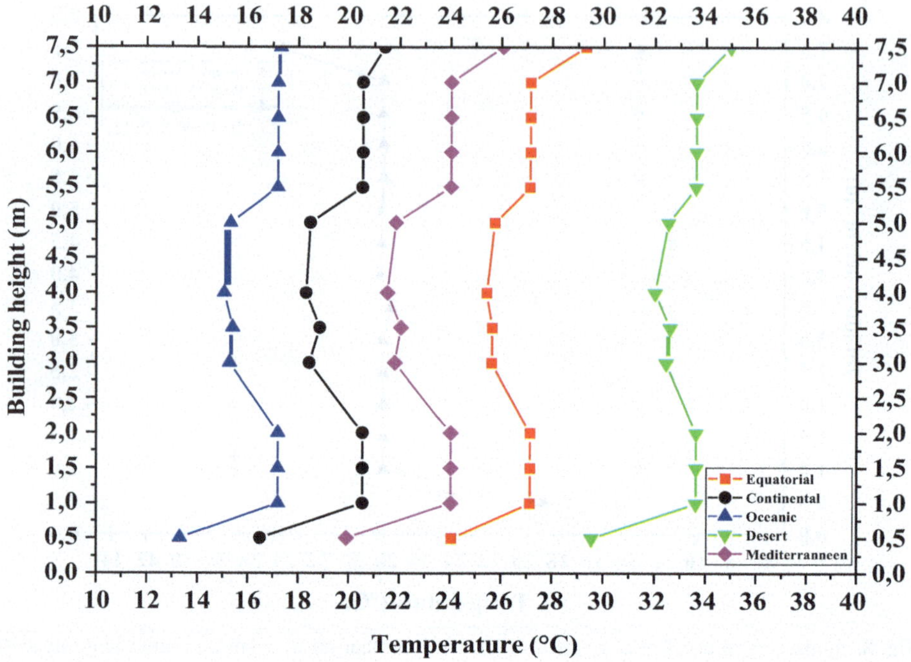

Fig. 9. Thermal stratification for the five climates studied during the pic summer period

4 Conclusion

In this study, we have undertaken a systematic and comprehensive investigation into the dynamics of thermal stratification in large-volume buildings, with a focus on the influence of building height, internal gains, and climatic conditions. Our approach involved the integration of these factors into Building Energy Simulation (BES) tools and the utilization of a three-dimensional floor model (Type49-TRNSYS). This model considers the thermophysical properties of the ground and its interaction with the building, leading to a more accurate prediction of indoor temperature variations, particularly in sub-volumes near the floor.

Through the incorporation of natural ventilation, we observed a temperature contrast of approximately 7 °C between 8 a.m. and 12 p.m., which expanded to around 9 °C by 4 p.m. and reached a peak of approximately 10 °C at midnight. Furthermore, our findings revealed significant deviations in vertical temperature variations within sub-volumes near the windows. However, these deviations showed similarities in the vertical profile of indoor air temperature across different simulation cases. Furthermore, our study has revealed that the variation of the number of sub-volumes within the mesh, ranging from 2 to 13, resulted in an approximately 0.5 °C increase in both hourly and mean annual temperatures. These findings underscore the efficacy of employing a finer mesh refinement in reducing the uncertainty associated with numerical results. In addition, the internal heat gains of 130 W per person, resulting from a total of 112 individuals engaged in light activity with a presence from 08:00 to 13:00 and from 14:00 to 19:00, led to a slight increase in temperature, primarily in the sub-volume in contact with the floor, while the other occupied sub-volumes and upper zones remained unaffected due to heat dissipation through open windows. It is evident that this vertical subdivision does not accurately predict the actual thermal distribution in large-volume buildings due to limitations arising from the vertical-only subdivision of the volume. These limitations pertain to aspects such as solar radiation distribution and horizontal temperature distribution, which have not been considered.

To address these limitations and offer practical insights, our research introduced a zonal thermo-aeraulic modeling approach that seamlessly integrated dynamic thermal simulation tools, specifically TRNSYS and CONTAM. This methodology not only enhances our understanding of thermal stratification in large buildings but also provides a platform for architects, engineers, and building operators to explore strategies for improving energy efficiency and occupant comfort.

References

1. Wilkinson, C., Court, H., Hill, J.: The architecture of long-span, large-volume buildings. In: Livre, p. 50 (1980)
2. Turner, S.: ASHRAE's thermal comfort standard in America: future steps away from energy intensive design. In: Proceedings of the Conference on Air Conditioned Low Carbon Cool. Challenge - Wind. 2008 Conference, pp. 1–8 (2008)
3. Gassar, A.A.A., Cha, S.H.: Energy prediction techniques for large-scale buildings towards a sustainable built environment: a review. Energy Build. **224**, 110238 (2020). https://doi.org/10.1016/j.enbuild.2020.110238
4. Santos, R.M., Bakhshoodeh, R.: Climate change/global warming/climate emergency versus general climate research: comparative bibliometric trends of publications. Heliyon **7**(11), e08219 (2021). https://doi.org/10.1016/j.heliyon.2021.e08219
5. U. S. IEA. International Energy Outlook 2017 Overview. In: International Energy Outlook, vol. IEO2017, p. 143 (2017)
6. Hernandez-Cruz, P., Giraldo-Soto, C., Escudero-Revilla, C., Hidalgo-Betanzos, J.M., Flores-Abascal, I.: Energy efficiency and energy performance gap in centralized social housing buildings of the Basque Country. Energy Build. **298**, 113534 (2023). https://doi.org/10.1016/j.enbuild.2023.113534
7. Lapisa, R., Bozonnet, E., Salagnac, P., Abadie, M.O.: Optimized design of low-rise commercial buildings under various climates – energy performance and passive cooling strategies. Build. Environ. **132**, 83–95 (2018). https://doi.org/10.1016/j.buildenv.2018.01.029
8. Laghmich, N., Romani, Z., Lapisa, R., Draoui, A.: The impact of internal gains on thermal stratification for public buildings. IOP Conf. Ser. Mater. Sci. Eng. Pap. (2019). https://doi.org/10.1088/1757-899X/609/4/042090
9. Saïd, M.N.A., MacDonald, R.A., Durrant, G.C.: Measurement of thermal stratification in large single-cell buildings. Energy Build. **24**(2), 105–115 (1996). https://doi.org/10.1016/0378-7788(95)00966-3
10. Mora, L.: Prédiction des performances thermo-aérauliques des bâtiments par association de modèles de différents niveaux de finesse au sein d'un environnement orienté objet, Ph.D. thesis, pp. 1–222 (2003)
11. Lu, Y., Dong, J., Liu, J.: Zonal modelling for thermal and energy performance of large space buildings: a review. Renew. Sustain. Energy Rev. **133**(73), 110241 (2020). https://doi.org/10.1016/j.rser.2020.110241. Elsevier Ltd.

12. Kaijima, S., Bouffanais, R., Willcox, K.: Computational fluid dynamics for architectural design. Open System - Proceedings of the 18th International Conference on Computer-Aided Architectural Design Research in Asia, CAADRIA 2013, pp. 169–178 (2013)
13. Flores-Abascal, I., et al.: A novel multicriteria methodology to assess the renovation of social buildings. J. Build. Eng. **77** (2023). https://doi.org/10.1016/j.jobe.2023.107505
14. Kummert, M.: TESS COMPONENT LIBRARIES General Descriptions (2014)
15. Hamdani, M., Bekkouche, S.M.A., Benouaz, T., Cherier, M.K., Benamrane, N., Benkhelifa, R.D.A.: Etude de l'Effet de la Ventilation Naturelle Transversale sur un Bâtiment Semi-Collectif dans la Région de Ghardaïa. In: 5th International Seminar New Renewable Energies, pp. 1–5 (2018)
16. Dols, B.J.P.W.S.: NIST Technical Note 1887 CONTAM User Guide and Program Documentation. (2007). https://doi.org/10.1071/PP98146
17. Musy, M.: Génération automatique de modèles zonaux pour l'étude du comportement thermo-aéraulique des bâtiments. These de l'Université de La Rochelle (1999)
18. Khalifa, I., Ernez, L.G., Znouda, E., Bouden, C.: Coupling TRNSYS 17 and CONTAM: simulation of a naturally ventilated double-skin façade. Adv. Build. Energy Res. **9**(2), 293–304 (2015). https://doi.org/10.1080/17512549.2015.1050694
19. Gilani, S., Montazeri, H., Blocken, B.: CFD simulation of stratified indoor environment in displacement ventilation: validation and sensitivity analysis. Build. Environ. **95**, 299–313 (2016). https://doi.org/10.1016/j.buildenv.2015.09.010
20. Perron, C., et al.: La turbulence en CFD : Modélisation et Simulation. Lab. Mécanique des Fluides Numérique Département génie mécanique, p. 2013 (2013)
21. Laghmich, N., Romani, Z., Lapisa, R., Draoui, A.: Numerical analysis of horizontal temperature distribution in large buildings by thermo-aeraulic zonal approach. Building Simulation, p. 17 (2021)
22. ASHRAE 55. Thermal Environmental Conditions for Human Occupancy, vol. 2004 (2004)
23. Nisrine, L., Romani, Z., Draoui, A.: Prédiction des performances thermo-aérauliques des bâtiments de grand volume par l'utilisation d'une approche de modélisation zonale intégrée dans des outils STD (2022)

Evaluate the Impact of Heating and Cooling on the Interface Shear Strength Between Clay and Concrete Surface

Murad Y. Abu-Farsakh[1](✉) and Abedalqader Idries[2]

[1] Louisiana Transportation Research Center, Louisiana State University, 4101 Gourrier Ave., Baton Rouge, LA 70808, USA
cefars@lsu.edu

[2] Graduate Research Assistant in Civil and Environmental Engineering, University of Illinois at Urbana-Champaign, Champaign, IL 61801, USA
aidries2@illinois.edu

Abstract. This work examined the impact of applying one cycle of heating and cooling on the interface behavior of clay soil to concrete surface as well as the possibility of using heating to enhance the pile's engineering capacity. To conduct the soil-concrete interface tests, a large-size direct shear test (DST) device with dimensions of 300 mm width by 300 mm length by 200 mm height was modified for this purpose. Low, medium, and high plasticity index (PI) clay soils with PIs of 12, 29, and 60, respectively, were used for the experimental tests. For every normal stress (σ_n) and clay soil type, two samples were tested at two selected temperatures (~22 °C, 70 °C) under various σ_n values that ranges from 30 kPa to 150 kPa. During the heating process, the temperature of the samples was gradually raised from room temperature to approximately 70 °C. Then the samples were allowed to return to room temperature. The interface friction angles at peak (δ_p) and at large-displacement (δ_{LD}) for samples exposed to one cycle of heating and cooling experienced substantial increase in magnitude of up to 35%.

1 Introduction

Numerous investigations were carried out to assess the impact of applying heating and cooling cycles on the characteristics of clayey soils. According to the majority of these investigations, when clay that is regularly consolidated is heated, it exhibits contraction behavior (Laloui, 2001); nevertheless, when clay that is somewhat over-consolidated is heated, it first expands before contracting. Moreover, heating causes a very over consolidated clay to expand (Laloui, 2001). Many studies (e.g., Abuel-Naga et al., 2006) reported permanent volumetric deformation for normally consolidated material during a heating and cooling cycle, whereas some invetigator (e.g., Do Donna, 2014) reported a partially recoverable expansion. Reversible (recoverable) volumetric deformation would occur in a substantially over-consolidated clay (OC) material (OC > 1.5–3) following a heating–cooling cycle (Towhata et al., 1993).

The apparent preconsolidation pressure (σ'_c) would decrease with an increase in thermal load, or temperature, while maintaining similar applied load and void ratio

A. S. Mosallam et al. (Eds.): ICATH 2023, SCI, pp. 227–234, 2024.
https://doi.org/10.1007/978-3-031-76557-5_18

(Laloui, 2001; Abuel-Naga et al., 2006). According to Tidfors and Sallfors (1989), there is a discernible decrease in the σ'_c value when temperature rises, and this decrease is more as clay content rises. Additionally, Laloui (2001) discovered that the value of σ'_c decreased nonlinearly as the temperature rose. It was discovered that the apparent preconsolidation pressure (σ'_c) and temperature behavior was not affected by viscosity of soil (Laloui, 2001).

Di Donna (2014) conducted direct shear tests (DSTs) at ambient 50 °C temperature, while discovered an increase in the clay to concrete interface shear strength of roughly 18% to 38%. Yavari et al. (2016) showed negligible variations in the parameters of interface shear strength during tests conducted at 5 °C and 40 °C temperatures. DSTs were carried out on the clay to concrete interface by Yazdani et al. (2019a) at different temperatures from 24.0 °C and 34.0 °C during heating and cooling cycles. They discovered a 5% decrease in shear strain at peak strength as well as an increase in peak shear strength. Under cyclic and monotonic heating conditions, the interface friction angle (δ_p) increased by 27.0% and 25.0%, respectively, while adhesion decreased by 17.0% to 20.0% (Yazdani et al., 2019b). DSTs were performed at 6.0 °C, 21.0 °C, and 38.0 °C on silt-concrete interface by Xiao et al. (2014). Through comparing the DST results, they discovered that the value of δ_p increased by 15.0% at 38.0 °C compared to δ_p value determined at 6.0 °C, and by 5.0% to 16.0% at 21.0 °C as compared to 6 °C (Xiao et al., 2014).

2 Properties of Soils

In this study, three different soil types with varying plasticity indices (PI): low PI = 12 soil, medium PI = 29 soil, and high PI = 60 soil, were employed. The corresponding liquid limits (LLs) are 33, 49, and 96, respectively. The values of σ'_c for the low PI soil, medium PI soil, and high PI soil are 24.0, 43.0, and 240.0 kPa, respectively. To assess shear behavior of the three types of soil, small-size DSTs were carried out. For both clay to clay and clay to concrete testing, normal stresses (σ_n) of 69.0 kPa, 110.0 kPa, and 150.0 kPa were applied to all soil types, with the exception of the low PI soil, which underwent an additional σ_n of 30.0 kPa. The internal friction angle (ϕ'), and cohesion (c), were computed for all soils. The values of ϕ' and c at peak were determined to be (31.1°, 8.33 kPa), (17.0°, 9.65 kPa) and (12.6°, 13.4 kPa) for low PI soil, medium PI soil, and high PI soil, respectively. The shear strength of the high PI soil was achieved at 15 mm displacement, while its peak strength occurred at much less displacement. The term "shear strength at large-displacement" will be used in this paper to refer to the shear strength at 15 mm.

3 Testing Setup

As shown in Fig. 1, a large-size direct shear device (LDSD) was adjusted to examine the impact of applying one cycle of heating and cooling on the interface shear parameters between clay and concrete. The authors believe that this approach yield more accurate results over the small-size direct shear device (Idries, 2020; Idries et al., 2020). The maximum temperature used in this study was determined to be 70 °C. The apparatus comprises of two segments: the top segment, measuring 300 mm × 300 mm × 100 mm, remains fixed, while the lower segment, measuring 400 mm × 300 mm × 100 mm, can be shifted in horizontal direction while being loaded. The upper section's top was 50 mm below the water level. A stepper motor was used to apply the vertical stress atop the plate, and a motor was used to move the bottom section horizontally at a specific shear rate. Three grooves were opened on one face of the LDSD to accommodate the thermocouple wires. After preparation and curing, a concrete block with a roughness comparable to the pile's surface was put inside the lower side of box. The top box was filled with clay up to 50 mm thickness.

In order to heat the water that circulate to maintain the required temperature, two heaters were immersed in the water path and controlled by a heat controller. In order to measure the temperature of the soil, an additional four thermocouples were installed; three of these were placed inside the clay sample at different locations (corner, center, and halfway) and the fourth was placed inside the water path. In order to guarantee saturation, speed up the heating process, and maintain a constant temperature throughout the entire system, a pump was lastly utilized to circulate the water.

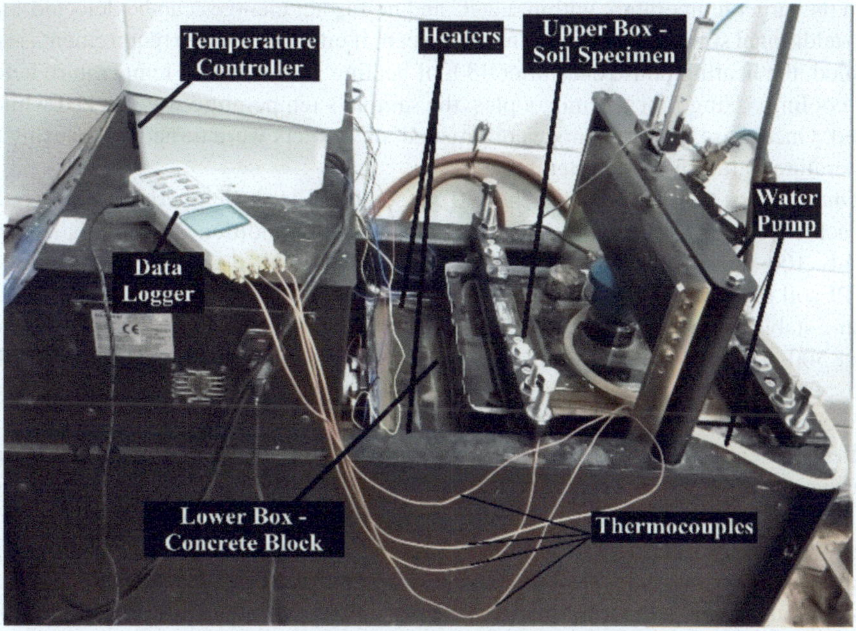

Fig. 1. Setup of LDSD used in this study.

4 Testing Procedure

Initially, the concrete piece was positioned inside the lower portion of LDSD. After the soil samples being prepared at water content higher than the optimum water content, the clay sample was positioned atop the shear box section. Using a modified proctor hammer, each soil layer received 70 blows to compact the top layer of soil to a final thickness of 25 mm. The clay sample had three thermocouples positioned at different locations. Similarly, the second layer of soil was added and compacted. The initial degree of saturation (S) for the soil samples ranged from 92% to 96.5%. Following the preparation of clay sample, the clay surface was covered with a saturated filter paper. After that, the filter was covered with the porous stone. To guarantee that saturation is satisfied, the shear box was positioned inside the water bath of LDSD, and left immersed for a full day at a typical sitting stress of 3.45 kPa. The value of σ_n was then raised from the sitting stress to the desired σ_n. Until the primary consolidation process was finished, the clay samples were left under load. Shearing begins right after consolidation ends for tests that without heating.

After completing consolidation, heating load was applied for samples that involve heating and cooling. Three steps comprised thermal loading: step 1 involved heating from room temperature (about 22 °C) to 70 °C; step 2 required maintaining at 70 °C; and step 3 involved cooling sample from 70 °C to room temperature. Before phase 3 began, the following two conditions were used to determine how long stage 2 would last to fully dissipate the induced excess porewater pressure: 1) All four thermocouples must reach the target temperature within ±2 °C and no further changes can be detected); and 2) no additional soil volumetric strain can be seen. If either of the two requirements is not satisfied, the heating phase ends after 48 h of heating at the desired temperature before start cooling. Using four thermocouples, the sample's temperature was tracked while it heated. Once these requirements were satisfied, the heaters were turned off, causing the temperature to return to room temperature.

Once the first thermocouple reached 70 °C (condition 1), the temperature stabilization was achieved in 60–80 min. For low and medium PI, condition 2 was satisfied after roughly 100 and 600 min, respectively. On the other hand, it took up to 400 min for low PI soil at 30 kPa and took up to 800 min for medium PI soil at 150 kPa to reach volume stabilization. However, for low PI soil and medium PI soil, the heating was kept up for 500–550 min and 1000 min, respectively, during the heating phase. On the other hand, after two days of heating at 70 °C, none of the high PI soil tests achieved volume stabilization, and the heating was halted.

Before shearing, a gap of about 2.0 mm was made between the two steel boxes. During shearing, the goal was to maintain a constant shearing rate. To maintain a drained state and enable the excess porewater pressure to dissipate, the shear rate was chosen to be very low at 0.005 mm/min. At failure, 1.5 cm of horizontal deformation was thought to have occurred. For every test, the shearing process took 3000 min, or roughly two days. Throughout the shearing phase, the heated water circulation was maintained to guarantee complete saturation.

Because of LVDT sensitivity and expansion of the shear device components from heating, the volume change from heating and cooling processes must be calibrated. This was accomplished by carrying out calibrated tests, including heating and cooling, on the system during consolidation at four distinct stresses of 30, 69, 110, and 150 kPa. The clay samples used in these tests were taken out of the system. In these experiments, temperature changes were noted in addition to volume changes. The system's overall volume change, excluding the clay sample that was eliminated from the initial test results, is what caused the observed volume change. Therefore, the value calibrated for only clay sample during heating and cooling cycle represents the final value of vertical displacement.

5 Tests Results and Discussion

The shear stresses (τ) against horizontal displacements (d_h) curves of clay to concrete interface for low PI soil, medium PI soil, and high PI soil are shown in Fig. 2–1 (a), 2–2 (a), and 2–3 (a), respectively. The figures demonstrate that, for all clay types, the interface strength between clay and concrete increases as the σ_n increases. Furthermore, the interface strength of all soil types tested under one cycle of heating was higher than those tested without a heating cycle. According to these findings, the maximum interface strength of low PI soil, medium PI soil, and high PI soil increased by an average of 20%, 28%, and 22%, respectively. During the interface DSTs, the matching vertical and horizontal displacements were observed. The vertical displacements (d_v) versus the horizontal displacements (d_h) for the three PI soils under all values of σ_n are shown in Fig. 2–1 (b) to 2–3 (b). The results of clay to clay DSTs of samples tested with and without a single heating cycle applied to the clay to concrete interface are included in the figures for comparison. The values of d_v without heating of the clay to clay are higher than those for interface between clay and concrete for all clay samples. The average reduction in d_v values ranges from 26% for low PI soil to 92% for high PI soil. Furthermore, compared to interface tests conducted with no heating cycle, the values of d_v for the interface tests conducted with one cycle of heating significantly decreased.

The value of δ_p for low PI soil is 27.6° for heated samples and 24.3° for not heated samples. However, the δ_{LD} value for samples with heating is 25.9° and 22.4° for samples without heating. The (δ_p, δ_{LD}) values for medium PI samples heated for one cycle of heating and those without heating are (16.8°, 14.9°) and (13.6°, 11.0°), respectively. For high PI soil, the values of (δ_p, δ_{LD}) are (10.8°, 6.3°) and (8.7°, 6.9°), respectively, for samples heated once and not heated samples. The values of peak adhesion (c_{ap}) are zero for all low and medium PI soil samples with and without heating. On the other hand, the adhesion at large-displacement (c_{aLD}) is measured as 8.07 kPa and 6.34 kPa for samples with heating and without heating, respectively, whereas c_{ap} for the high PI soil is 11.0 and 9.38 kPa for heated samples and not heated samples, respectively. For the high PI soil, the cohesion values are 12.13 kPa at large-displacement and 13.44 kPa at peak.

Studies by Laloui (2001) and Abuel-Naga et al. (2006) showed that heating saturated clay is anticipated to cause excess porewater pressure because of greater thermal water expansion as compared to clay and high heating/draining rate. According to Campanella and Mitchell (1968), the variation in physico-chemical interaction and dissipation of excess porewater pressure results in decreasing the volumetric strain. When the soil samples were heated, there was a decrease in vertical displacement, and hence increase in volume reduction, until the volume is stabilized. For all tests, an additional reduction in the vertical displacement was noted during cooling.

Thermally-induced volumetric strains (TIVS) was calculated for all the heated interface tested samples. The values of TIVS for low PI soil ranges from 0.5% to 0.8%. The TIVSs for clay to concrete interface of medium PI soil ranges from 0.89% to 1.0%.

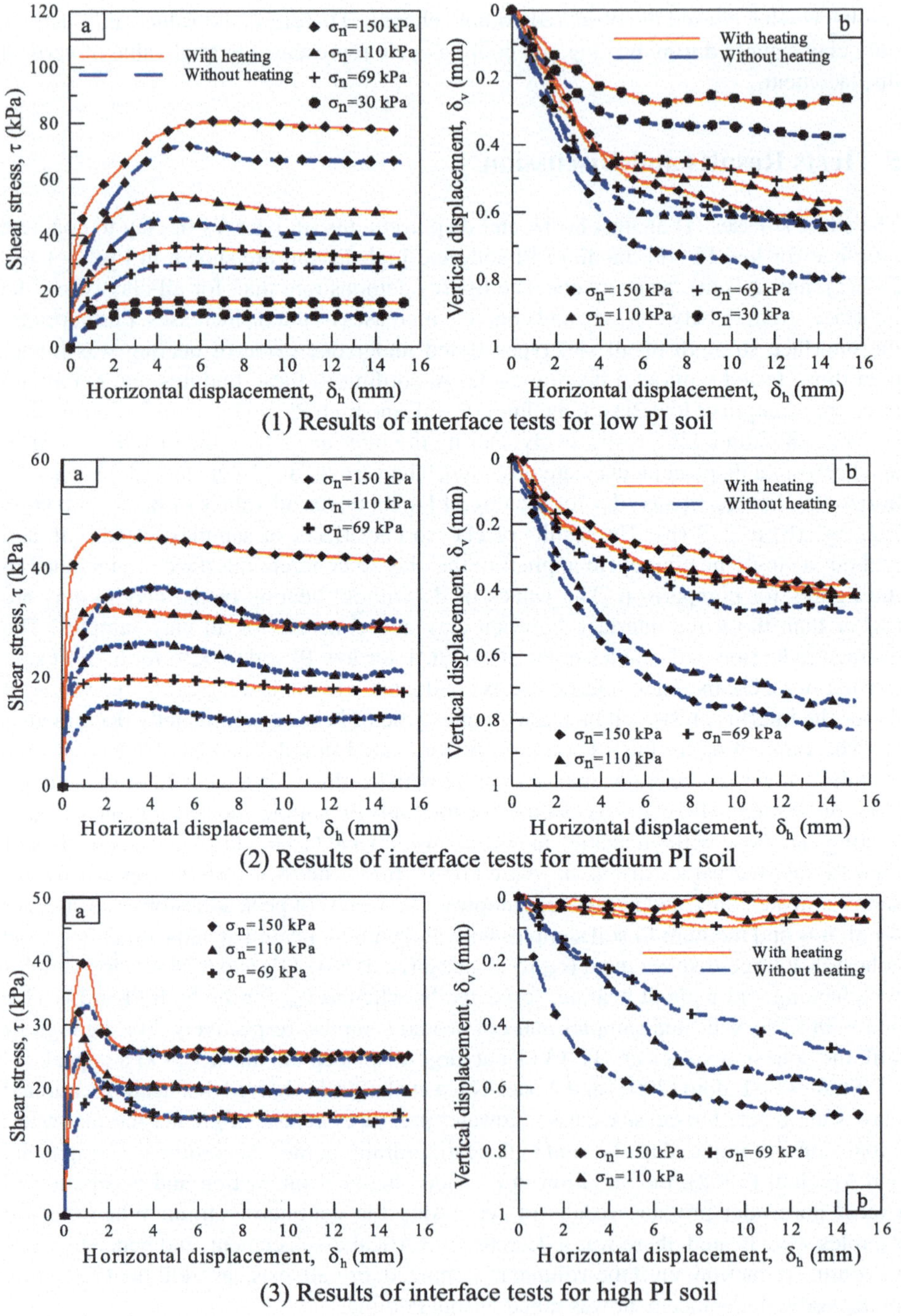

Fig. 2. (1) Results of interface tests for low PI soil. (2) Results of interface tests for medium PI soil. (3) Results of interface tests for high PI soil

In the meantime, the TIVSs for clay to concrete interface of high PI soil ranges from 2.5% to 3.35%. The outcomes show that the TIVS values following the cooling stage are unrelated to the applied σ_n value, which is consistent with the study by Abuel-Naga et al. (2007). The rate and amount of porewater pressure dissipation, which are by the properties of the soil (i.e., thermal performance), is the main cause of TIVS behavior. Thus, it is expected that the same clay under different σ_n values and same thermal load will exhibit the same TIVS values. Yet, variations in the conditions of preparation may be the cause of the slight variation in TIVS values (Yazdani et al. 2019a). The average values of TIVS for the low PI soil, medium PI soil, and high PI soil are 0.65%, 0.93%, and 2.98%, respectively. This indicates that the TIVS value rises as the soils' PI increases, as demonstrated by Abuel-Naga et al. (2006).

6 Conclusions

- The test results demonstrated a substantial enhancement in the interface strength between clay soil and concrete for samples that underwent a single heating and cooling cycle. The peak shear strength as well as the interface shear strength increased, with the exception of the OC high PI soil, which only showed an increase in the peak strength.
- Samples that underwent a single heating and cooling cycle had higher interface friction angles (δ_p and δ_{LD}) at peak and at large-displacement than samples that did not receive any heating. The percentage increase in at peak interface angle (δ_p) ranges from 13.6% for low PI soil to 25.0% for medium PI soil. Concurrently, the percentage increase in at large-displacement interface angle (δ_{LD}) ranged from 5% for high PI soil to 35.5% for medium PI soils.
- For the samples that underwent a single cycle of heating and cooling, the vertical displacement during shear was significantly reduced by 26% for low PI soil to 92% for high PI soils.
- It was discovered that the TIVS increased independently of the applied σ_n but increased with temperature loading and PI. This can be explained by the change in the physico-chemical interaction and dissipation of thermally-induced porewater pressure that was created during heating.

Acknowledgements. This study was funded by the Louisiana Transportation Research Center (LTRC) and the Louisiana Department of Transportation and Development. The authors would like to thank the engineers at LA DOTD for their support and valuable assistance during this study.

References

Abuel-Naga, H.M., Bergado, D.T., Lim, B.F.: Effect of temperature on shear strength and yielding behavior of soft Bangkok clay. Soils Found. **47**(3), 423–436 (2007)

Abuel-Naga, H.M., Bergado, D.T., Ramana, G.V., Grino, L., Rujivipat, P., Thet, Y.: Experimental evaluation of engineering behavior of soft Bangkok clay under elevated temperature. J. Geotech. Geoenviron. Eng. **132**(7), 902–910 (2006)

Campanella, R.G., Mitchell, J.K.: Influence of temperature variations on soil behavior. J. Soil Mech. Found. Div. **94**(3), 709–734 (1968)

Di Donna, A.: Thermo-mechanical aspects of energy piles. Ph.D. thesis, Swiss Federal Institute of Technology in Lausanne (EPFL), Lausanne (2014)

Idries, A., Ghaaowd, I., Abu-Farsakh, M.: Effect of one cycle of heating and cooling on the clay to concrete pile interface behavior. E3S Web Conf. **205**, 05012 (2020). EDP Sciences

Idries, A.A.: Study of the Effect of Heating - Cooling Cycles on the Clay - Concrete Pile Interface Characteristics. LSU Master's Theses. 5148 (2020). https://digitalcommons.lsu.edu/gradschool_theses/5148

Laloui, L.: Thermo-mechanical behavior of soils. In: Environmental Geomechanics, pp. 809–843. EPFL Press, Lausanne (2001)

Tidfors, M., Sallfors, G.: Temperature effect on preconsolidation pressure. Geotech. Test. J. **12**(1), 93–97 (1989)

Towhata, I., Kuntiwattanakul, P., Kobayashi, H.: A preliminary study on heating of clays to examine possible effects of temperature on soil-mechanical properties. Soils Found. **33**(4), 184–190 (1993)

Xiao, S., Suleiman, M.T., McCartney, J.S.: Shear behavior of silty soil and soil-structure interface under temperature effects. In: Geo-Congress 2014: Geo-characterization and Modeling for Sustainability, pp. 4105–4114 (2014)

Yavari, N., Tang, A.M., Pereira, J.M., Hassen, G.: Effect of temperature on the shear strength of soils and the soil–structure interface. Can. Geotech. J. **53**(7), 1186–1194 (2016)

Yazdani, S., Helwany, S., Olgun, G.: Influence of temperature on soil–pile interface shear strength. Geomech. Energy Environ. **18**, 69–78 (2019a)

Yazdani, S., Helwany, S., Olgun, G.: Experimental evaluation of shear strength of kaolin clay under cyclic and noncyclic thermal loading. Geotech. Test. J. **42**(6), 1518–1548 (2019b)

Application of MASW and SRT Surveys in Engineering Geotechnical Evaluation for Safe Urban Extension: A Case Study in Eastern Agadir, Morocco

Ismaail Khadrouf[1](✉), Ouafa El Hammoumi[1], and Najib El Goumi[2]

[1] Laboratory of Applied Geology, Geoinformatics and Environment, Department of Geology, Faculty of Sciences Ben M'sick, Hassan II University, Casablanca, Morocco
ismaail.khadrouf-etu@etu.univh2c.ma

[2] Soil Experimental Center, Public Laboratory for Trials and Studies, Casablanca, Morocco

Abstract. Despite the strategic and economic importance of the Agadir area, many buildings in the city were constructed without a thorough understanding of the subsoil. The 1960 Agadir earthquake, even though it had a moderate magnitude (Mw = 5.8), led to the destruction of more than 75% of the city, revealing the vulnerability of its structures. Due to population growth, the Eastern Agadir area was chosen as a planned urban expansion area. To ensure the durability of structures and safe urban development, extensive geotechnical investigations are needed. However, these investigations are expensive and time-consuming. Our main goal is to determine the geotechnical, petrophysical, and elastic-dynamic properties of the subsoil using seismic surveys, which are cost-effective and efficient. We used MASW and SRT to calculate S-waves and P-waves, respectively and then derived the mentioned properties. Our results show three layers: (1) The first layer, with a thickness of 5 m, has V_P and V_S values ranging from 382 to 3757 m/s and 220 to > 850 m/s. (2) The second layer, at a depth of -5 to -10 m, has V_P and V_S values ranging from 432 to 3494 m/s and 215 to > 850 m/s. (3) The third layer, located at -10 to -20 m, has V_P and V_S values ranging from 432 to 432 m/s and 215 to > 850 m/s, respectively. V_{S30} was calculated for near-surface geologic materials by applying the EUROCODE 8 classification. The study area was categorized into three classes: class A, class B, and class C. We have provided maps and tables containing the calculated parameters.

Keywords: Agadir · MASW · SRT · Urban extension · Geotechnical · petrophysical · and elastic-dynamic soil properties

1 Introduction

The demographic expansion, urban growth, and planned city extension areas in many Moroccan cities, including Agadir, necessitate the provision of essential parameters related to geotechnical soil conditions, rock/soil petrophysical properties, elastic-dynamic moduli, and seismic soil classes. This is crucial for successful, cost-effective

A. S. Mosallam et al. (Eds.): ICATH 2023, SCI, pp. 235–244, 2024.
https://doi.org/10.1007/978-3-031-76557-5_19

territorial planning and safe and durable construction. Therefore, geotechnical and seismic site characterization plays an important role in identifying suitable areas for any proposed city extensions and thus preventing construction and human life from future natural hazards.

Most structures in Agadir city were constructed without comprehensive subsoil characterization, which resulted in the destruction of more than 75% of the city during the 1960 Agadir earthquake, despite its moderate magnitude (Mw = 5.8). The main reasons for this extensive destruction were the construction vulnerability, site-specific conditions, nature/type of soil, and the city's proximity to the epicenter [1]. Due to demographic growth and urban development, the eastern Agadir area was selected to accommodate the city's urban expansion.

Therefore, it is vital to have a construction plan to ensure the longevity of superstructures and reduce human and economic losses. Having a detailed understanding of soil and rock characteristics, geological structures, and seismic site properties is invaluable for geotechnical and earthquake engineers. However, detailed site investigations require invasive in- situ techniques such as coring boreholes, trial pits, SPT, and laboratory tests. These techniques are high-cost, time-consuming, and offer limited spatial information. To address these challenges, several studies used near-geophysical surveys in geotechnical and seismic site characterization and evaluation [2–4] and have yielded effective results.

The Multichannel Analysis of Surface Waves (MASW) method is a recent and highly regarded technique for estimating subsoil shear wave (S-wave) velocity. This technique is considered the most effective for determining engineering and elasticity parameters [5]. Seismic refraction tomography (SRT) is a highly precise in-situ seismic method for resolving subsurface inhomogeneities with greater resolution in both lateral and vertical gradients [6].

The study area is located in the eastern part of Agadir. This zone was selected to receive the urban land expansion of the city due to the large amount of state-owned land, land use diversity, infrastructure availability, and economic opportunities and impacts. On the other hand, this area is prone to floods, earthquakes, and soil engineering problems such as liquefaction, soil heterogeneities, groundwater contamination, and soil erosion [7–10].

The aim of this study was to investigate the geotechnical, elastic-dynamic, and seismic properties of subsurface geological materials in the Agadir area by interpreting S-waves from MASW and P-waves from SRT. This information will be used to identify suitable areas for safeguarding structures, thus contributing to safe urban extension.

2 Geologic and Seismotectonic Setting

The Agadir region is situated in the northeastern section of the Souss Basin, with geographic coordinates spanning from approximately 30.443°N to 30.327°N in latitude, and from approximately − 9.574°W to − 9.354°W in longitude. The Souss Basin is characterized as a tertiary synclinorium structure with two depocenters located in its western and eastern parts [11]. The basin subsurface consisted of Plio-Quaternary sediments, which included sands, silt, gravels, and lacustrine limestone. These sediments covered a Cretaceous syncline in the northern part of the basin and a Paleozoic schist basement in the southern part [12] (Fig. 1B). The Plio-Quaternary sediments in the study area exhibited heterogeneity both vertically and laterally. The tectonic evolution of the Souss basin is related to the High Atlas Mountains (HA), which were controlled by the Neogene-Quaternary tectonic inversion. As a result, several fault tracts were developed and reactivated [13]. The southern boundary of the Souss Basin was defined by the South Atlas Front, a fault zone that stretched for approximately 150 km in a WSW–ENE direction. This fault zone was created by four primary fault tracts that formed from west to east. These fault tracts include the Agadir–Tagragra fault tract, which strikes in a WNW direction and spans 30 km in length; the Oued Issen fault tract, which strikes in a WSW direction and spans 3 km in length; the Northern Taroudant fault tract, which has an east–west strike and spans 45 km in length; and the Oulad Berhil fault tract, which has a strike ranging from E–W to WNW and spans 40 km in length [14]. The seismicity of this area is linked to the High Atlas Mountains, primarily affected by the South Atlas Front. It is characterized by a compressional regime resulting from interactions between the Arabian and Eurasia plates, which began during the Late Cretaceous [15] (Fig. 1B). Historically, this area has witnessed several destructive earthquakes, such as the 1731 Santa Cruz earthquake with Mw 8.7 and the 1755 Lisbon earthquake, which affected most Moroccan coastal cities [1]. The Agadir earthquake occurred in 1960 with a magnitude of 5.9, claiming over 12,000 lives and destroying approximately 75% of the city. More recently, the 2023 Al'Haouez earthquake, with a magnitude of 6.8, claimed over 2,946 lives.

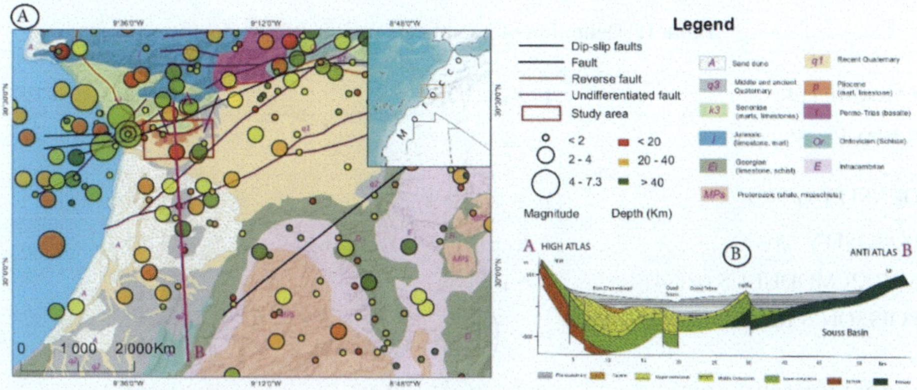

Fig. 1. Geological and seismotectonic setting of the study area. (A) Seismotectonic map of the study area. (B) Cross section of the study area [16].

3 Data and Methodology

Two seismic surveys (MASW and SRT) were used to calculate the seismic velocities. Both the MASW and SRT methods were implemented at the same locations. Vp-waves were calculated from thirty-three lines, while Vs-waves were calculated from fifteen

1D-MASW lines (Fig. 2). To characterize the suitability of subsurface materials, various parameters were derived from the Vp and Vs velocities. Geotechnical characteristics, including Rock Quality Designation (RQD), ultimate bearing capacity (Qult) and allowable bearing capacity (Qa), were assessed, while V_{s30} were used to determine seismic soil classes according to Eurocode 8. Petrophysical parameters, such as density and porosity, were also considered, alongside elastic-dynamic properties like Poisson's ratio, shear modulus, and Bulk's modulus. Definitions of these parameters can be found in Table 1. In cases where only SRT surveys were conducted, V_s velocities were estimated using formula listed in Table 1.

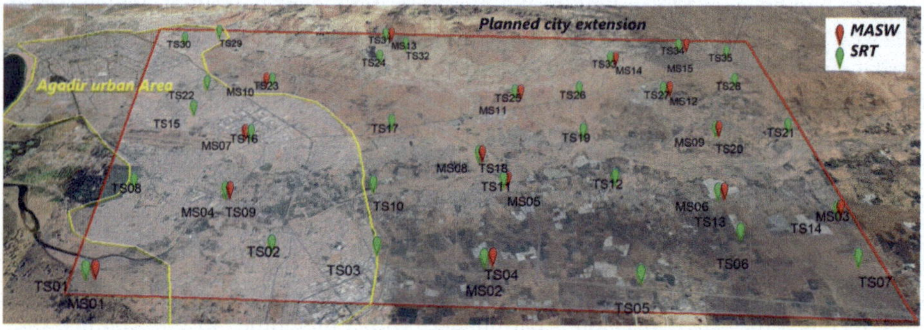

Fig. 2. Locations of MASW and SRT surveys within the study area.

Table 1. Definition of parameters used in this work.

PARAMETER	EQUATION	REFERENCES
S-WAVES (**V_S**) (1)	$V_s = \frac{V_p}{\sqrt{3}}$	[17]
DENSITY (**ρ**) (2)	$\rho = 0.31 * V_p^{0.25}$	[18]
POROSITY (**ϕ**) (3)	$\phi = -0.175\ln(V_p) + 1.65$	[19]
SHEAR MODULUS (**μ**) (4)	$\mu = \rho\, V_s{}^2$	[20]
POISSON'S RATIO (**σ**) (5)	$\sigma = \left(1 - 2\left(\frac{V_s^2}{V_p^2}\right)\right)/2\left(1 - \left(\frac{V_s^2}{V_p^2}\right)\right)$	[21]
BULK MODULUS (**K**) (6)	$K = \rho(V_p{}^2 - \frac{3V_s}{4})$	[22]
ROCK QUALITY DESIGNATION (RQD) (7)	V_p= 21.951 * RQD + 0.1368	[17]
ULTIMATE BEARING CAPACITY (Q_{ult}) (8)	$Q_{ult} = 10^{2.932(\log V_s - 1.45)}$	[23]
AVERAGE SHEAR WAVE VELOCITY OF THE UPPER 30 m (**V_{S30}**) (10)	$V_{S30} = \frac{30}{\sum_{i=1}^{N}\left(\frac{n_i}{c_i}\right)}$	[4]

3.1 MASW Acquisition and Processing

Fifteen 1D-MASW profiles were conducted using 24-channel ABEM seismograph covering the study site. Profile length is 69 m with 3m offset distance for all surveyed lines. The recorded length was 1s with sampling interval of 0.5 ms. To generate seismic energy, a metal plate was employed to get the 20 kg sledgehammer attacks at least five shots. Several shots were recorded at each profile. In these surveys, 4.5 Hz vertical geophones were used in data acquisition. The stacked 1D models were processed using SeisImager/SW software package. The process involved two primary stages: first, calculating dispersion curves (Fig. 3A), and second, inverting these curves using the WaveEq module (Fig. 3B), which employs mathematical transformation process to derive the shear wave velocity profile.

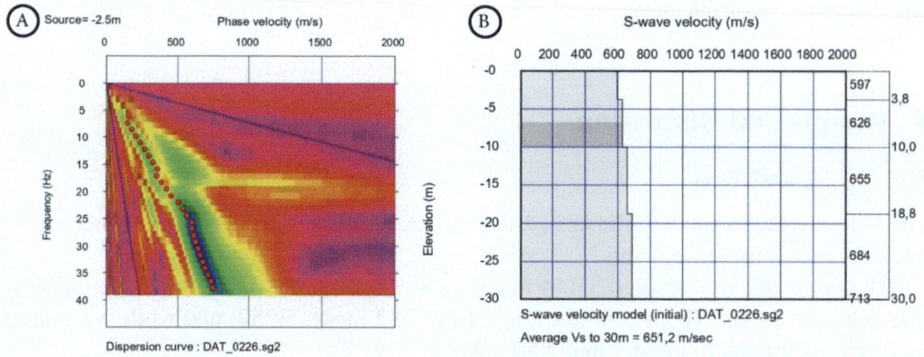

Fig. 3. Example of S-wave interpretation. (A) is the dispersion curve, (B) is model of 1D VS.

3.2 SRT Acquisition and Processing

In this study, the SRT procedure was conducted using an ABEM TERRALOC Pro 2 seismograph equipped with 24 channels and a 1 ms sensitivity. This seismograph system is designed for efficient fieldwork, boasting 24-bit accuracy and a dynamic range exceeding 100 dB. Data were digitally recorded on its built-in hard disk. Seismic waves were generated by striking an iron plate (30 × 30 cm2) with a 15 kg sledgehammer, ensuring vertical impact for optimal signal quality. Three shots were recorded in forward, inline, mid-point, and reverse positions. Sensitive geophones captured both direct and refracted waves as they propagated through the ground. Data quality depended on equipment accuracy, topography, geophone placement, ambient noise, and source characteristics. Vertical geophones (4,5 Hz) were used to convert ground motion into electrical signals and required secure coupling to the ground. Data acquisition involved a 24-channel seismograph with 115 m profile lengths and 5 m geophone spacings. P waves data were processed using SeisImager/2D software. The initial steps involved reading raw field data, assigning geometry, and applying geophone elevations. Subsequently, waveforms were analyzed by identifying first arrivals, establishing travel time-distance curves (Fig. 4A), and ultimately calculating average P-wave velocities for sub-soil layers (Fig. 4B).

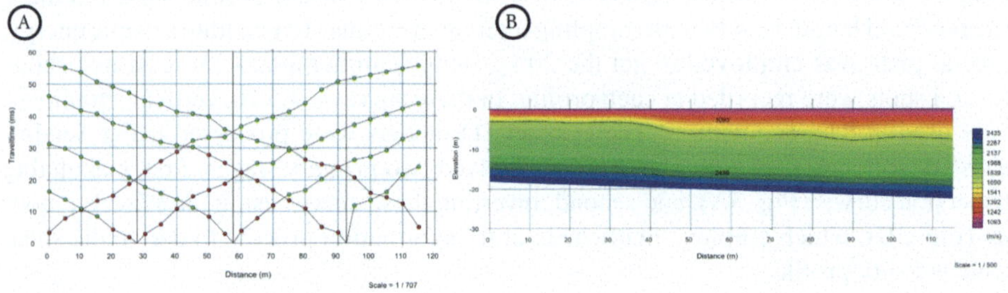

Fig. 4. Example of P-wave interpretation. (A) Time distance (T–D) curves of P-wave (B) and its associated velocity-depth model.

4 Results and Discussion

4.1 V_p, V_s and V_{S30}

The velocity-depth models illustrated in Fig. 5A and Fig. 5B, reveal the existence of three distinct subsurface layers.

The first layer is characterized by heterogenous sediments, exhibiting different seismic velocity values. The V_P values range from 382 m/s to 3757 m/s, while V_S values range from 220 m/s to greater than 850 m/s.

Beneath the first layer, the second layer is observed, featuring higher and homogenous seismic velocities values. The V_P values range from 432 m/s to 3494 m/s, while V_S values range from 215 m/s to greater than 850 m/s. This layer corresponds to compacted soils and rocks, such as limestone, sandstone and conglomerates.

The third layer, which can be considered a geotechnical bedrock, represents a substantial and dense stratum and has a highest seismic velocity. Within this layer, V_P values span from 456 m/s to 3800 m/s, while Vs values range from 250 m/s to greater than 850 m/s.

The V_{s30} map shows V_s values ranging from 179 m/s to 830 m/s, classifying the study area into soil classes A (rock or other geological formations including 5m of weak materials), B (deposits of dense sand, gravel or stiff clay), and C (deep deposits of dense or medium-dense sand, gravel or stiff clay) according to Eurocode 8.

4.2 Geotechnical, Elastic-Dynamic and Petrophysical Parameters

Table 2 provides statistical summary of the findings obtained from the analysis of 33 profiles conducted within the study area.

- The Ultimate Bearing Capacity (Qult) values range between 27 and 2345 kN/m^2, 38.6 and 7454.1 kN/m^2, and 60.2 and 34144.6 kN/m^2 for the first, second, and third layers, respectively. The Qult values increase with depth, with the highest values observed in the third layer.

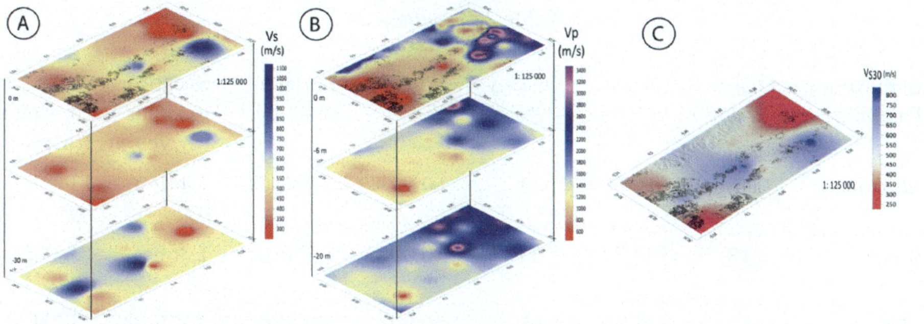

Fig. 5. 3D illustration maps: (A) Vp-waves velocities, (B) Vs-waves velocities, (C) Vs30 map

- RQD values range between 17.3 and 100%, 19.7 and 100%, and 20.7 and 100% for the first, second, and third layers, respectively.
- Density values for the first, second, and the third layers are between 1.37 KN/m^3 and 2.42 KN/m^3, 1.4–2.4 KN/m^3, 1.5–2.4 KN/m^3, respectively.
- Porosity values for the first, second, and the third layers are between 20.9% and 60.9%, 22,2–58,8%, 20,9–57,8%, respectively.
- Poisson ratio values for the first, second, and the third layers are between-7.58E-03 and 4.96E-01, 2.32E-02-4.90E-01, -6.16E-02-4.87E-01, respectively.
- Shear modulus values range between 7.83E + 05 and 1.44E + 07 dyn/cm^2, 8.92E + 05–3.50E + 07 dyn/cm^2, and 1.36E + 06–1.12E + 08 dyn/cm^2 for the first, second, and third layers, respectively.
- Bulk modulus values range between 1.96E + 03 and 3.36E + 05 Mpa, 2.58E + 03–2.85E + 05 Mpa, and 2.92E + 03–3.37E + 05 dyn/cm^2 for the first, second, and third layers, respectively.

The examination of the calculated parameters reveals soil heterogeneities in both vertical and horizontal directions. In general, the first and second layers show fairly competent materials, while the third layer shows good competent materials and can be considered as geotechnical bedrock.

Table 2. Statistical analysis of the geotechnical, petrophysical, and elastic-dynamic parameters at the study area.

Engineering Parameters	Qult (kN/m^2)	RQD (%)	Density (KN/m3)	Porosity (%)	Poisson ratio	Shear modulus (dyn/cm^2)	Bulk modulus (Mpa)
Layer 1							
Minimum	27	17.30	1.37	20.90	-7.58E-03	7.83E + 05	1.96E + 03

(*continued*)

Table 2. *(continued)*

Engineering Parameters	Qult (kN/m^2)	RQD (%)	Density (KN/m3)	Porosity (%)	Poisson ratio	Shear modulus (dyn/cm^2)	Bulk modulus (Mpa)
Maximum	2345	100.00	2.42	60.90	4.96E-01	1.44E + 07	3.36E + 05
Range	2318	82.70	1.05	40.00	5.03E-01	1.36E + 07	3.34E + 05
Mean	361	41.80	1.68	46.90	2.00E-01	3.17E + 06	2.71E + 04
Standard deviation	461.40	19.80	0.21	8,50	1.00E-01	2.86E + 06	6.08E + 04
Layer 2							
Minimum	38.60	19.7	1.4	22.20	2.32E-02	8.92E + 05	2.58E + 03
Maximum	7454.10	100	2.4	58.80	4.90E-01	3.50E + 07	2.85E + 05
Range	7415.50	80.3	1	36.60	4.67E-01	3.41E + 07	2.83E + 05
Mean	1349.80	64.5	1.9	39.20	2.00E-01	8.66E + 06	4.62E + 04
Standard deviation	1671.40	26.2	0.17	6.60	1.41E-01	7.88E + 06	5.24E + 04
Layer 3							
Minimum	60.20	20.70	1.50	20.90	-6.16E-02	1.36E + 06	2.92E + 03
Maximum	34144.60	100.00	2.40	57.80	4.87E-01	1.12E + 08	3.37E + 05
Range	34024.20	79.30	0.90	36.90	5.48E-01	1.11E + 08	3.35E + 05
Mean	3357.50	79.30	1.90	35.80	1.90E-01	1.64E + 07	7.34E + 04
Standard deviation	6344.60	32.40	0.20	7.10	1.53E-01	2.12E + 07	7.32E + 04

5 Conclusion

Due to demographic growth and economic development, the eastern Agadir area was selected to accommodate the urban expansion of the city. However, this area is characterized by heterogeneous and poorly compacted sediments, making it prone to several geotechnical problems. To address these challenges, site characterization using in-situ and laboratory tests is required to measure geotechnical, seismic, petrophysical, and elastic-dynamic parameters. However, these investigations are costly and time-consuming. Seismic surveys are nowadays acceptable tools that provide a clear picture of the subsurface characteristics. For these reasons, MASW and SRT methods were used to calculate S and P wave velocities, respectively. The calculated data were then used to estimate the aforementioned parameters.

The results show that the study area is mainly composed of three geoseismic layers. The first layer is characterized by heterogeneous sediments and exhibits different seismic velocity values (V_P between 382 and 757 m/s, V_S between 220 and 850 m/s). The second layer is more competent, with V_p and V_s ranging between 432 m/s and 3494 m/s, and 215 m/s to greater than 850 m/s, respectively. The third layer, which can be considered as a geotechnical bedrock, has V_p and V_s values spanning from 456 m/s to 3800 m/s and 250 m/s to greater than 850 m/s, respectively.

Seismic soil classification based on V_{s30} shows that the study area is composed of soil classes A, B, and C according to Eurocode 8. The examination of the calculated parameters reveals soil heterogeneities in both vertical and horizontal directions.

References

1. Cherkaoui, T.-E., Hassani, A.E.: Seismicity and Seismic Hazard in Morocco 1901-2010, 12 (2012)
2. Karastathis, V.K., Karmis, P., Novikova, T., Roumelioti, Z., Gerolymatou, E., Papanastassiou, D., et al.: The contribution of geophysical techniques to site characterisation and liquefaction risk assessment: case study of Nafplion City Greece. J. Appl. Geophys. **72**, 194–211 (2010). https://doi.org/10.1016/j.jappgeo.2010.09.003
3. Pegah, E., Liu, H.: Application of near-surface seismic refraction tomography and multichannel analysis of surface waves for geotechnical site characterizations: a case study. Eng. Geol. **208**, 100–113 (2016). https://doi.org/10.1016/j.enggeo.2016.04.021
4. Abudeif, A.M., Fat-Helbary, R.E., Mohammed, M.A., Alkhashab, H.M., Masoud, M.M.: Geotechnical engineering evaluation of soil utilizing 2D multichannel analysis of surface waves (MASW) technique in New Akhmim city, Sohag, Upper Egypt. J. Afr. Earth Sc. **157**, 103512 (2019). https://doi.org/10.1016/j.jafrearsci.2019.05.020
5. Park, C.B., Shawver, J.: MASW survey using multiple source offsets. In: Symposium on the Application of Geophysics to Engineering and Environmental Problems 2009, Environment and Engineering Geophysical Society, pp. 9–15 (2009). https://doi.org/10.4133/1.3176691
6. Butchibabu, B., Jha, P.C., Sandeep, N., Sivaram, Y.V.: Seismic refraction tomography using underwater and land based seismic data for evaluation of foundation of civil structures. J. Appl. Geophys. **210**, 104934 (2023). https://doi.org/10.1016/j.jappgeo.2023.104934
7. Bouaakkaz, B., El Morjani, Z.E.A., Bouchaou, L.: Social vulnerability assessment to flood hazard in Souss basin Morocco. J. African Earth Sci. **198**, 104774 (2023). https://doi.org/10.1016/j.jafrearsci.2022.104774
8. El Morjani, Z.E.A., Seif Ennasr, M., Elmouden, A., Idbraim, S., Bouaakaz, B., Saad, A.: Flood hazard mapping and modeling using GIS applied to the souss river watershed. In: Choukr-Allah, R., Ragab, R., Bouchaou, L., Barceló, D. (eds.) The Souss-Massa River Basin, Morocco. THEC, vol. 53, pp. 57–93. Springer, Cham (2016). https://doi.org/10.1007/698_2016_69
9. Elmouden, A., Alahiane, N., Faskaoui, M., El Morjani, Z.E.A.: Dams Siltation and Soil Erosion in the Souss–Massa River Basin, pp. 1–26 (2016). https://doi.org/10.1007/698_2016_70

10. Malki, M., Choukr-Allah, R., Bouchaou, L., Hirich, A., Ait-Brahim, Y., Krimissa, S., et al.: Assessment of groundwater quality: impact of natural and anthropogenic contamination in Souss-Massa River Basin, pp. 1–20 (2016). https://doi.org/10.1007/698_2016_72
11. Hssaisoune, M., Boutaleb, S., Benssaou, M., Bouaakkaz, B., Bouchaou, L.: Physical geography, geology, and water resource availability of the Souss-Massa River Basin. The Souss-Massa River Basin, Morocco, pp. 27–56 (2016)
12. Mustaphi, H., Medina, F., Jabour, H., Hoepffner, C.: Le bassin du Souss (Zone de Faille du Tizi n'Test, Haut Atlas occidental, Maroc): résultat d'une inversion tectonique contrÎée par une faille de détachement profonde. J. Afr. Earth Sc. **24**, 153–168 (1997). https://doi.org/10.1016/S0899-5362(97)00033-X
13. Mridekh, A., Toto, E.R., Hafid, M., Ouataoui, A.: Structure sismique de la plate-forme Atlantique au large d'Agadir (Maroc sud-occidental). Comptes Rendus De L Academie Des Sciences Serie Ii Fascicule A-Sciences De La Terre Et Des Planetes - C R ACAD SCI SER II A **331**, 387–392 (2000). https://doi.org/10.1016/S1251-8050(00)01422-1
14. Sébrier, M., Siame, L., Zouine, E.M., Winter, T., Missenard, Y., Leturmy, P.: Active tectonics in the Moroccan High Atlas. C.R. Geosci. **338**, 65–79 (2006). https://doi.org/10.1016/j.crte.2005.12.001
15. Dewey, J., Helman, M., Knott, S., Turco, E., Hutton, D.: Kinematics of the western Mediterranean. Geological Soc. London, Special Publ. **45**, 265–283 (1989). https://doi.org/10.1144/GSL.SP.1989.045.01.15
16. Ambroggi, R.: Etude géologique du versant méridional du Haut Atlas occidental et de la plaine du Souss. Ed. de la division de la géologie. Rabat (1963)
17. Castagna, J., Batzle, M., Eastwood, R.: Relationship between compressional and shear-wave velocities in classic silicate rocks. Geophysics **50**, 571–581 (1985). https://doi.org/10.1190/1.1441933
18. Gardner, G.H.F., Gardner, L.W., Gregory, A.R.: Formation velocity and density—the diagnostic basics for stratigraphic traps. Geophysics (1974). https://doi.org/10.1190/1.1440465
19. Watkins, J.S., Walters, L.A., Godson, R.H.: Dependence of in-situ compressional-wave velocity on porosity in unsaturated rocks. Geophysics **37**, 29–35 (1972). https://doi.org/10.1190/1.1440249
20. Toksoz, M.N., Cheng, C.H., Timur, A.: Velocities of seismic waves in porous rocks. Geophysics **41**, 621–645 (1976)
21. Telford, W.M., Geldart, L.P., Sheriff, R.E., Keys, D.A.: Applied Geophysics, pp. 1–860. Cambridge University Press, NY (1976)
22. Mott, P.H., Dorgan, J.R., Roland, C.M.: The bulk modulus and Poisson's ratio of "incompressible" materials. J. Sound Vib. **312**, 572–575 (2008). https://doi.org/10.1016/j.jsv.2008.01.026
23. Parry, R.H.G.: Estimating bearing capacity in sand from SPT values. J. Geotech. Eng. Div. **103**, 1014–1019 (1977). https://doi.org/10.1061/AJGEB6.0000484

The Flexural Performance of Geopolymer Reinforced Concrete Beam: A Review

Ashraf M. Ali[1], Nehal M. Ayash[1], Mostafa A. Osman[1], Ahmed F. Deifalla[2(✉)], and Hala Mamdouh[1]

[1] Civil Engineering Department, Faculty of Engineering at Mataria, Helwan University, Cairo 11718, Egypt
{nehal82ayash,Hala_mamdouh}@m-eng.helwan.edu.eg

[2] Structural Engineering and Construction Management, Future University in Egpt, New Cairo, Egypt
Ahmed.deifalla@fue.edu.eg

Abstract. Geopolymer concrete (GC) is an environmentally beneficial choice that may be used in place of conventional concrete (CC). When compared to GC, the manufacturing process for Portland cement results in the production of harmful elements. The bulk of the research that were encountered were focused on examining the behavior of GC with regard to its mechanical characteristics and durability. On the other hand, the flexure behavior of GC reinforced beams has not been well explored in the published literature. The flexure behavior of GC beams is the subject of this review, which aims to improve current knowledge of the subject. After conducting a comprehensive research on the flexure behavior of GC, it was discovered that there are a number of gaps in the existing literature. Beams made of geopolymer concrete were shown to have a greater flexural capacity, despite the fact that several aspects have not yet been thoroughly examined. In addition, geopolymer beams exhibited greater deformations by virtue of their lower elastic modulus in comparison to that of conventional concrete. Beams made of geopolymer concrete exhibited satisfactory behavior when compared to traditional concrete components, which suggests that this material has potential applications in the construction industry.

Keywords: Geopolymer concrete (GC) · Conventional Concrete (CC) · Flexural behavior · Fly Ash (FA) · Alkali activation

1 Introduction

The request for concrete increases as a material for construction, which results in an increase in cement production. Figure 1 shows the increase in global cement production from 1990 to 2050. We noticed from the figure that this industry is growing rapidly, especially in developing countries that have a high demand for construction [1]. Thus, the purchase orders for Portland cement is rising, where the average Portland cement utilization is about 1 ton per year for living individual [2]. Moreover, the cement production has a harmful influence on the environment, where around 80% of CO2 sum

A. S. Mosallam et al. (Eds.): ICATH 2023, SCI, pp. 245–256, 2024.
https://doi.org/10.1007/978-3-031-76557-5_20

radiated from Portland cement concrete manufacturing is due to Portland cement, which emitted 7% of global CO2 [3].

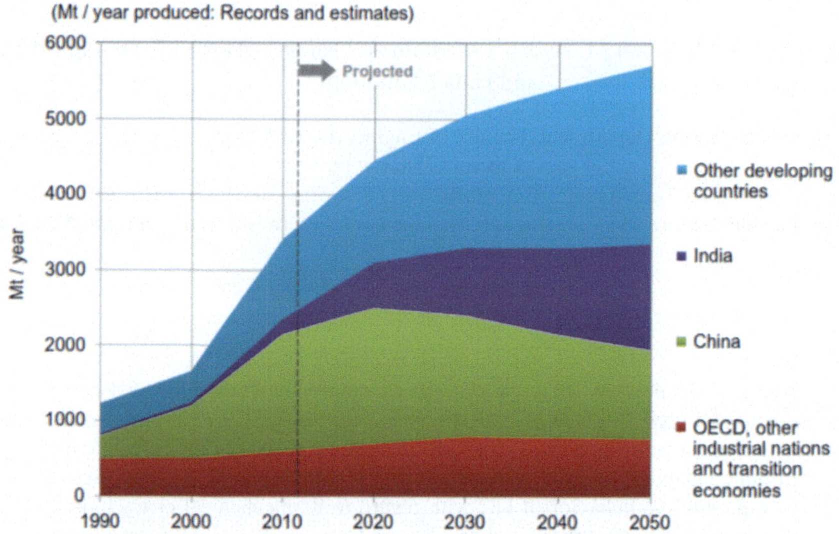

Fig. 1. World Portland cement production form 1990 – 2050 [1].

Furthermore, Portland cement manufacturing requires energy, which is about 50–60% of its production cost [4]. Thus, it consumes a large amount of electrical and thermal energy, for example, to produce one ton of clinker we need about 110 – 120 kWh and about 3000–6500 MJ [5].

In recent times, many research works showed that geopolymer concrete (GC) is a viable arrangement that can be served as a compelling elective to Portland cement concrete (PC) [6]. Thus, the usage of GC reduces CO_2 emissions compared to using PC. Geopolymer system is composed of one or more rich aluminum and silicon material such as (FA), slag, silica fume and metakaolin [7–9]. The main binder reacts with an alkaline activator solution (AAS) that could be a mix of sodium hydroxide solids (SHS) and sodium silicate solution (SSS) and this combination is very important for the polymerization process [10]. The usage of GC may decrease the amount of emissions of CO_2 by up to 64% compared to the use of cement [11]. Moreover, from an economic view, the price of source materials is lower than cement, for example, the price of GC which depended on FA as aluminosilicate material is cheaper than PC by 10–30% after taking into consideration the price of alkaline activator [12]. In addition, many researchers found using GC instead of using CC similar and even better strength, stiffness, and other mechanical properties compared to CC [13, 14]. Castel et al. [15] observed that GC appears exceptionally little drying shrinkage. Fernandez-Jimenez et al. [16] reported that the bond between steel and GC is exceptionally strong and the failure mode was through break of the steel bar. Amer et al. [17] completely supplanted Portland cement by (GGBFS) in a concrete mix that accomplished a compressive strength (CS) of 43 MPa at ambient curing condition after 28 days. Achieving sufficient CS in ambient curing is a desirable advantage compared to CC that might need water curing to obtain comparable CS.

In contrast to the CC, the GC was able to achieve 95% of the CS in only seven days, as highlighted by Ashour et al. [18]. This discovery suggests that the GC will be more effective in rapidly pushing projects forward. In addition, the tensile strength of GC was often greater than that of CC at the same CS [19–21] as shown in Fig. 2 [22]. Moreover, the mix design has a considerable impact on the bending behaviors of GC beams during fabrication [23]. However, most investigated mixes are not practical either due to extremely high slump loss, or high required heat curing to achieve adequate performance. Thus, fibers are recommended for geopolymer concrete [24, 25].

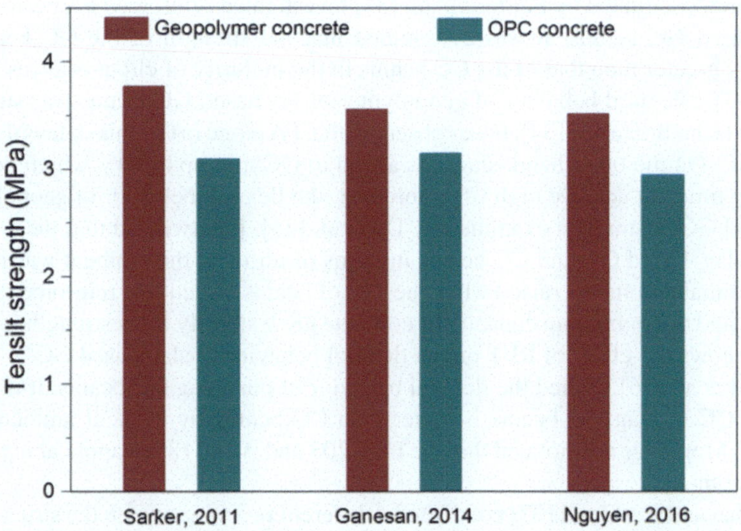

Fig. 2. Comparison between the tensile strength of GC and CC [22].

Over the last several years, there has been an increase in the amount of research conducted on GC. The scope of this methodology encompasses not only the research of micro-scale investigations, but also the study of the behavior of structures. In addition, the behavior of certain structural components, such as beams, among other things. This review focused on the results of previous research on the flexural behavior of GC beams and emphasized those findings. In addition, it was discovered that the beams produced by GC were determined to be sufficiently equivalent to those produced by CC.

2 Flexural Behavior of GC Beams

Sumajouw et al. [26] studied the flexure behavior of fly-ash GC reinforced beams with various flexure RFT ratios (0.64–2.69%). The flexural capacity of fly ash GC beams increases with the raise of tensile reinforcement ratio. AS 3600 [27] predicts the strength in reasonable manner with average ratio between measured and calculated resistance ranged from 0.98 to 1.28.

A study by Dattatreya et al. [28] also investigated that the FA based GC beams had lower post-peak ductility, which was confirmed by Yost et al. [29], which is because failure was more brittle due to concrete smashing compared to CC.

Moreover, Antony et al. [30] observed that reinforced GC beams had a higher value of deflection than CC beams. It is noticed that it had maximum capacity as well as lower crack width. Also, Prachasaree et al. [31] developed concrete stress block values for FA based GC beams that were similar to experimental results. It was noticed that the nominal moment capacity differed by an average of 13%, which resulted in a roughly 1.4 times reduction in the disparity between the results of the experimental testing and the predictions made by the researchers. In addition to it was found that that AS 3600 [27] and ACI 318 [32] were applicable with the design plan strategies.

Research conducted by Ferdous et al. [33] investigated reinforced FA beams as well as slag-based GC beams. It was determined that the strength of the GC beams was marginally greater than that of the CC beams in the majority of circumstances.

Also, The flexural behavior of geopolymer of FA based GC beams was studied by Hutagi and Khadiranaikar [34]. It was observed that FA alone in GC mix delays the setting time of GC. On the other hand, slag was added to GC mix up to 40% which decreased the setting time but achieve high CS. Moreover, the flexural behavior of geopolymer of slag–based GC beams was examined by Du et al. [35]. It is twigged that the difference between slag–based GC and CC beams in terms of ultimate the moment was minimal, and the ultimate moments raised when the size of beams or bending reinforcement ratio increased and the maximum capacity of concrete grew slightly as its strength increased. Figure 3 shows the effect of RFT on the flexural behavior of slag based GC.

Ashraf et al. [36] studied the flexural behavior of pure slag GC beams. It is twigged that the LCC of slag GC beams is better than CC beams by 7.4% at ambient curing condition. Moreover it is noticed that the ECP 203 and ACI 318 are applicable to design slag GC beams.

Gopalan and Kathirvel [37] conducted a different investigation on the structural performance of slag GC beams. In this study, the tensile reinforcement ratio was investigated at three different levels: 1%, 1.5%, and 2%. According to their findings, the ultimate load bearing capacity of the beam improves as the tensile reinforcement ratio and the deflection at the ultimate load both increase.

Pham et al. [38] used advanced FE-based modeling to compare the results of GC beams to results of numerical analysis using ABAQUS. The numerical findings were similar to the experimental results for the qualitative and quantitative findings (both M-phi and crack pattern results). It was observed that the discrepancies ranged from 1% to 5%. The created FE models were recommended to be used as a precise design tool for GC beams. The structural behavior of GC beams containing various concrete ingredients was also investigated. For example, the influence of recycled aggregate has been studied by Zhang et al. and Akduman et al. [39, 40]. The substitution of recycled aggregate in geopolymer recycled aggregate concrete beams has been demonstrated to have no impact on the failure mechanism or flexural capacity of the beams. Recycling aggregates resulted in a considerable reduction in the amount of energy that was lost by GC beams. Moreover, Kalaivani et al. [41] investigated the effect of adding lime into GC mix. It is observed a heat production, which aids in the polymerization process. When

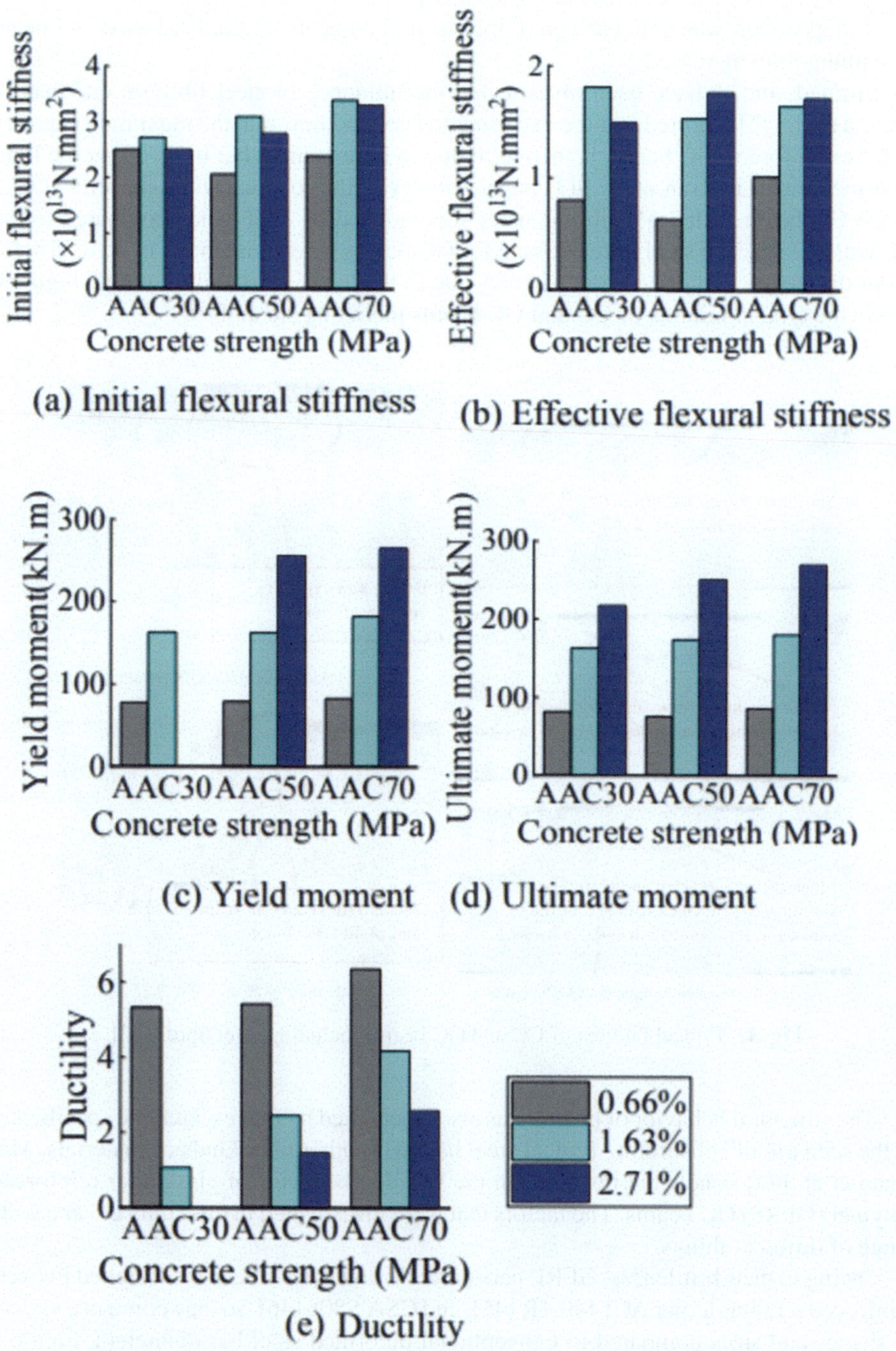

Fig. 3. Influence of RFT ratio on the performance of slag-based GC beams [35].

the test specimen was ambient cured, the flexural capacity of the FA based GC beam containing lime increased.

Limited studies have been investigated the influence of steel fibre on GC beams. Hassan et al. [42] reported that the experimental results show that the maximum capacity of fibre-reinforced GC beams is increased due to fibre content but up to a specific limit as it mentioned in Tran et al. [43] paper. However, the load-carrying capacity (LCC) of the GC beam with 1.5% fibre content was reduced by 18% when compared to the GC with 0.75%. The steel fibres dosage for GC beams is recommended to be 0.75% for better dispersion and orientation of fibres due to the lower workability of GC. Figure 4 shows the typical failures of CC and GC beams including steel fibres.

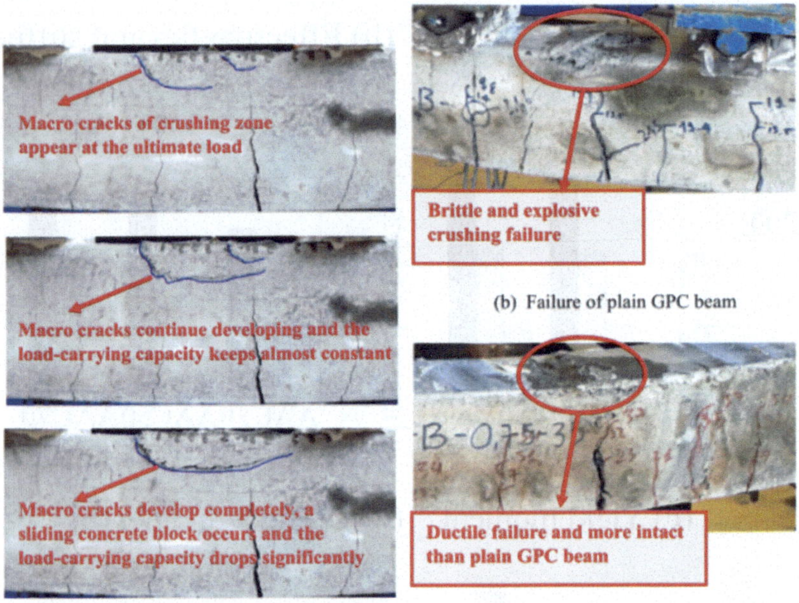

(a) A progressive failure of OPC beam (b) Failure of plain GPC beam (c) Failure of Beam GPC-0.75-35

Fig. 4. Typical failures of CC and GC beams including steel fibres [43].

The structural behavior of beams was also researched by the researchers, specifically in the scenario of substituting typical steel bars with other fiber kinds of materials. Maranan et al. [44] conducted research on the bending behavior of glass fiber reinforced polymer (GFRP) GC beams. The factors that were investigated in this study cover a wide range of different things.

Owing to their brittleness, GFRP bars (straight and headed) can be designed in over-reinforced sections as per ACI 440.1R [45], and CSA S806 [46], so they come in a variety of shapes and sizes compared to conventional deformed steel bar diameters. Figure 5 showed that the influence of GFRP-RFT on the LCC response of GC beams differed from ordinary RFT. Increasing the reinforcement ratio of GFRP was beneficial for serviceability and crack control of GC beams. Straight and hooked GFRP bars were equivalent to regular steel bars in terms of moment capacity. However, they were substantially different to ordinary steel bars because of the varying in elastic moduli. GFRP-reinforced beams had higher flexural capability than that of ordinary steel-reinforced beams for equal reinforcement ratios because of yielding of steel. Comparing the LCC of GFRP-reinforced GC beams to different codes of practice. The test-to-prediction ratios of 1.31 and 1.23, respectively.

Moreover, Ahmed et al. [47] reported that the stiffness of the beams was affected by the RFT ratio. As a result, the beams with a low RFT ratio had significant deformation. While beams with a higher rebar ratio noticed more load enhancement (73.3%–302.9%) in terms of load-deflection behavior and it is observed that increase in the rebar ratio had a greater impact on GC beams than CC beams.

El-Sayed et al. [48] made experiments and nonlinear investigations on ultra-high-performance (UHP) with GFRP bars as a RFT. It is twigged that the ductility of UHP geopolymer CC beams reinforced with GFRP bars was found to be greater. Ahmed et al. [49] compared the flexural behavior of FA based GC beams reinforced with different types of bars. CFRP bars were preferred over GFRP bars as the LCC for CFRP-RGC beams was higher for the same RFT ratio. By achieving the same RFT ratio with a smaller cross-section of CFRP-RGC beams, the dead load was reduced and the CFRP-RGC beams had the smallest crack width. Few studies are made on replacing steel bars with basalt bars. Fan and Zhang [50] reported that basalt rebar reinforced GC beam had 4 times the deflection of ordinary reinforced concrete beam. The ACI 440.1R [45] was found applicable to estimating the flexural capacity of basalt fibre-reinforced GC beams. Moreover, Rashid et al. [51] compared between the bending behavior of steel, CFRP, and BFRP reinforced GC beams. They found that CFRP-reinforced GC beams had the

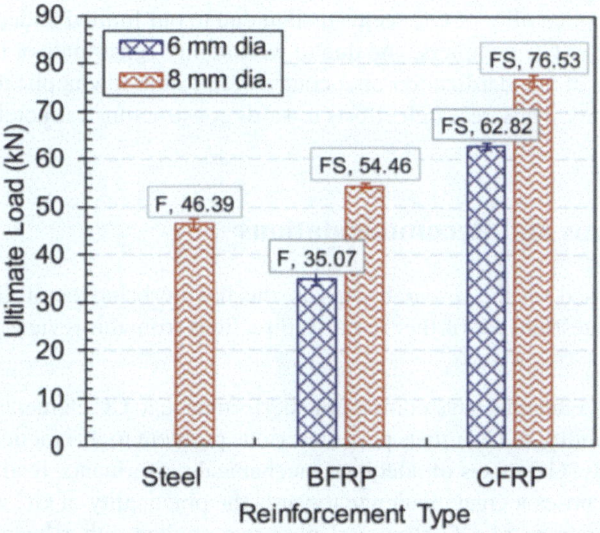

Fig. 5. Ultimate loads and their corresponding failure mode for different rebar reinforcement [51].

highest ultimate load. Figure 5 shows a comparison between ultimate loads and their corresponding modes of failure.

Furthermore, the flexural behavior of FA based GC beams exposed to elevated temperature was investigated by Mathew et al. [52]. It is twigged that the deformation behavior was similar to CC beams. On the other hand, it was noticed the ductility of FA based GC beams was reduced rapidly.

3 Discussion

This paper investigated past studies of the flexural behavior of GC beams. As the GC consists of one or more rich aluminum and silicon material such as (FA), slag, silica fume and metakaolin [7, 8]. The main binder reacts with an AAS that could be a blend of SHS and (SSS) and this combination is very important for the polymerization process.Research on the flexure behavior of GC elements remains very limited. First, Research that discussed the behavior of other structural elements is sparse. Moreover, GC has a lot of variability and a wide range of alternatives, unlike ordinary concrete. However, the bulk of the literature mostly utilized FA based GC mixes. Metakaolin mixes, for example, were never used in the literature for studying structural behavior. Furthermore, the influence of certain mixed parameters on the flexure behavior of elements is not properly studied –such as the effect of molarity of NaOH and the content of water glass.

Generally, the behavior of reinforced GC members was found to be equivalent to ordinary concrete members in most cases. It is accepted that GC members may be constructed in a manner like CC members. Since their general flexure behavior - such as LCC, cracking behavior, and failure mode– are identical to CC members. However, the Structural applicability of GC seems distant due to our limited understanding of the flexure behavior of GC members, and due to workability constraints of the geopolymer mixes. The lack of standardization also contributed to the low applicability of GC in structural elements. Thus, research efforts towards constructing a structural code for GC are essential.

4 Conclusions and Recommendations

This paper focused on the research gaps in the flexure behavior of GC beams. The findings and suggestions for further research directions from this review can be summed up as follows:

- Structural GC elements have comparable performance to CC elements. However, the loss of workability of GC mixes poses a serious problem to its practical implementation. Currently, GC mixes of adequate mechanical performance tend to rapidly lose slump which poses a great challenge towards the practicality of GC in construction.
- Most investigations of GC structural elements studied either beams or columns. Other structural members –e.g. frames, slabs, and walls– are rarely discussed in the literature, or discussed only is special cases of loading –e.g. impact loading on slabs.

- There is not much variation in the binder type used in structural geopolymer studies since they mostly use FA based geopolymer concrete, while other binder types –such as metakaolin– are never used. Moreover, the effect of some mix design parameters –e.g. NaOH molarity, water glass content– on the structural behavior of elements is neglected.
- The shear behavior of plain GC beams is not sufficiently studied, even though the effect of fibers on the shear behavior is investigated in the literature. Moreover, many aspects of the structural behavior of GC elements, beside the plain shear behavior, remain yet to be properly investigated.
- Experimental and numerical investigations are required on multiple points including:

  - serviceability, especially in deflection and crack propagation.
  - fire resistance of geopolymer concrete.
  - long-term behavior of GC structural elements.

Acknowledgement. In particular, Drs. Mohamed Kohail, Ismail Amer, Sayed El-Araby, and Abdalla Talaat from the structural department at Ain Shams University are respectfully acknowledged by the authors for their support.

Declaration of Competing Interest. The authors claim that they have no known financial conflicts of interest or close personal relationships that would have affected the research provided in this paper.

References

1. Imbabi, M.S., Carrigan, C., McKenna, S.: Trends and developments in green cement and concrete technology. Int. J. Sustain. Built Environ. **1**, 194–216 (2012). https://doi.org/10.1016/J.IJSBE.2013.05.001
2. Lippiatt, B.C., Ahmad, S.: Measuring the life-cycle environmental and economic performance of concrete: the bees approach (2004)
3. Pal, A.: Developing low-clinker ternary blends for Indian cement industry. J. Inst. Eng. Ser. A. 99 (2018). https://doi.org/10.1007/s40030-018-0309-4
4. Wang, J., Dai, Y., Gao, L.: Exergy analyses and parametric optimizations for different cogeneration power plants in cement industry. Appl. Energy **86**, 941–948 (2009). https://doi.org/10.1016/J.APENERGY.2008.09.001
5. Mejeoumov, G.G.: Improved Cement Quality and Grinding Efficiency by Means of Closed Mill Circuit Modeling A Dissertation (2007)
6. Hassan, A., Arif, M., Shariq, M.: Use of geopolymer concrete for a cleaner and sustainable environment – a review of mechanical properties and microstructure. J. Clean. Prod. **223**, 704–728 (2019). https://doi.org/10.1016/J.JCLEPRO.2019.03.051
7. Part, W.K., Ramli, M., Cheah, C.B.: An overview on the influence of various factors on the properties of geopolymer concrete derived from industrial by-products. Constr. Build. Mater. **77**, 370–395 (2015). https://doi.org/10.1016/J.CONBUILDMAT.2014.12.065
8. Amer, I., Kohail, M., El-Feky, M.S., Rashad, A., Khalaf, M.A.: A review on alkali-activated slag concrete. Ain Shams Eng. J. **12**, 1475–1499 (2021). https://doi.org/10.1016/J.ASEJ.2020.12.003

9. Amer, I., Kohail, M., Elfeky, M., Rashad, M., Khalaf, M.: Evaluation of using cement in alkali-activated slag concrete. Int. J. Sci. Technol. Res. **9**, 245 (2020)
10. Palomo, A., Grutzeck, M.W., Blanco, M.T.: Alkali-activated fly ashes: a cement for the future. Cem. Concr. Res. **29**, 1323–1329 (1999). https://doi.org/10.1016/S0008-8846(98)00243-9
11. McLellan, B.C., Williams, R.P., Lay, J., Van Riessen, A., Corder, G.D.: Costs and carbon emissions for geopolymer pastes in comparison to ordinary portland cement. J. Clean. Prod. **19**, 1080–1090 (2011). https://doi.org/10.1016/J.JCLEPRO.2011.02.010
12. Lloyd, N., Rangan, V.: Geopolymer concrete with fly ash, In: Proceeedings of the Second International Conference Sustain. Construction Material Technology, UWM Center for By-Products Utilization, pp. 1493–1504 (2010)
13. Farooq, F., et al.: Geopolymer concrete as sustainable material: a state of the art review. Constr. Build. Mater. **306**, 124762 (2021). https://doi.org/10.1016/j.conbuildmat.2021.124762
14. Manish, S., Banerjee, R.: Comparison of biohydrogen production processes. Int. J. Hydrogen Energy **33**, 279–286 (2008). https://doi.org/10.1016/J.IJHYDENE.2007.07.026
15. Castel, A., Foster, S.J., Ng, T., Sanjayan, J.G., Gilbert, R.I.: Creep and drying shrinkage of a blended slag and low calcium fly ash geopolymer Concrete, Mater. Struct. **49**(5), 1619–1628 (2015). https://doi.org/10.1617/S11527-015-0599-1
16. Fernandez-Jimenez, A.M., Palomo, A., Lopez-Hombrados, C.: Engineering properties of alkali-activated fly ash concrete. Mater. J. **103**, 106–112 (2006). https://doi.org/10.14359/15261
17. Amer, I., Kohail, M., El-Feky, M.S., Rashad, A., Khalaf, M.A.: Characterization of alkali-activated hybrid slag/cement concrete. Ain Shams Eng. J. **12**, 135–144 (2021). https://doi.org/10.1016/J.ASEJ.2020.08.003
18. Ashour et al.: Using metakaolin and ground granulated blast furnace slag in production of geopolymer concrete (2021). https://doi.org/10.17605/OSF.IO/7Z8CU
19. Ganesan, N., Abraham, R., Deepa Raj, S., Sasi, D.: Stress–strain behaviour of confined Geopolymer concrete, Constr. Build. Mater. **73**, 326–331 (2014). https://doi.org/10.1016/J.CONBUILDMAT.2014.09.092
20. Nguyen, K.T., Ahn, N., Le, T.A., Lee, K.: Theoretical and experimental study on mechanical properties and flexural strength of fly ash-geopolymer concrete. Constr. Build. Mater. **106**, 65–77 (2016). https://doi.org/10.1016/J.CONBUILDMAT.2015.12.033
21. Sarker, P.: Bond strength of reinforcing steel embedded in fly ash-based geopolymer concrete. Mater. Struct. **44**, 1021–1030 (2011). https://doi.org/10.1617/S11527-010-9683-8
22. Zhang, P., Wang, K., Li, Q., Wang, J., Ling, Y.: Fabrication and engineering properties of concretes based on geopolymers/alkali-activated binders - A review. J. Clean. Prod. **258**, 120896 (2020). https://doi.org/10.1016/J.JCLEPRO.2020.120896
23. Mo, K.H., Alengaram, U.J., Jumaat, M.Z.: Structural performance of reinforced geopolymer concrete members: a review. Constr. Build. Mater. **120**, 251–264 (2016). https://doi.org/10.1016/J.CONBUILDMAT.2016.05.088
24. de Azevedo, A.R.G., et al.: Natural fibers as an alternative to synthetic fibers in reinforcement of geopolymer matrices: a comparative review. Polym. **13**, 2493 (2021). https://doi.org/10.3390/POLYM13152493
25. Yanou, R.N., Kaze, R.C., Adesina, A., Nemaleu, J.G.D., Jiofack, S.B.K., Djobo, J.N.Y.: Performance of laterite-based geopolymers reinforced with sugarcane bagasse fibers. Case Stud. Constr. Mater. **15**, e00762 (2021). https://doi.org/10.1016/J.CSCM.2021.E00762
26. Sumajouw, D., Hardjito, D., Wallah, S.E., Rangan, B.V.: Behaviour and strength of reinforced fly ash-based geopolymer concrete beams. Civ. Eng. Dimens. (2005)
27. AS, AS. AS 3600. Concr. Struct. Stand. Aust. Sydney (2004)
28. Dattatreya, J.K., Rajamane, N., Sabitha, D., Ambily, P., Nataraja, M.C.: Flexural behaviour of reinforced Geopolymer concrete beams. Int. J. Civ. Struct. Eng. (2011)

29. Yost, J.R., Radlińska, A., Ernst, S., Salera, M., Martignetti, N.J.: Structural behavior of alkali activated fly ash concrete. Part. Structural testing and experimental findings, Mater. Struct. Constr. **46**, 449–462 (2013). https://doi.org/10.1617/S11527-012-9985-0/FIGURES/8
30. Antony, J.C., Saravanan, G., Salahuddin, M., Thirugnanasambandam, S.: Development of fly ash based geopolymer precast concrete elements, ASIAN J. Civ. Eng. (BUILDING HOUSING) (2013)
31. Prachasaree, W., Limkatanyu, S., Hawa, A., Samakrattakit, A.: Development of equivalent stress block parameters for fly-ash-based geopolymer concrete. Arab. J. Sci. Eng. 2014 3912. 39, 8549–8558 (2014). https://doi.org/10.1007/S13369-014-1447-2
32. ACI Committee 318, ACI 318. Building code requirements for structural concrete (ACI 318-05) and commentary (ACI 318R-05) (2005)
33. Ferdous, W., Manalo, A., Khennane, A., Kayali, O.: Geopolymer concrete-filled pultruded composite beams – concrete mix design and application. Cem. Concr. Compos. **58**, 1–13 (2015). https://doi.org/10.1016/J.CEMCONCOMP.2014.12.012
34. Hutagi, A., Khadiranaikar, R.B.: Flexural behavior of reinforced geopolymer concrete beams. Int. Conf. Electr. Electron. Optim. Tech. ICEEOT **2016**, 3463–3467 (2016). https://doi.org/10.1109/ICEEOT.2016.7755347
35. Du, Y., Wang, J., Shi, C., Hwang, H.J., Li, N.: Flexural behavior of alkali-activated slag-based concrete beams. Eng. Struct. **229**, 111644 (2021). https://doi.org/10.1016/J.ENGSTRUCT.2020.111644
36. Mamdouh, H., Ali, A.M., Osman, M.A., Deifalla, A.F., Ayash, N.M.: Effects of size and flexural reinforcement ratio on ambient-cured geopolymer slag concrete beams under four-point bending. Build **12**, 1554 (2022). https://doi.org/10.3390/BUILDINGS12101554
37. Gopalan, R., Kathirvel, P.: Effect of tensile and shear reinforcement on the flexural behaviour of reinforced geopolymer concrete beams. Int. J. Civ. Eng. Technol. **8**, 1025–1033 (2017). https://iaeme.com/Home/article_id/IJCIET_08_05_110. Accessed 9 Feb 2024
38. Pham, D.Q., Nguyen, T.N., Le, S.T., Pham, T.T., Ngo, T.D.: The structural behaviours of steel reinforced geopolymer concrete beams: an experimental and numerical investigation. Structures. **33**, 567–580 (2021). https://doi.org/10.1016/J.ISTRUC.2021.04.077
39. Zhang, H., Wan, K., Wu, B., Hu, Z.: Flexural behavior of reinforced geopolymer concrete beams with recycled coarse aggregates **24**, 3281–3298 (2021). https://doi.org/10.1177/13694332211026224
40. Akduman, Ş., et al.: Experimental investigations on the structural behaviour of reinforced geopolymer beams produced from recycled construction materials. J. Build. Eng. **41** (2021). https://doi.org/10.1016/J.JOBE.2021.102776
41. Kalaivani, M., Shyamala, G., Ramesh, S., Angusenthil, K., Jagadeesan, R.: Performance evaluation of fly ash/slag based geopolymer concrete beams with addition of lime. Mater. Today Proc. **27**, 652–656 (2020). https://doi.org/10.1016/J.MATPR.2020.01.596
42. Hassan, A., Arif, M., Shariq, M.: Structural performance of ambient-cured reinforced geopolymer concrete beams with steel fibres. Struct. Concr. **22**, 457–475 (2021). https://doi.org/10.1002/SUCO.202000191
43. Tran, T.T., Pham, T.M., Hao, H.: Experimental and analytical investigation on flexural behaviour of ambient cured geopolymer concrete beams reinforced with steel fibers. Eng. Struct. **200**, 109707 (2019). https://doi.org/10.1016/J.ENGSTRUCT.2019.109707
44. Maranan, G.B., Manalo, A.C., Benmokrane, B., Karunasena, W., Mendis, P.: Evaluation of the flexural strength and serviceability of geopolymer concrete beams reinforced with glass-fibre-reinforced polymer (GFRP) bars. Eng. Struct. **101**, 529–541 (2015). https://doi.org/10.1016/J.ENGSTRUCT.2015.08.003
45. A.C. 440, Guide for the Design and Construction of Concrete Reinforced with FRP Bars. ACI 440.1R-03, (2003)

46. CSA Standard, CAN/CSA-S806–02. Design and Construction of Building Components with Fibre-Reinforced Polymers, Canadian Standards Association, Mississauga, Canada (2002)
47. Ahmed, H.Q., Jaf, D.K., Yaseen, S.A.: Flexural capacity and behaviour of geopolymer concrete beams reinforced with glass fibre-reinforced polymer bars. Int. J. Concr. Struct. Mater. **14**, 1–16 (2020). https://doi.org/10.1186/S40069-019-0389-1/FIGURES/15
48. El-Sayed, T.A., Algash, Y.A.: Flexural behavior of ultra-high performance geopolymer RC beams reinforced with GFRP bars, Case Stud. Constr. Mater. **15** (2021). https://doi.org/10.1016/J.CSCM.2021.E00604
49. Ahmed, H.Q. Jaf, D.K., Yaseen, S.A.: Comparison of the flexural performance and behaviour of fly-ash-based geopolymer concrete beams reinforced with CFRP and GFRP bars. Adv. Mater. Sci. Eng. 2020 (2020). https://doi.org/10.1155/2020/3495276
50. Fan, X., Zhang, M.: Experimental study on flexural behaviour of inorganic polymer concrete beams reinforced with basalt rebar. Compos. Part B Eng. **93**, 174–183 (2016). https://doi.org/10.1016/J.COMPOSITESB.2016.03.021
51. Rashid, K., Li, X., Xie, Y., Deng, J., Zhang, F.: Cracking behavior of geopolymer concrete beams reinforced with steel and fiber reinforced polymer bars under flexural load. Compos. Part B Eng. **186**, 107777 (2020). https://doi.org/10.1016/J.COMPOSITESB.2020.107777
52. Mathew, G., Joseph, B.: Flexural behaviour of geopolymer concrete beams exposed to elevated temperatures. J. Build. Eng. **15**, 311–317 (2018). https://doi.org/10.1016/J.JOBE.2017.09.009

Flexural Behavior of Over-Reinforced Concrete Beams Strengthened with Steel Fiber

Waleed Abdallah[1], Ahmed Farouk Deifalla[2](✉), and Amal. H. Ibrahim[1]

[1] Civil Department, Faculty of Engineering at, Mataria Helwan University, Cairo, Egypt
{Waleed_abdallah,Aml_Hassanin}@m-eng.helwan.edu.eg
[2] Future University in Egypt, Cairo 11835, Egypt
ahmed.deifalla@fue.edu.eg

Abstract. Adding steel fibers for the reinforced concrete **(SFRC)** increase some properties of concrete such as the tensile strength, shear strength, toughness and the flexural strength. Most specifications limit the reinforcement ratios in the beams to avoid the brittle failure. However, sometimes a high reinforcement ratio is used in order to minimize beam depth and still provide adequate stiffness. The objective of this work is to investigate and evaluate the using of steel fiber in improving the behavior of over-reinforced beams and its effect on the enhancement of strength and ductility of such beams was presented. The present study aims to investigate the effect of steel fiber on the behavior of the over reinforced concrete beam in terms of crack formation, displacement, and deformation. Six over-reinforced concrete (RC) beams were tested in this study to use steel fibre at various depths to improve the flexure behaviour and ultimate strength capabilities. One is a control beam without steel fibers, three beams of steel fibre ratios 0.5%, 1%, and 2% by volume of concrete were applied to reinforce them to half depth. The final two beams were strengthened completely with the two steel fibre ratios of 1% and 2%. The beams were tested in flexure under one-point loading system. The ultimate load, First cracking load, mid span deflections and steel reinforcement strains were investigated and compared, using the outcomes for the control beam as a benchmark, the flexural behaviour of the SFRC beams was examined and contrasted. SFRC beams were discovered to display ductile behaviour, good fracture management, efficient flexure strengthening, and noticeably decreased mid-span deflection based on the examination of the data. However, they are inversely proportional to the steel strain, crack width, fracture spacing, ductility, and mid-span deflection. The steel to fibre ratio and distributed volume are directly related to the initial stiffness, cracking load, and failure load.

Keywords: Over Reinforced concrete beams · Flexure capacity · Steel fiber · strengthening **Non-linear** · Finite element · Analysis

1 Introduction

The reinforcement ratio of the flexural reinforced concrete beams is classified into three categories, which may be balanced, under-reinforced, and over-reinforced Several of international codes of practice are not allow for the design of flexural beam in the brittle

A. S. Mosallam et al. (Eds.): ICATH 2023, SCI, pp. 257–273, 2024.
https://doi.org/10.1007/978-3-031-76557-5_21

failure of concrete due to the low ductility of the members. The over-reinforced section is one of the most significant aspects is in the structure. However, To avoid the brittle failure and excessive ageing. The development of the over reinforcement sections is essential for the strengthening and improvement of them and exhibit better performance. With increase the section capacity the beams possess higher stiffness, ductility and have high energy absorption to carry extra loads [1, 2]. Several methods were used to develop the over reinforcement sections with various degrees of success, as short stirrups with/without steel fiber [3], helical reinforcement [4] and an external bonding with a steel plate in the side of the RC beam [5]. Overall, these methods were used to avoid brittle behavior and crushing failure. The use of helical reinforcements at mid span of the beams in the critical zone was one of the most important methods to develop the over-reinforced beam [6]. A linear decrease in ultimate load and increase in the spalling loads as the helical pitch increased. However, the decrease in helical pitch was due to the increase in displacement [7]. In order to change the brittle mode of composite into a strong cement matrix with superior crack resistance, steel fiber is added to concrete. This is because steel fiber has the ability to limit the propagation of cracks. It enhances specific concrete element attributes such controlling crack propagation, preventing wide cracks, increasing ultimate flexural strength and toughness, preventing brittle failure and stiffness, and reducing concrete beam deflection [8, 9 and 10]. The compressive strength in concrete does not significantly improve with the adding of small amounts of steel fibers to concrete, however, post-cracking ductility of the composite may be improved [8]. The influence of steel fibers on the flexural strength of concrete is much greater than for direct tension and compression [11]. Steel fibers can transfer the tension stresses across the cracked section, providing residual strength. This is dependent on the interfacial activity between the fiber and the concrete matrix and fiber dispersion [12]. The bond property between the steel fiber and the concrete matrix, in particular, performs better when the fibers are distributed uniformly [13]. Steel fibers improve the resistance of impact and the toughness characteristics of the high strength concrete and increase the first cracking load, ultimate load, stiffness and ductility of the concrete beams [14, 15]. The deflection can be reduced by using the steel fiber which inhibits the crack growth. As a result, adding fibers causes decrease in the neutral axis depth of the beam cross-section, especially at lighter loads [16, 17]. Steel fiber improves the flexural stiffness after cracking [18]. Using steel fiber with reinforcing bars may provide conflicting outcomes, where depending on the relative reinforcing bar ratio, the instance a deflection softening fiber behavior may or may not lead to crack localization [19]. The flexural strength of steel fiber reinforced concrete increased with the inclusion of steel fiber, 1.5% & 2% was the maximum steel fiber content for compressive & flexural strength respectively and for aspect ratio of 70. Also, the cracking/failure pattern of SFRC specimens showed improved ductility over that of plain concrete [20].

2 Experimental Methodology

2.1 Materials

The concrete mix was designed to achieve 25 MP a compressive strength after 28 days. Ordinary Portland cement which has initial and final setting times of 75 min and 216 min respectively was used. The fine aggregate used was natural river sand. The coarse aggregate was crushed dolomite which has specific gravity and a fineness modulus of 2.85 and 3.28 respectively. Normal tap water was used for mixing and curing procedure. The mix proportions of the concrete used in this study are summarized in Table 1. Standard cubes of 150 × 150 × 150 mm were taken from each concrete batch to ensure obtaining the required compressive strengths. High-tensile steel bars are used in the tested specimens, 16 and 10 diameters for tension and compression steel reinforcement respectively and mild smooth steel 8 diameters for stirrups. The properties of the used steel reinforcing bars are given in Table 2.

Table 1. Proportions of the concrete's mix

Ingredients	Grade 25 MP
w/c ratio	0.56
Mixed water	230
Portland cement	410
Fine aggregate (sand)	902
Coarse aggregate (max size = 16 mm)	833

Table 2. Properties of the used steel reinforcing bars

Properties	Ø 8	Φ 10	Φ 16
Yield stress (MP)	280	395	392
Ultimate stress (MP)	410	590	582
Actual area (mm^2)	50	78	201

2.1.1 Steel Fiber

Hooked-ended macro steel fiber was used in this study as shown in Fig. 1. The length and the diameter of the used steel fiber are 50 mm and 0.75 mm respectively with 1100 N/mm^2 tensile strength. The steel fiber was added with three different volume fractions V_f (0.5%, 1% and 2% of the total volume of the concrete).

Fig. 1. Hooked-ended shaped steel fiber used in this study

2.2 Laboratory Specimens

Six over reinforced concrete beams having the cross-sectional dimensions of 150 mm × 200 mm and 1600 mm effective span were casted and tested until failure. Al specimens have longitudinal bottom steel reinforcement of 4Ø16 in tension, 2Ø10 in compression and twelve stirrups of Ø8 mm @ 200 mm distributed along the beam length as shown in Fig. 2. All tested beams were over-reinforced with a 3.1% reinforcement ratio.

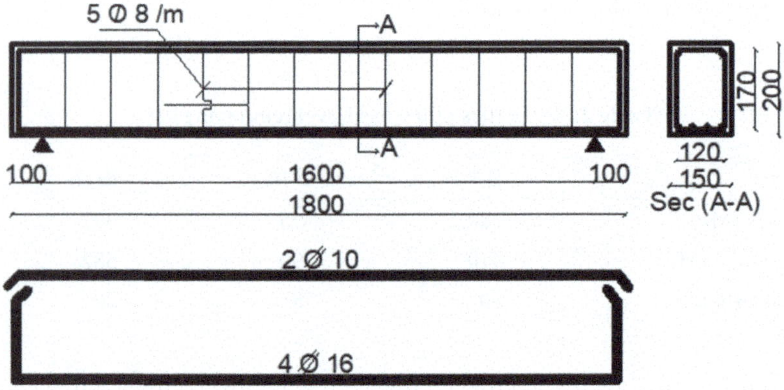

Fig. 2. Typical beam specimen dimension and reinforcement details

2.3 Experimental Program

The main objectives in this study are to assess how the steel fibre position and ratio affect the over reinforced beam. This study evaluated six over-reinforced concrete beam specimens until failure. Prior to strengthening, the structural performance of an over-reinforced beam was compared using the control beam (B1), the other five specimens were strengthening with steel fibers. Steel fibres in the amounts of 0.5%, 1%, and 2% by volume were added to the bottom at the middle third region till half depth for specimens (B2, B3, and B4) respectively. Steel fibres of 1% and 2% by volume were used to reinforce specimens (B5 and B6) respectively in the same region to the all depths.

The details of the specimens are shown in Fig. 3 and Table 3.

Table 3. Test program and specimen details

Beams	Tension RFT	Compression RFT	Stirrups	SF Ratio	SF Location
B1	4Ø16	2Ø10	5Ø8/m	0%	
B2	4Ø16	2Ø10	5Ø8/m	0.5%	0.5d
B3	4Ø16	2Ø10	5Ø8/m	1%	0.5d
B4	4Ø16	2Ø10	5Ø8/m	2%	0.5d
B5	4Ø16	2Ø10	5Ø8/m	1%	d
B6	4Ø16	2Ø10	5Ø8/m	2%	d

2.4 Test Setup and Instrumentation

The beams specimens were simply supported and tested under one central point loading using a hydraulic oil jack till failure. All beams were tested with 1600 mm effective length. Two Linear Variable Displacement transducer (LVDTs) were placed on the bottom of the beam at the mid and third-span to record the deflection as shown in Fig. 4. One electrical steel strain gauge was attached at the middle of the beam span in the longitudinal direction of the tensile bars on the bottom to measure the tensile strain. Load, displacement, location and shape of cracks on the beams and mode of failure were recorded after each increment until failure.

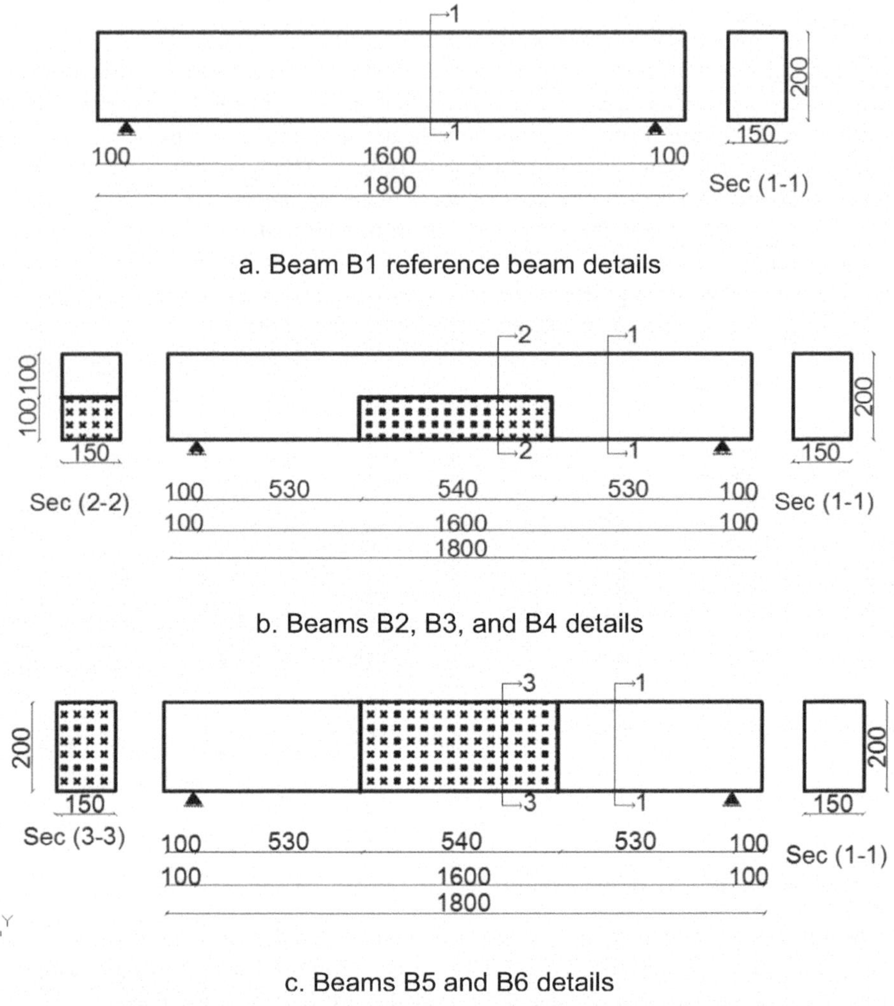

Fig. 3. Details of the steel fiber locations in the tested specimens

3 Test Results and Discussion

3.1 Crack Pattern and Failure Mode

The over-reinforced tested beams used in this study were designed to fail in brittle failure, but the observed crack patterns and failure modes for all tested beams were varying according to the various steel fibre ratios, the various distributed depth regions, as shown in Fig. 5 and Table 4. The flexural cracks were initiated and slowly propagated from the mid-span of the beam at the bottom fibers in the moment zone. By increasing the load, new cracks were created and the formed cracks propagated towards the loading point. A typical brittle failure was observed in beams (B1, B2 and B3) which could be noticed from the crashing of the compressive zone under the point load while a typical

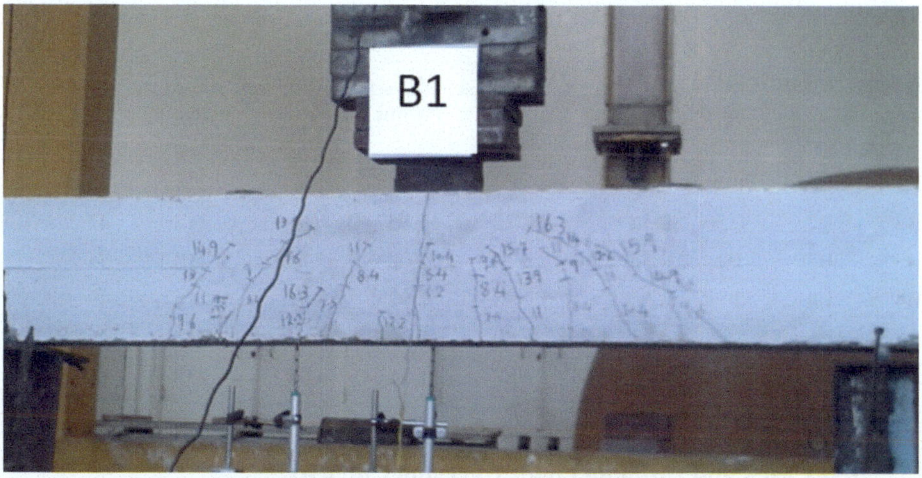

Fig. 4. Test setup and LVDT arrangement

ductile failure was observed in beams (B4, B5 and B6). The rate of crack propagation for all tested beams was smaller than that of the control beam B1 (without any steel fiber). The cracks height was greater for beam B1 more than that of all tested beams (about 0.66 of total depth (d)), the beams B3, B4 and B5 have cracks height about (0.6d), and the beams B5 and B6 have cracks height about (0.5d) as shown in the figure.

It can be concluded that the compression zone increased by increasing the steel fiber ratio and the presence of steel fiber in the concrete reduces the width and intensity of cracks.

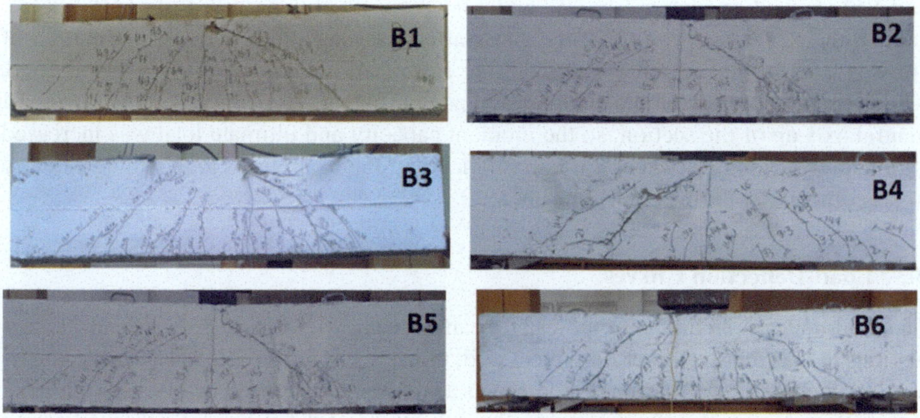

Fig. 5. Failure Mode of all tested specimen beams.

Table 4. Test results

Beams	Beams	SF ratio	SF location	P_{Crack} (KN)	Δ_{Crack} (mm)	*ΔUltimate (mm)* At 130 KN	$P_{u\,MAX}$ (KN)	Initial stiffness (KN/mm)
B1	B1	0%	Control beam	62	0.94	11.7	218.5	42.55
B2	B2	0.5%	0.5d	72	1.057	6.9	217	43.24
B3	B3	1%	0.5d	76.3	1.083	5.856	218.1	44.23
B4	B4	2%	0.5d	78.9	1.02	5.1	216	49
B5	B5	1%	d	80.1	1.05	4.5	233.7	46.48
B6	B6	2%	d	88	1.155	4.71	250	49.18

3.2 First Crack and Ultimate Loads Carrying Capacity

Loads at first cracking and failure from the experimental results are illustrated in Table 4. The effect of the steel fibre ratio at various distribution depths on the cracking and failure loads is examined. Is illustrated in Figs. 6 and 7. It can be seen from the figure that the load at the first crack is increased with the addition of steel fiber compared to control concrete beam B1. For cracking loads, Beams (B3 and B4) containing 1% and 2% steel fiber distributed at half depth showed an increase by 23% and 29% respectively compared to B1 (control beam). While for beams (B5 and B6) containing 1% and 2% steel fiber distributed at full depth showed an increase by 27% and 42% respectively compared to control beam. This may be due to the improvement in the bond of concrete component of steel fibers. The failure load for beams (B3 and B4) was not influenced considerably with the presence of steel fiber compared to control beam while for (B5 and B6) showed an increase by 7% and 14% respectively compared to control beam. In fact, the steel fibers can resist the stress after cracking stage, some redistribution of stresses was formed which approached the full plastic condition in the tension zone, while remaining elastic in the compression zone. This mechanism caused the raising of neutral axis up of the section, so the moment capacity and ultimate load was increased significantly. Using steel fiber through full depth of the beam leads to the highest failure loads compared with the control beam.

3.3 Load-Deflection Curves

The load-deflection curves for all tested beams are shown in Figs. 8, 9, 10 and 11. It is obvious that as the steel fiber ratio increased, the deflection decreased. The curve can be divided into three stages, stage (1) represents a pre-cracking portion nearly a straight line, stage (2) represents the after cracking where the slope of the curve decreases indicating a reduction in the beam stiffness until the crushing of the concrete in compression zone or tension steel yields. It is important to mention that, although the steel fiber decreased the deflection, but the behavior of beams still considered ductile. Figures 12, 13, 14 and 15 show the deflection distribution along the beam length at 100 KN represents the after cracking stage. It is observed that the reduction in the deflection was 10% and 18%

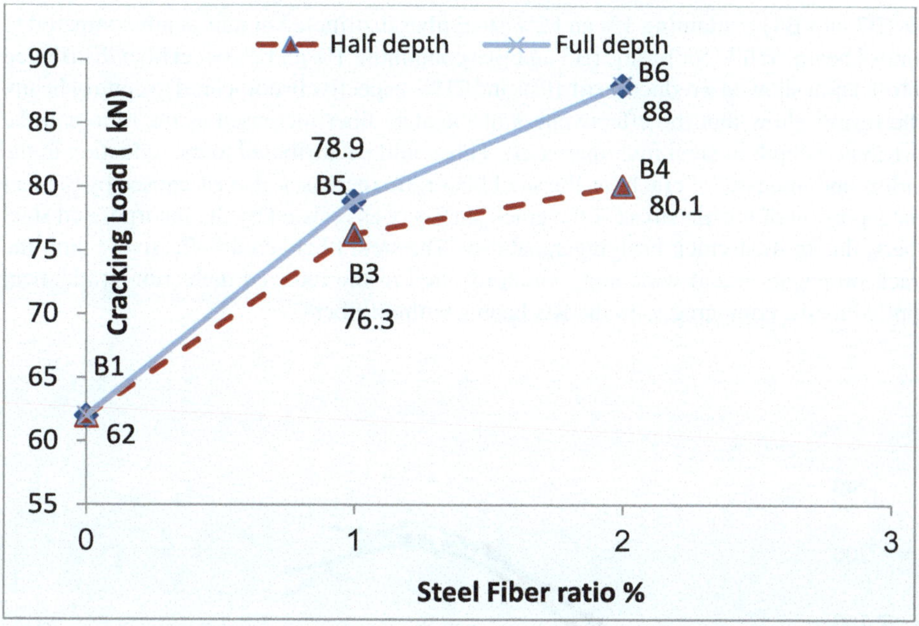

Fig. 6. The effect of steel fibre distribution depth and ratios on cracking load.

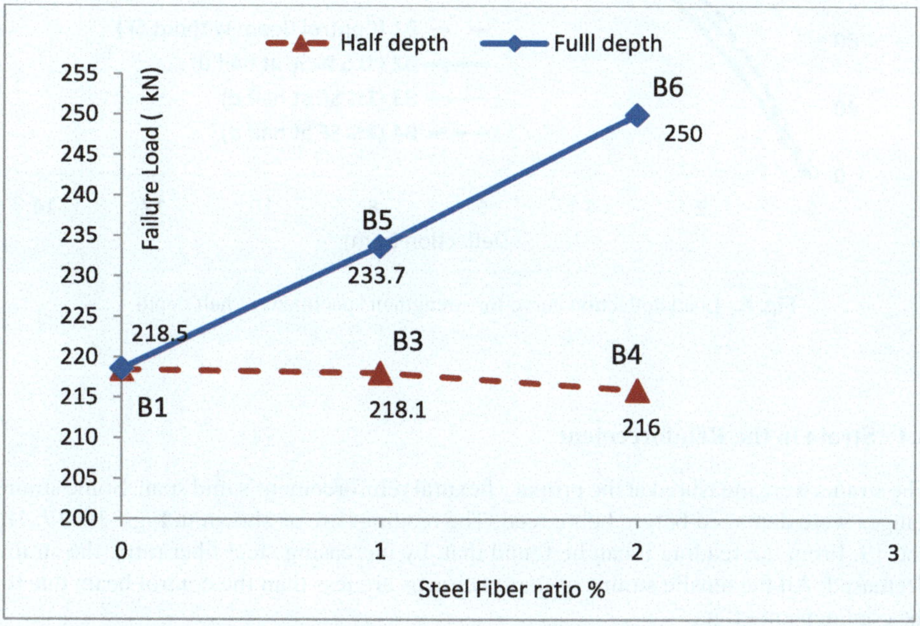

Fig. 7. The effect of steel fibre distribution depth and ratios on the failure load.

for (B3 and B4) containing 1% and 2% steel fiber distributed at half depth compared to control beam. While for beams (B5 and B6) containing 1% and 2% steel fiber distributed at full depth showed a reduction of 19% and 21% respectively compared to control beam. The results show that the effectiveness of the steel fiber increased as the ratio and the distributed depth of steel fiber increased. This could be attributed to the reduction in the widths and intensity of cracks as the steel fiber ratio increases. This is caused by the fact that a portion of tensile stress at the crack surface was resisted by the incorporated steel fibers, due to their crack bridging capability. The steel fibers could effectively limit the crack propagation and widening, which resulted in the reduced deflection at identical applied loads, compared with the RC beams without fibers.

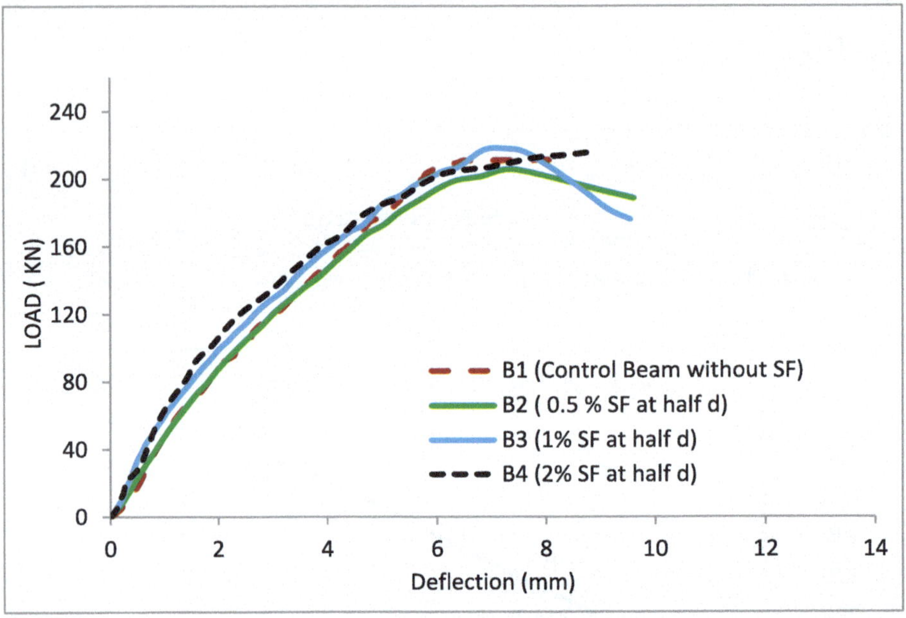

Fig. 8. Load deflection curve for strengthen specimens at half depth

3.4 Strain in the Reinforcement

The strains were measured at the primary flexural reinforcement's mid span. Some strain gauges were damaged before being read. The readings are as shown in Figs. 16, 17, 18 and 19. From the reading it can be found that, by increasing steel fiber ratio, the strain decreased. All the tensile strains of the specimens are less than the control beam due to the confinement effect.

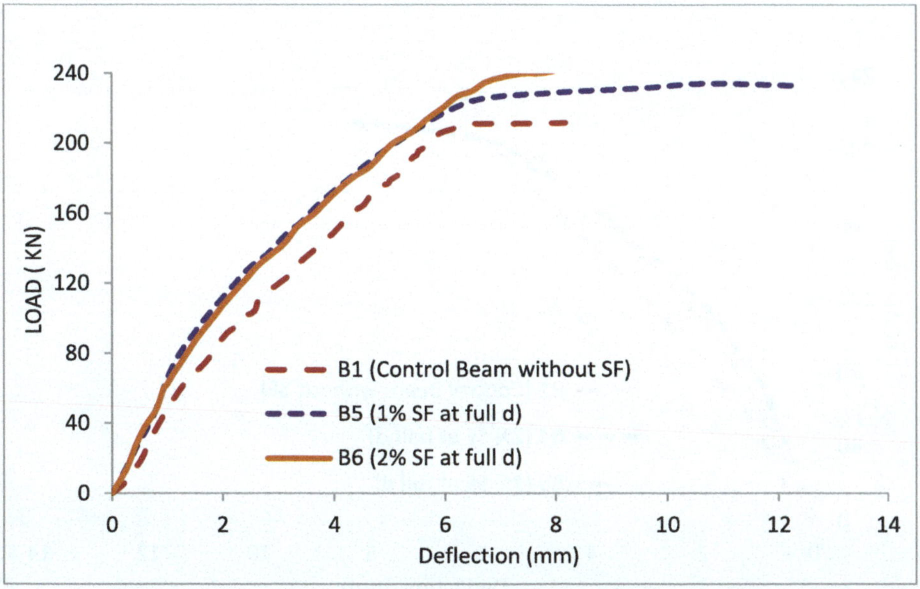

Fig. 9. Load deflection curve for strengthen specimens at full depth

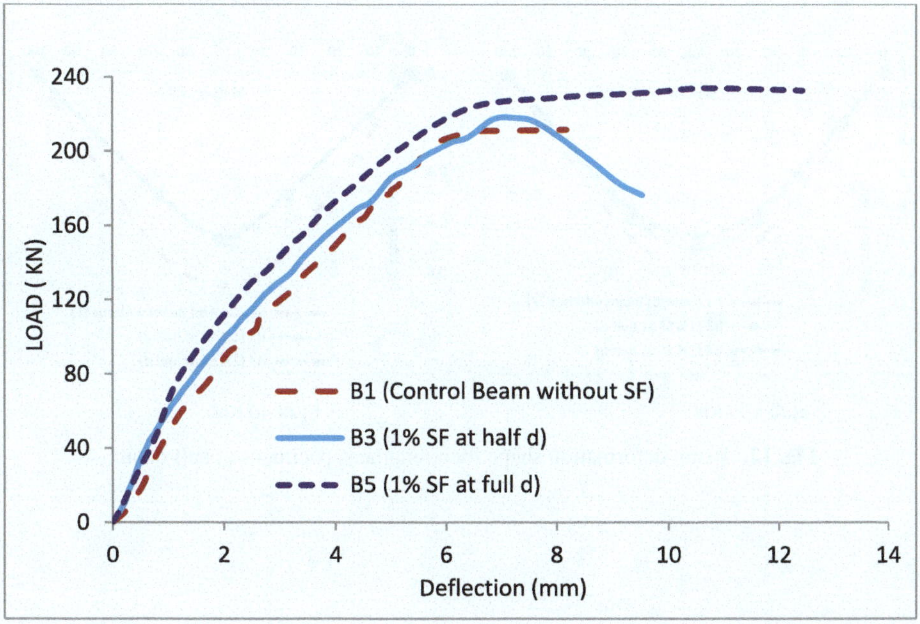

Fig. 10. Load deflection curve for 1% strengthen specimens at half and full depths

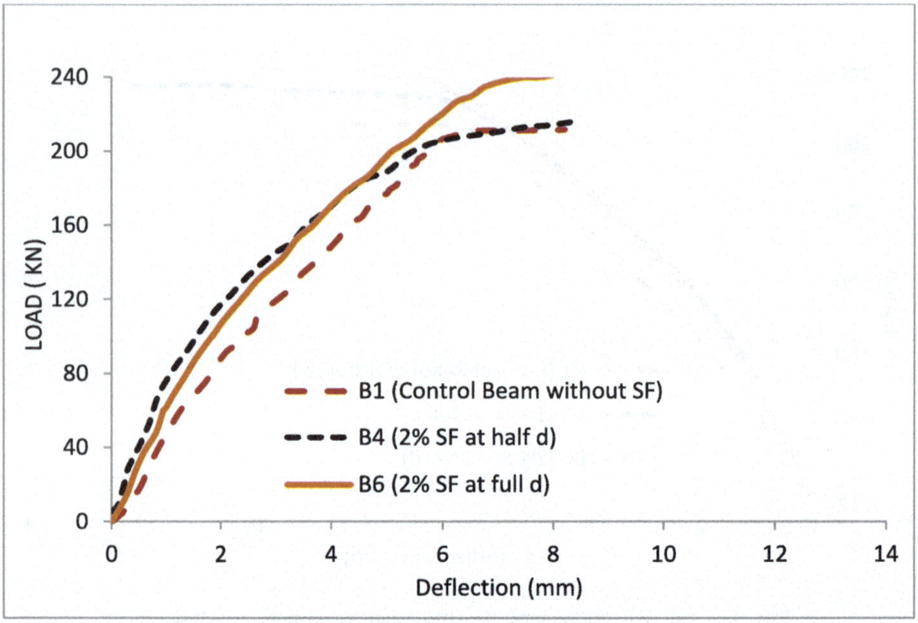

Fig. 11. Load deflection curve for 2% strengthen specimens at half and full depths

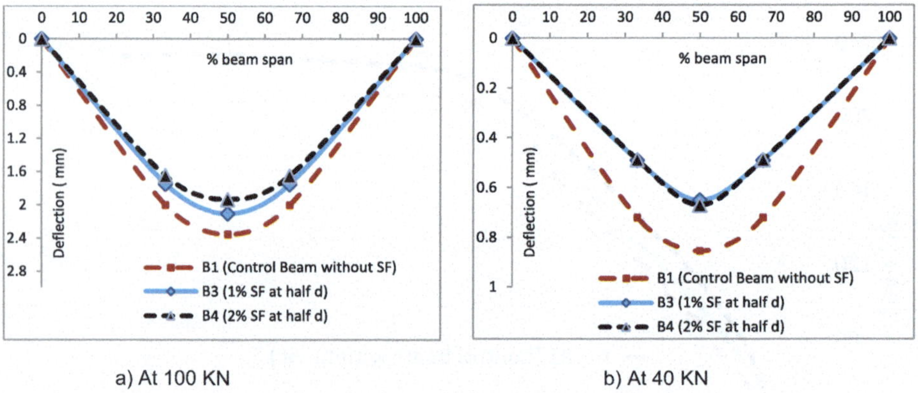

Fig. 12. Beam deformation shape for strengthen specimens at half depth

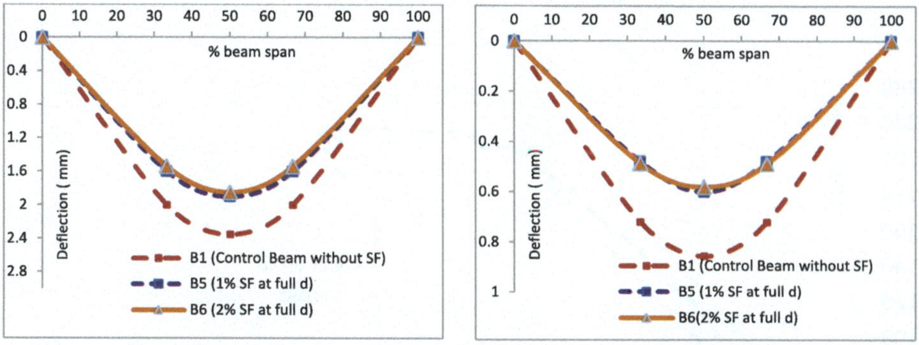

a) At 100 KN b) At 40 KN

Fig. 13. Beam deformation shape for strengthen specimens at full depth

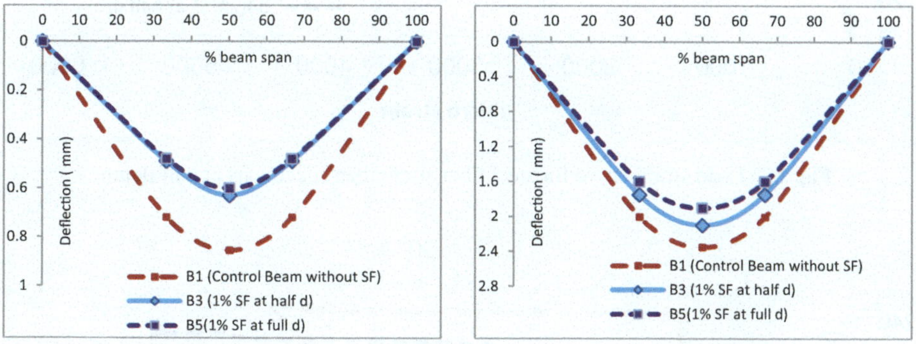

a) At 100 KN b) At 40 KN

Fig. 14. Beam deformation shape for specimens strengthen with 1% steel fiber at half and full depths

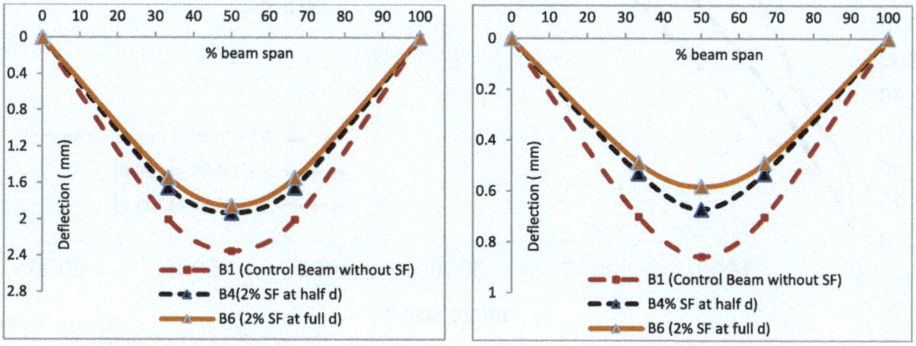

a) At 100 KN b) At 40 KN

Fig. 15. Beam deformation shape for specimens strengthen with 2% steel fiber at half and full depths

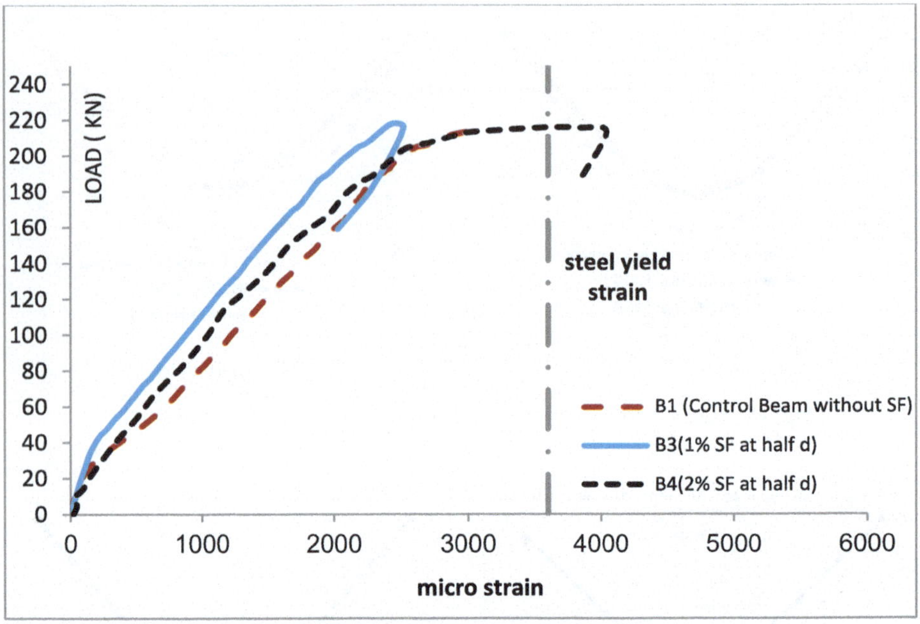

Fig. 16. Load strain curve for steel fiber strengthen specimens at half depth.

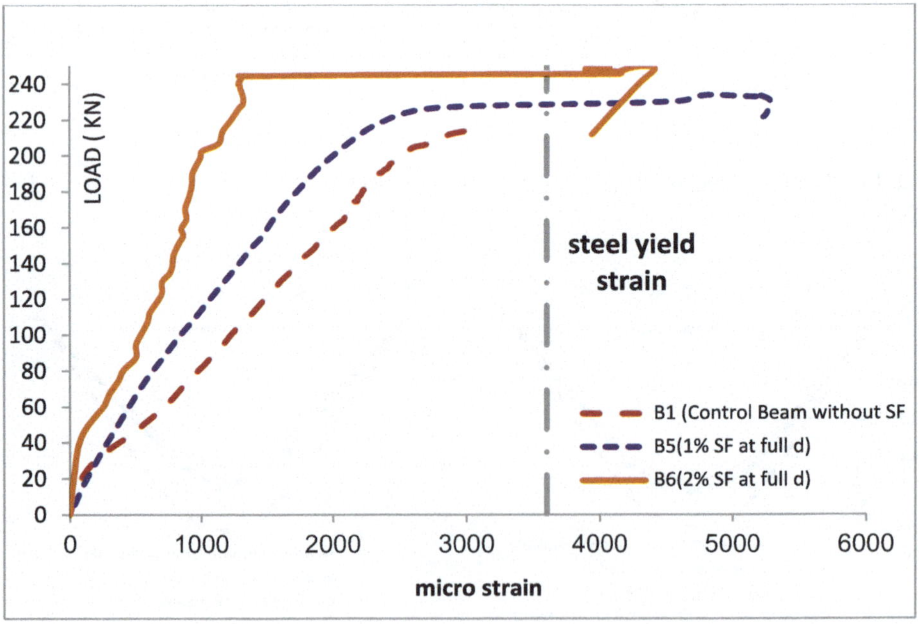

Fig. 17. Load strain curve for steel fiber strengthen specimens at full depth.

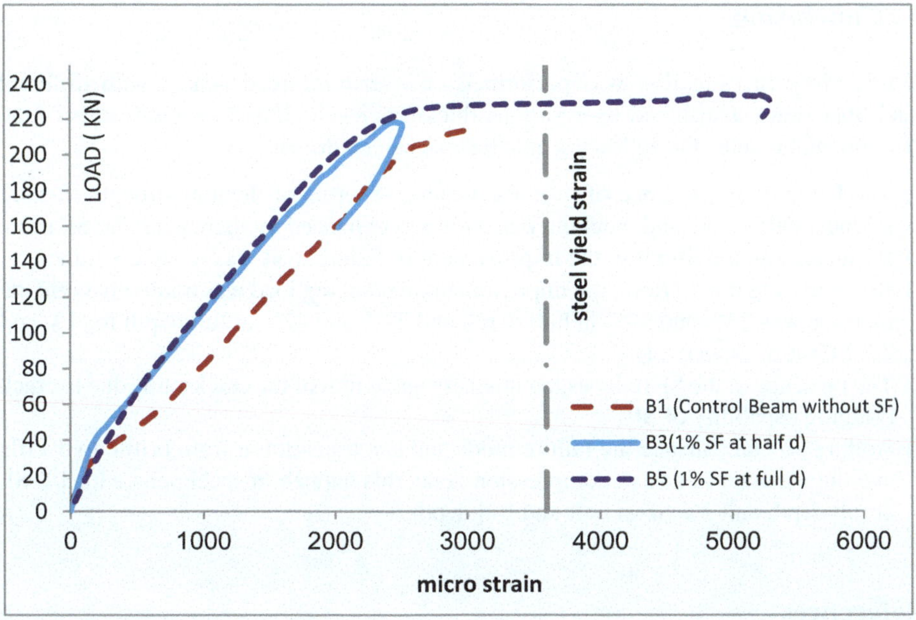

Fig. 18. Load strain curve for 1% steel fiber strengthen specimens at half and full depth.

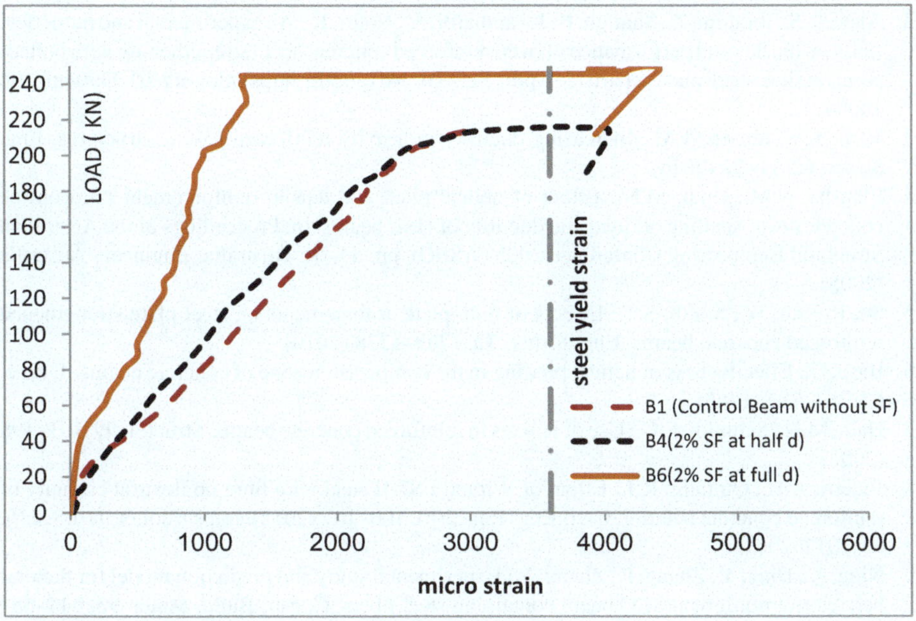

Fig. 19. Load strain curve for 2% steel fiber strengthen specimens at half and full depth.

4 Conclusion

A laboratory investigation was performed on over-reinforced beams with different steel fiber ratios distributed over various thickness levels. Based on the results of the experimental results, the following conclusions can be drawn:

1) The flexural performance of over RC beams, in terms of flexural strength, flexure stiffness deflection, and cracking behavior was enhanced by increasing the SF ratio.
2) By increasing the SF ratio, the improvement in failure load was relatively minor.
3) By increasing the SF ratio, the improvement in cracking load was relatively high; the increase was 23% and 29% at half depth and 27% and 42% at full depth for 1% and 2% SF ratio respectively.
4) The presence of the SF reduces the intensity and width of the cracks, this due to crack bridging capability of SF.
5) Adding SF can enhance the failure mode and can transition it from brittle to ductile avoiding sudden failure in compression zone, this happen in specimens with 1% SF at full depth and 2% SF at half and full depth.

References

1. Ahmed, M., Farghal, O., Nagah, A., Haridy, A.: Effect of confining method on the ductility of over-reinforced concrete beams. J. Eng. Sci. **35**, 617–633 (2007)
2. Alasadi, S., Ibrahim, Z., Shafigh, P., Javanmardi, A., Nouri, K.: An experimental and numerical study on the flexural performance of over-reinforced concrete beam strengthening with bolted-compression steel plates: part II. Appl. Sci. **10**, 94 (2020). https://doi.org/10.3390/app10010094
3. Verli, J.A., Jensen, T.M.: Increasing ductility in heavily reinforced LWAC structures. Eng. Struct. **62**, 11–22 (2014)
4. Elbasha, N.M., Hadi, M.N.: Effect of helical pitch and tensile reinforcement ratio on the concrete cover spalling off load and ductility of HSC beams. In: Proceedings of the Australian Structural Engineering Conference 2005 (ASEC), pp. 54–64. Australia: Engineers Australia (2005)
5. Su, R., Siu, W., Smith, S.T.: Effects of bolt–plate arrangements on steel plate strengthened reinforced concrete beams. Eng. Struct. **32**, 1769–1778 (2010)
6. Base, G.: Effectiveness of helical binding in the compression zone of concrete beams. J. Proc. **62**, 763–782 (1965)
7. Hadi, M.N., Schmidt, L.C.: Use of helixes in reinforced concrete beams. Struct. J. **99**, 191–198 (2002)
8. Jodeiri, A.H., Quitalig, R.J.: Effect of Wirand FS7-II steel wire fibre on flexural capacity of reinforced concrete beam. J. Civil Eng. Res. **2**(6), 100–107 (2012). https://doi.org/10.5923/j.jce.20120206.06
9. Ning, X., Ding, Y., Zhang, F., Zhang, Y.: Experimental study and prediction model for flexural behavior of reinforced SCC beam containing steel fibers. Constr. Build. Mater. **93**, 644–653 (2015)
10. Küsel, F., Kearsley, E.: Effect of steel fibres in combination with different reinforcing ratios on the performance of continuous beams. Constr. Build. Mater. **227**, 116553 (2019)

11. Padmarajaiah, S.K., Ramaswamy, A.: Finite element assessment of flexural strength of fiber reinforced high strength concrete prestressed beams. J. Cement Concrete Compos. **24**(2), 229–241 (2002)
12. Chi, Y., Xu, L., Zhang, Y.: Experimental study on hybrid fiber-reinforced concrete subjected to uniaxial compression. J. Mater. Civ. Eng. **26**(2), 211–218 (2014)
13. Khaloo, A., Molaei Raisi, E., Hosseini, P., Tahsiri, H.: Mechanical performance of self-compacting concrete reinforced with steel fibers. Constr. Build Mater. **51**, 179–186 (2014)
14. Marar, K., Eren, Ö., Celik, T.: Relationship between impact energy and compression toughness energy of high-strength fiber-reinforced concrete. Mater. Lett. **47**, 297–304 (2001)
15. Jodeiri, A., Quitalig, R.: Effect of Steel Fibre on Flexural Capacity of Reinforced Concrete Beam,|| Journal of Civil Engineering Res. **2**, 100–107 (2012
16. Dancygier, A.N., Savir, Z.: Flexural behavior of HSFRC with low reinforcement ratios. Eng. Struct. **28**(11), 1503–1512 (2006)
17. Altun, F., Haktanir, T., Ari, K.: Effects of steel fiber addition on mechanical properties of concrete and RC beams. Constr. Build. Mater. **21**(3), 654–661 (2007)
18. Ackermann, F.P., Schnell, J.: Steel fibre reinforced continuous composite slabs. Compos. Constr. Steel Concr. **VI**, 125–137 (2011)
19. Yoo, D., Moon, D.: Effect of steel fibers on the flexural behavior of RC beams with very low reinforcement ratios. Constr. Build. Mater. **188**, 237–254 (2018)
20. Mohamed, A.K., Elamary, A.S., Ahmed, M.M.: Flexural behavior of over reinforced HSC beams confined by rectangular ties. J. Eng. Sci., Assiut Univ. **36**(6), 1379–1398 (2008)
21. Rjoub, M.I.M.: Moment capacity of steel fiber reinforced concrete beams. J. Eng. Sci. Assiut Univ. **34**(2), 413–422 (2006)
22. Sofi, A., Phanikumar, B.R.: An experimental investigation on flexural behavior of fibre-reinforced pond ash-modified concrete. Ain Shams Eng. J. **6**, 1133–1142 (2015)
23. Taha, A.: El-Sayed, Flexural behavior of RC beams containing recycled industrial wastes as steel fibers. Constr. Build. Mater. **212**, 27–38 (2019)
24. Alasadi, S., Shafigh, P., Ibrahim, Z.: Experimental study on the flexural behavior of over reinforced concrete beams bolted with compression steel plate: part I. Appl. Sci. **10**, 822 (2020). https://doi.org/10.3390/app10030822
25. Mindess, S., Chen, L., Morgan, D.R.: Determination of the First-Crack Strength and Flexural Toughness of Steel Fiber-Reinforced Concrete, Elsevier Science Inc. Received September 14, 1993 ISSN 1065–7355/94/$7.00
26. Abdalkader, A., Elzaroug, O., Abubaker, F.: Flexural cracking Behavior of steel fiber reinforced concrete beams. Int. J. Sci. Technol. Res. **6**(08) (2017)
27. Ahmed, M.M., Farghal, O.A., Nagah, A.K., Haridy, A.A.: Effect of confining method on the ductility of over-reinforced concrete beams. J. Eng. Sci. Assiut Univ. **35**(3), 617–633 (2007)
28. Lam, T.Q.K., Do, T.M.D., Ngo, V.T., Nguyen, T.C., Huynh, T.P.: Numerical simulation and experiment on steel fiber concrete Beams, Modelling and Methods of Structural Analysis IOP Conf. Series: Journal of Physics: Conf. Series **1425**, 012007 (2020)

The Behavior of High-strength Geo-polymer Columns with Ferro-cement

Taha A. El-Sayed[1] and Ahmed F. Deifalla[2(✉)]

[1] Structural Engineering Department, Faculty of Engineering at Shoubra, University of Benha, Benha, Egypt

[2] Department of Structural Engineering and Construction Management, Future University in Egypt, New Cairo 11835, Egypt

ahmed.deifalla@fue.edu.eg

Abstract. The focus of the study was on the structural response of geopolymer ferrocement columns made of high-strength concrete when subjected to axial loading. Thus, an experimental program was executed. Nine geopolymer ferrocement columns, each having 150 mm width × 150 mm depth × 1600 mm height, were tested up to failure while being subjected to axial compression forcese. The investigated variables included the number of wire meshes layers and steel reinforcement. The study assessed the performance of the geopolymer columns based on the cracking load, the cracking pattern, the maximum displacement, the maximum load, the ductility index and the energy absorption index. The results showed that columns with steel welded wire-mesh performed better with respect to the maximum load, the energy absorption index, and the ductility index.

Keywords: Rice-Straw-Ash (RSA) · Axial behavior · Geopolymer Concrete (GC) · Ferrocement · Wire meshes

1 Introduction

Industrial by-products can be effectively utilized in construction to reduce costs and safely dispose of manufacturing waste [1, 13, 14]. One such example is the geo-polymer cement developed from both GGBS and alkaline fly ash, which is used in special construction concrete manufacturing [2, 3]. Geopolymer cement production requires less energy compared to OPC, which is a significant source of greenhouse gases. Where alternative concrete solutions are necessary due to OPC being not suitable; thus, geo-polymer concrete is particularly useful.

In most construction projects, OPC is the primary binder used in concrete. As a civil engineer, it is well-known that the cement industry has significant impact on the environment. The cement production generates a significant amounts of the carbon dioxide, while require high energy. Thus, resulting in the consumption of fossil fuels [4, 5]. However, when partially replacing cement with fly ash in the presence of both polymerizes and water; thus, the hydration process occurs to produce a (C-S-H) gel, where the calcium hydroxide reacts with the fly ash. The environmental impact of cement production

A. S. Mosallam et al. (Eds.): ICATH 2023, SCI, pp. 274–297, 2024.
https://doi.org/10.1007/978-3-031-76557-5_22

can be mitigated by using the fly ash, which could replace up to 60% of the cement in concrete [4].

Extensive studies has been conducted over the years to investigate the feasibility of geopolymer concrete (GC) being used in construction of buildings. Numerous studies have been carried out to examine various aspects of GC, including its production, stiffness, mix design, durability and mechanical properties. These researchs have confirmed the potential of GC as a viable alternative to traditional concrete, particularly in applications where high chemical resistance is required.

The use of GC is widely spreading, especially in industries where structures are required to be chemical-resistant, and research is being conducted to explore its potential for a variety of applications. Numerous experimental studies have been conducted in many countries across the globe in order to investigate the durability, the stiffness, the mechanical properties, the mix design, the production of GC. Various researchers have proposed suitable starting materials and methods for producing GC with desirable properties. As a result, in specific applications, GC is becoming increasingly recognized as a promising alternative to traditional concrete [11–13].

For columns and other structural elements, in the use of ferro-cement as a structural material, several studies have been conducted on. For example, Mansur and Paramasivan and Kaushik et al. [15, 16] found that ferrocement columns could be used as structural columns under both concentric and eccentric compression, where ferrocement columns were examined.

Numerous other studies have investigated the behavior of structural elements under centric and eccentric compressive load, including beams, slabs, and walls with the use of ferro-cement [17, 18]. Overall, these studies have demonstrated that, in certain applications, ferrocement as a viable alternative for traditional RC. However, the different aspect inluding and not limited to the long-term durability, the design and the construction of ferro-cement structures require much more works in order to be optimum.

The main objective of this study is to investigate the behavior HSC columns under centric compression load with different types and numbers of wire meshes, which are RSA-based geo-polymer ferro-cement. To achieve this objective, an experimental plan was designed. Nine geopolymer ferrocement columns of dimensions 150 × 150 × 1600 mm were tested under axial compression until failure. The investigated variables were the type and number of wire meshes used in the columns. The study outcome will provide valuable insights into the behavior of geopolymer ferrocement columns under centric load and to help optimize their design and construction for specific applications. By identifying the most effective types and numbers of wire meshes, this research can help provide a cost-effective and a more efficient geopolymer ferrocement structures.

2 Experimental Work

In this study, the machine used for testing columns can apply up to 500 tons and test columns with height up to 6 m. The primary objective of the testing was to determine the failure mode, the maximum load, and the maximum displacement of the columns with GC. These parameters are critical in evaluating the performance of the columns and determining whether they meet the design requirements for specific applications.

By measuring the maximum displacement and load of the columns, researchers can determine the columns stiffness and capacity. The failure mode of the columns can help identify any potential weaknesses or failure mechanisms that need to be addressed in the construction and design of geopolymer ferro-cement structures and also provide insights into the material performance under various loading conditions.

2.1 Materials-Used

1. Fine aggregate:
   The specific gravity (SG) of sand value of 2.55 and the sand bulk density (BD) value of 1780 kg/m^3.
2. Coarse aggregate:
   The crushed aggregate shown in Fig. 1 has a size of 10 mm, a SG value of 2.60, and a BD value of 1750 kg/m^3.

Fig. 1. 10 mm aggregate size

3. Recycled Rice-Straw-Ash "RSA":
   The specific surface area of the RSA has a value of 5200 cm^2/g and its SG has a value of 2.91 g/cm^2 as shown in Fig. 2.

4. For curing and mixing purposes, clean water was used.
5. Alkaline activator: Fig. 3 shows the presence of sodium hydroxide (NaOH) and sodium meta-silicate (Na_2SiO_3).

6. Reinforcments:
   Two steel types were utilized: 6 mm smooth bars grade (24/35) and 12 deformed bars grade (42/60).
7. Wire mesh:
a) Both expanded and welded wire-mesh:
   Table 1 shows the mechanical properties of the expanded and welded steel wire-meshes used in the study, while Fig. 4 displays the ferrocement mesh types.

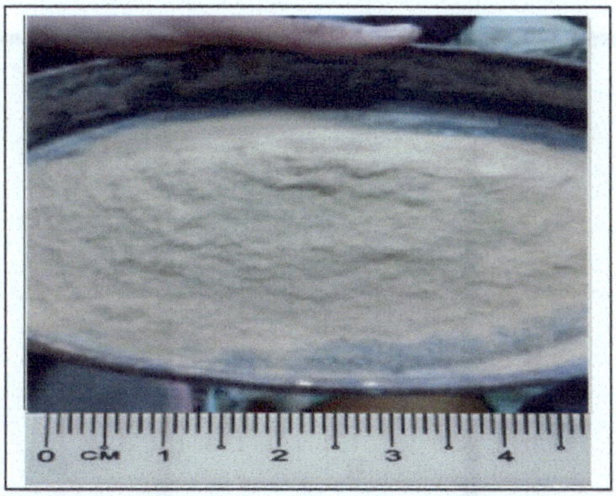

Fig. 2. Rice straw ash

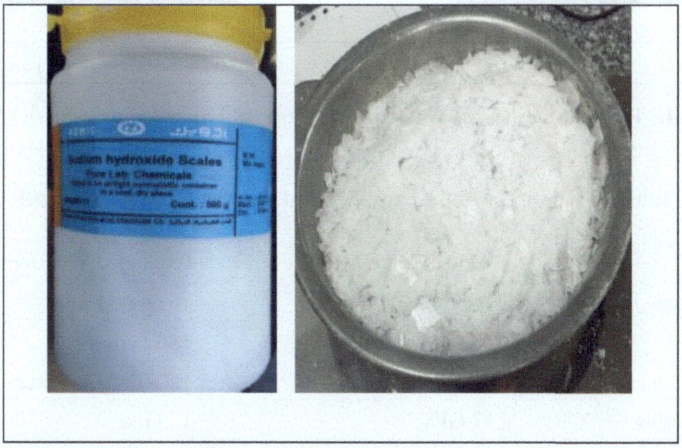

Fig. 3. Sodium-hydroxide [20]

b) Tensar meshes

The meshes depicted in Fig. 5 is constructed using high density polyethylene and is known as "Geogrid CE 121". It has a 6 × 8 mm in size, a 725 gm/m^2 in weight, a 2.04% in volume fraction, and a 3.3 mm in thickness.

c) Fiber glass meshes

The Gavazzi "V3-133-A" mesh, as seen in Fig. 6, has a 12.5 × 11.5 mm for opening dimension and a 1.66 × 0.66 mm for cross-section dimensions and 1.0 × 0.5 mm in the transverse direction. It has a 123 gm/m^2 in weight and a volume fraction value of 0.535%.

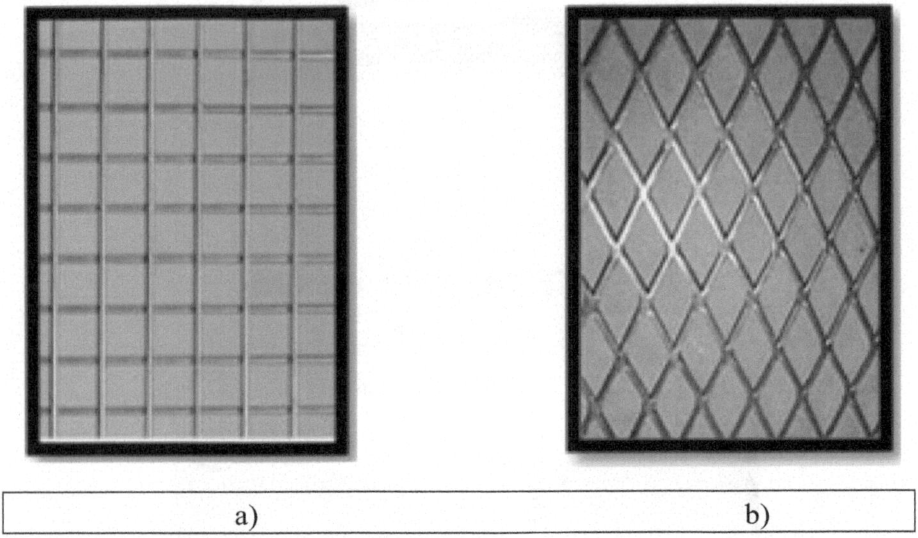

Fig. 4. Types of meshes; a) Welded wire-mesh, b) Expanded wire-mesh [20].

Table 1. The specs of the expanded and the welded wire mesh [20]

	Welded Wire Mesh	Expanded Wire mesh
Dimension size	12.5 × 12.5 mm	16.5 × 31 mm
Weight	600 gm/m^2 sheet	1660 gm/m^2
Thickness/Diameter	0.7 mm thickness	1.25 mm diameter
Young's modulus	17 GPa	12 GPa
Yield strength	400 MPa	250 MPa
Yield strain	1.17×10^{-3}	9.7×10^{-3}
Ultimate strength	600 MPa	380 MPa
Ultimate strain	58.8×10^{-3}	59.2×10^{-3}

2.2 Mix Design

A mix design was employed to create high-strength concrete (HSC) with a 60 MPa for the 28 days target strength. Table 2 outlines the design mix used to achieve this concrete strength.

2.3 Columns Specimens

The experimental work examined the behavior of columns with HSC geopolymer. Nine columns were tested in total, each with dimensions of 150 × 150 × 1600 mm and reinforced by 4φ12 steel bars. Using a machine with a capacity of 5000 kN, the columns

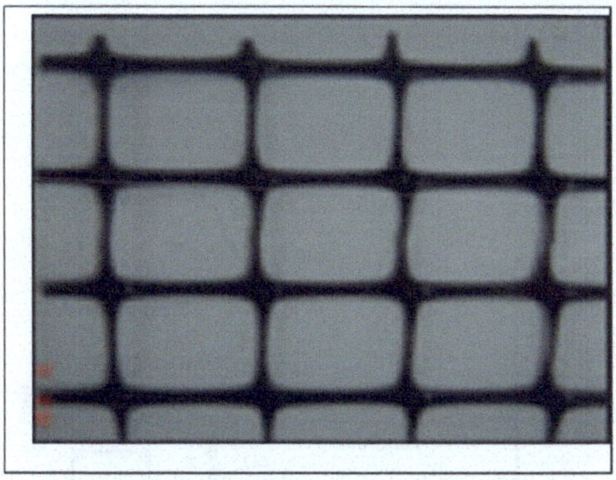

Fig. 5. Polyethylene (Tensar) mesh

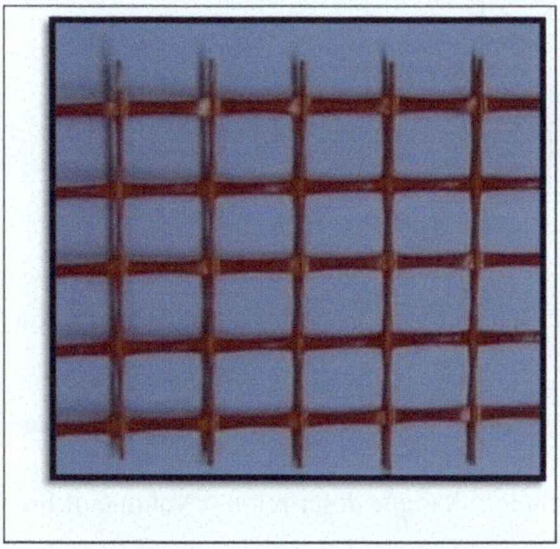

Fig. 6. Fiber glass mesh

Table 2. HSC Mix design

Item	RSA (kg/m^3)	Coarse aggregate (kg/m^3)	Fine aggregate (kg/m^3)	NaOH (kg/m^3)	Na_2SiO_3 (kg/m^3)	Water (kg/m^3)
Per m^3 of concrete	400	1150	650	50	150	47

were tested under compression. All columns were designed with a concrete mix of $f_{cu} = 60$ MPa at 28 days, and their concrete dimensions and RFT details are presented in Fig. 7. Table 3 provides information on all of the column samples, while Table 4 displays the reinforcement configurations for each of the tested columns.

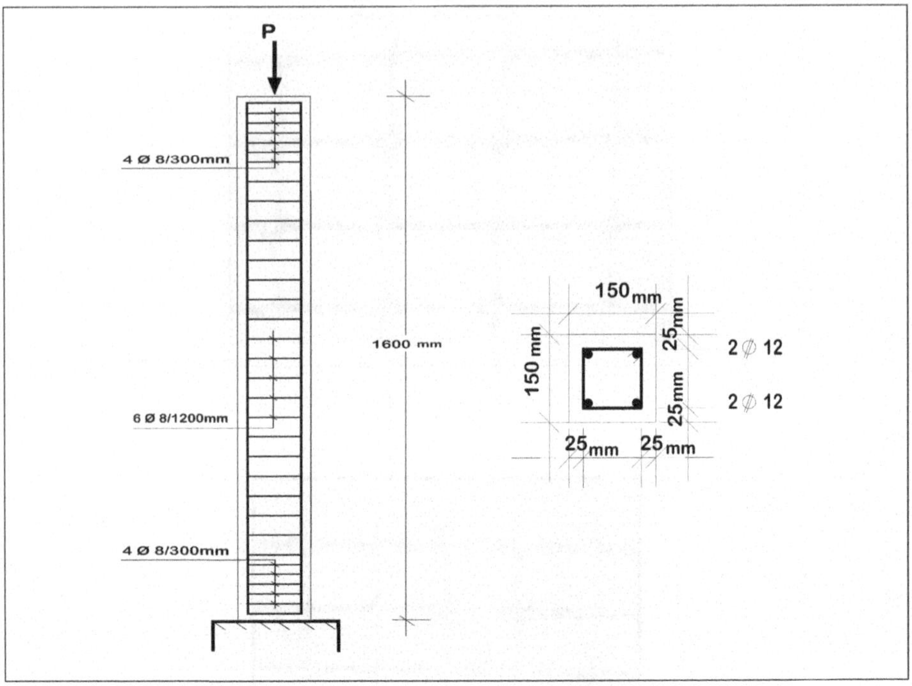

Fig. 7. The dimensions and the details of the columns.

Table 3. Explanation of the columns that were examined in the study

Series	Sample ID	Sample description	Volume of Fraction	RFT	Stirrups
Control	C1	Control	---------	4φ12	6 φ 8/’
Group A: Welded wire-mesh	C1-A	1-layer welded	0.00270	4 φ12	------
	C2-A	2-layers welded	0.00540	4 φ12	------
	C3-A	3-layers welded	0.00810	4 φ12	------
Group B: Expanded wire-mesh	C4-B	1-layer expanded	0.00753	4 φ12	------
	C5-B	2-layers expanded	0.01510	4 φ12	------
Group C: Tensar-mesh	C6-C	1-layer tensar	0.02040	4 φ12	------
Group D: Fiber glass-mesh	C7-D	1-layer fiber glass	0.00535	4 φ12	------
	C8-D	2-layers fiber glass	0.01070	4 φ12	------

2.4 Test Setup

Figure 8 depicts the typical column test setup. During the testing, LVDT were used to measure displacement, and all columns were tested until they failed.

Table 4. The reinforcement configurations of the columns that were tested in the study.

Col. ID	The reinforcement configurations of the columns that were tested in the study.
C1	
C1-A	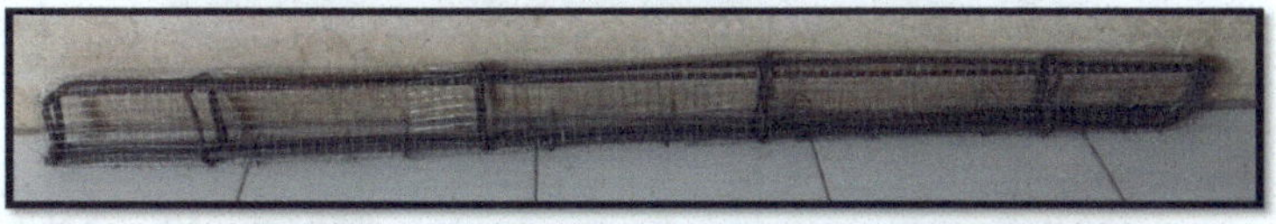
C2-A	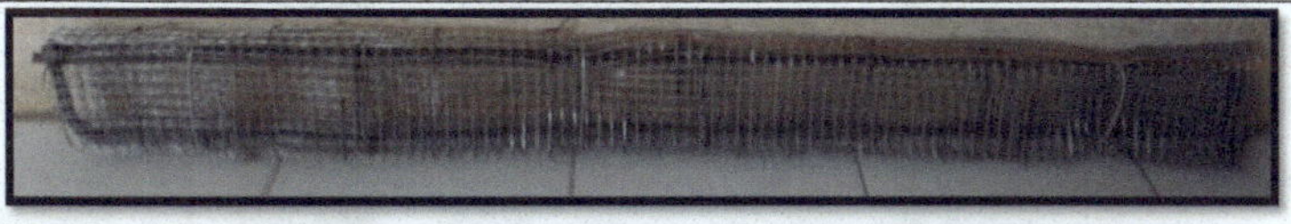
C3-A	
C4-B	
C5-B	
C6-C	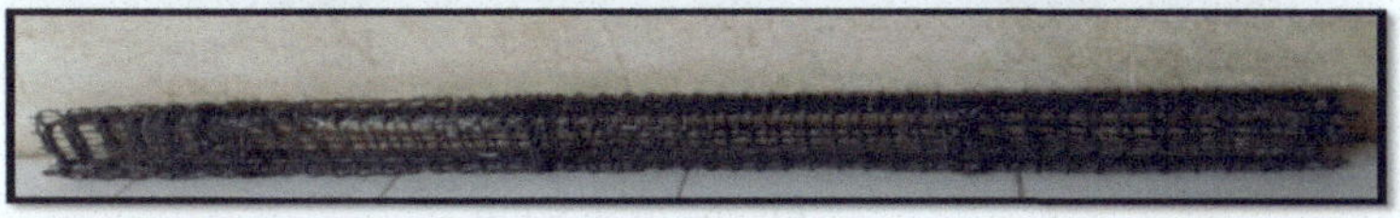
C7-D	
C8-D	

Fig. 8. Test Setup

3 The Experimental Test Results Were Analyzed and Discussed in Detail

The performance of the HSC geopolymer columns, as observed in terms of cracking behavior, failure mode, load-displacement relationship, maximum load, and maximum displacement will be discussed in detail.

3.1 Maximum Loads

Table 5 displays the tested columns maximum load. Column C1, which served as the control, had an maximum load value of 738,700 N., For columns C3-A, C2-A, and C1-A (group A), the maximum loads ranged from 771,800 N to 955,600 N. For columns C5-B and C4-B (group B) had maximum loads value of 821,700 N and 813,700 N, respectively. The use of both fibers in the modeled columns resulted in an increase in the maximum load. Column C6-C (group C) had an maximum load of 1027,200 N. Columns C8-D and C7-D (group D), the maximum loads were 841,300 N, and 793,400 N, respectively.

3.2 Maximum Displacement

For all the tested columns, Table 5 presents the values of the maximum displacement. Column C1 had a displacement of 12.02 mm. The maximum displacement at maximum load decreased from 14.50 mm to 12.72 mm for columns C3-A to C1-A, respectively, which were higher than that of the control column C1. The maximum displacement values of 16.96 mm and 15.45 mm were observed for columns C5-B and C4-B had, respectively, which is also higher than that of the control column C1. A maximum displacement value of 13.55 mm was observed for column C6-C. A maximum displacement values of 14.07 mm and 12.08 mm was observed for columns C8-D and C7-D, respectively.

Table 5. The results of the experimental tests

Col ID	First crack load (kN)	Serviceability Load (kN)	Ultimate load (kN)	Def. at first crack load (mm)	Def. at Ult. Load (mm)	Ductility ratio ---	Energy Absorption (kN.mm)
C1	296.00	460.59	738.70	3.80	12.02	3.16	6320.08
C1-A	325.00	481.27	771.80	4.30	12.72	2.96	9201.48
C2-A	380.00	525.46	842.50	4.70	14.36	3.06	9353.37
C3-A	455.00	596.15	955.60	5.20	14.50	2.79	10380.91
C4-B	443.00	507.46	813.70	5.90	15.45	2.62	10122.72
C5-B	449.00	574.96	921.70	6.30	16.96	2.69	12008.52
C6-C	515.00	640.90	1027.20	4.40	13.55	3.08	10414.93
C7-D	391.00	496.65	796.40	3.90	12.08	3.10	6983.16
C8-D	378.00	524.71	841.30	4.50	14.07	3.13	9179.36

3.3 The Load-Displacement Relationship

For the tested columns, Figs. 9, 10, 11, 12, 13, 14, 15, 16, and 17 display the relationship between the load and displacement. It is evident that the load-displacement relationship can be split into three phases:

1. Before the first onset of cracking (Elastic behavior), where a linear load-displacement relationship is observed. The slope of the load-displacement curve in this stage varies with different types of test specimens. The end of this stage is marked by a deviation from linearity.
2. In the second stage, the load-displacement curve slope changes slowly as a result of the stiffness reduction in the samples due to multiple cracking.
3. In the third stage, large plastic deformation occurs as a result of the yielding of reinforcing bars and the large extension in the reinforcing mesh of the ferrocement columns. This stage is terminated by failure of the test columns.

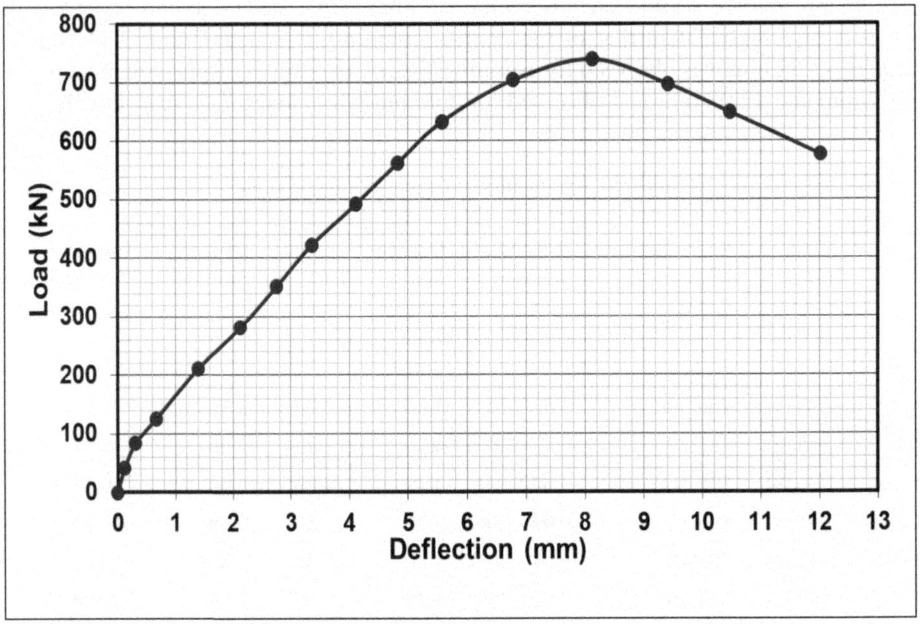

Fig. 9. C1

From Fig. 18, it can be concluded that column C6-C, which was reinforced with a single layer of tensar-mesh, had the highest first crack load and maximum load, while the control column C1 had a lower maximum load carrying capacity.

Table 5 displays the results obtained from all of the tested specimens. The table shows the first crack load, maximum load, and first crack and maximum load displacement that were obtained during testing. The ductility ratio, service load, and energy absorption were calculated from the load-displacement curve for each column sample. The first

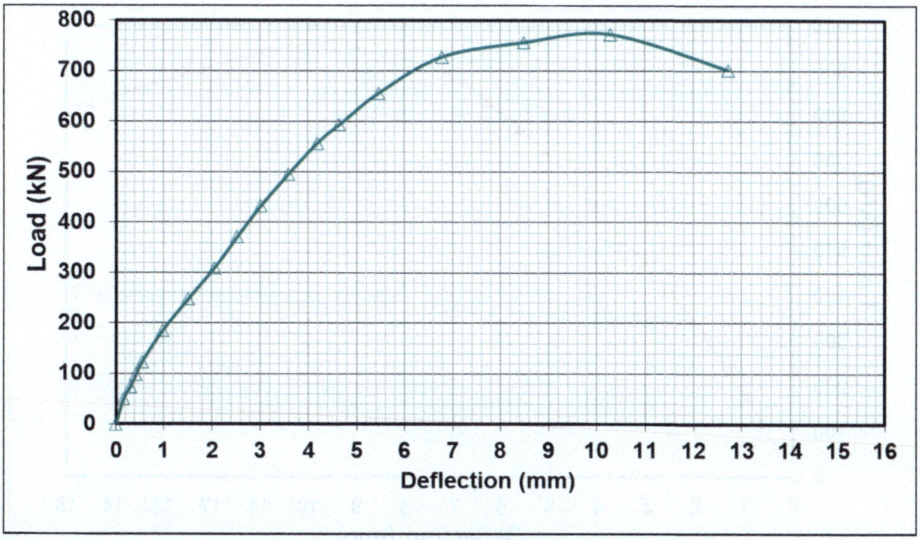

Fig. 10. C1-A

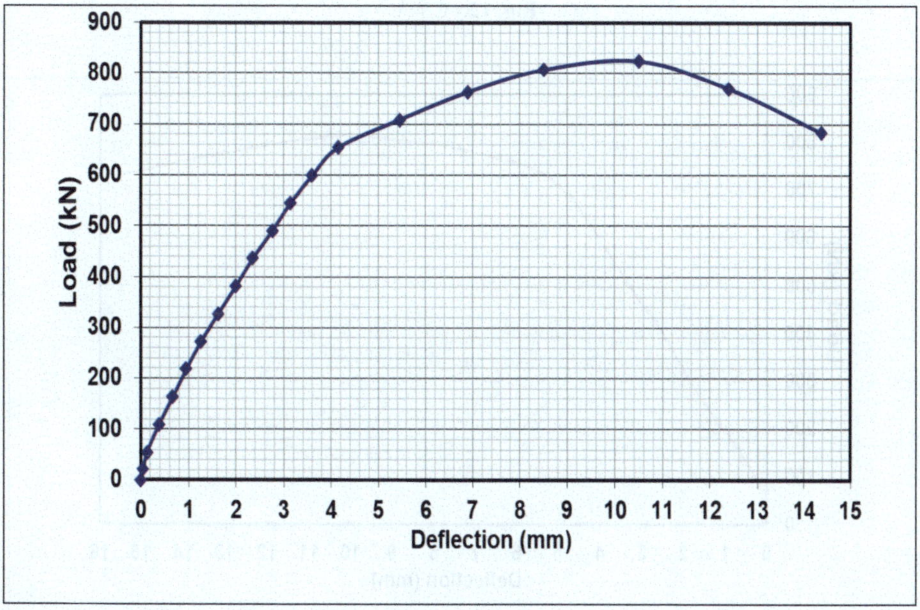

Fig. 11. C2-A

crack load was determined at the point where the load-displacement curve began to deviate from the linear relationship. The service load was calculated using Eq. 1.

$$P_{ser} = (P_{ult.} - 1.4 * D.L.)/1.6, \quad D.L. = \text{own weight of column} \quad (1)$$

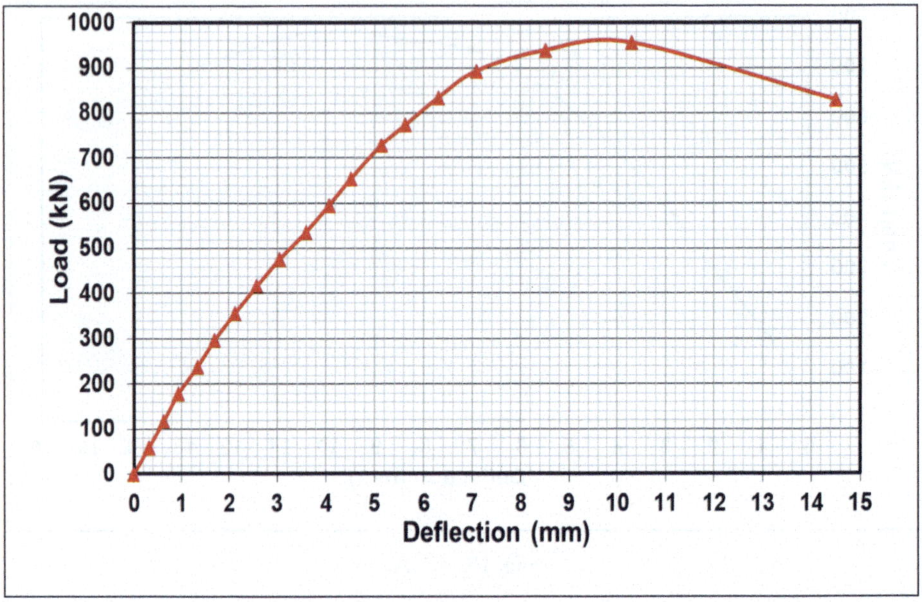

Fig. 12. C3-A

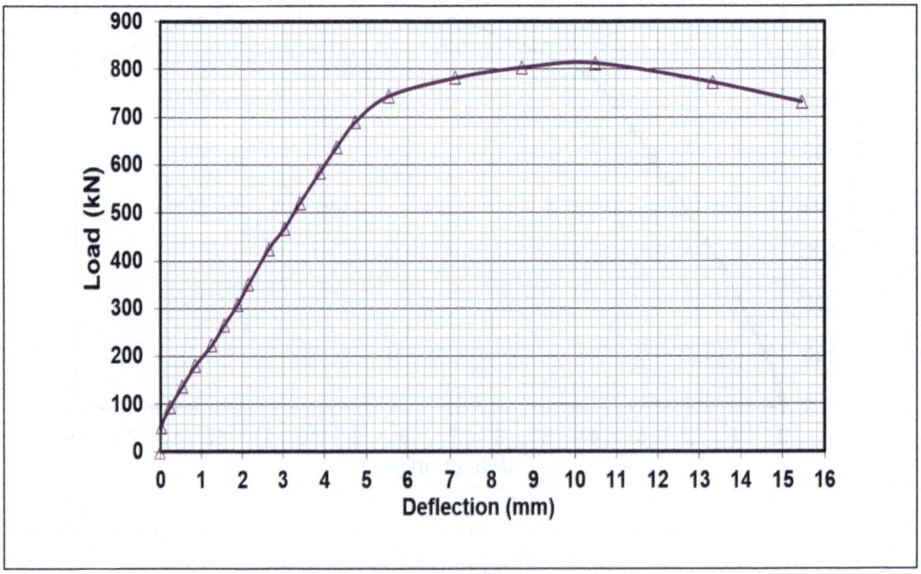

Fig. 13. C4-B

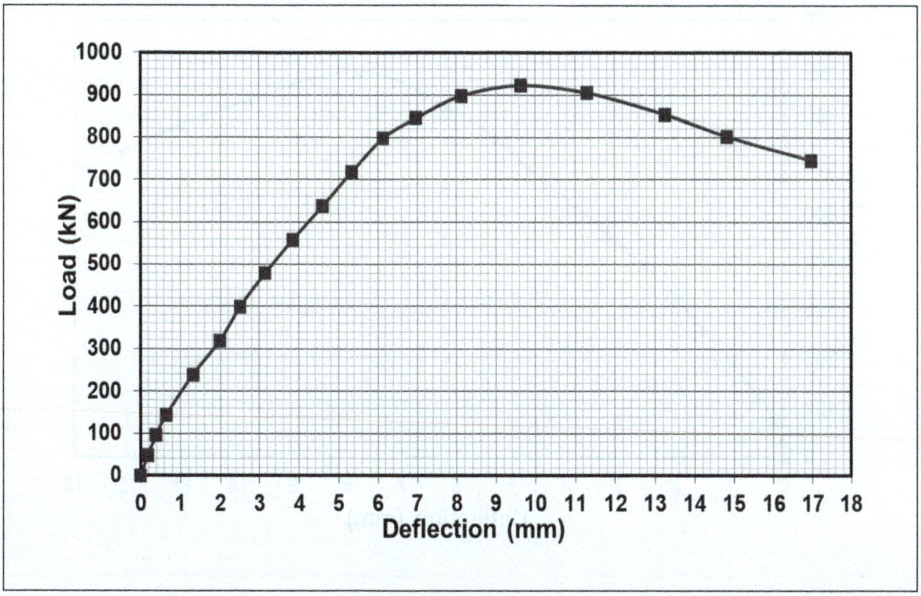

Fig. 14. C5-B

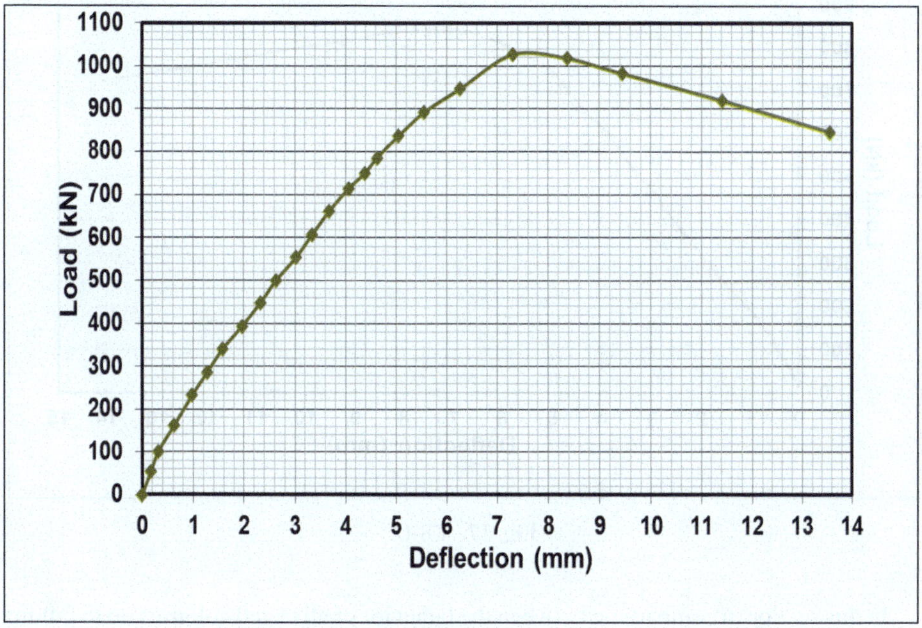

Fig. 15. C6-C

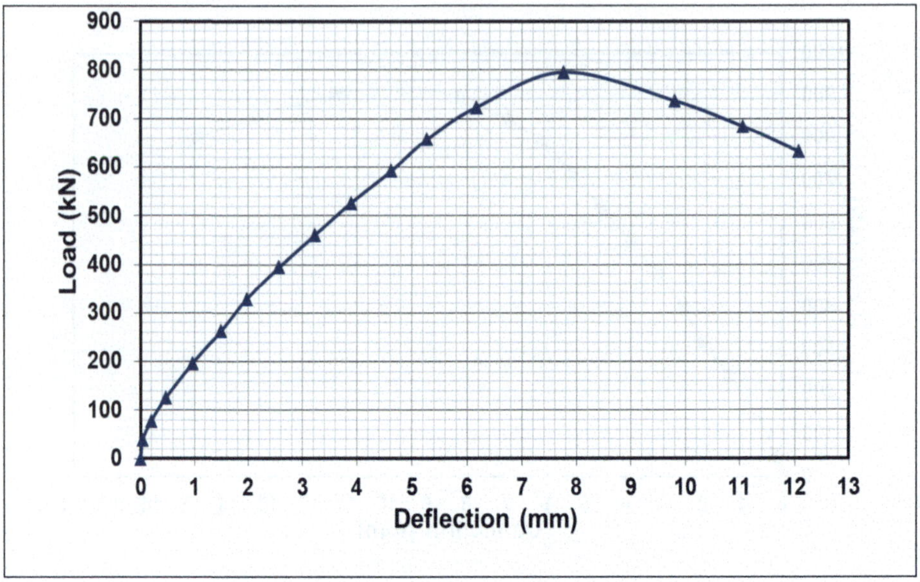

Fig. 16. C7-D

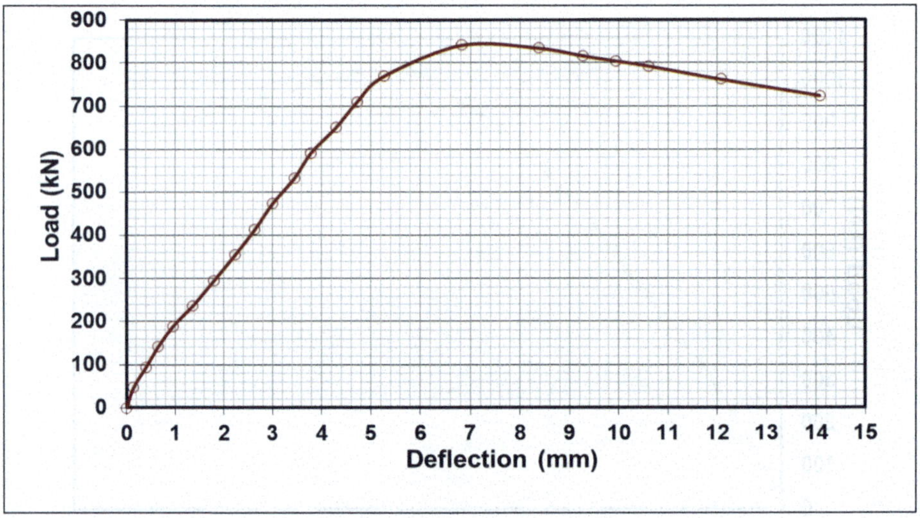

Fig. 17. C8-D

In this section, a comparison between the behavior of all tested columns up to failure, as obtained from the experimental results, is presented. The comparisons between the tested columns are shown in Figs. 19, 20, 21, 22, 23, and 24, which display the first crack load, serviceability load, maximum load, displacement at maximum load, ductility ratio, and energy absorption for all tested columns, respectively. These figures provide

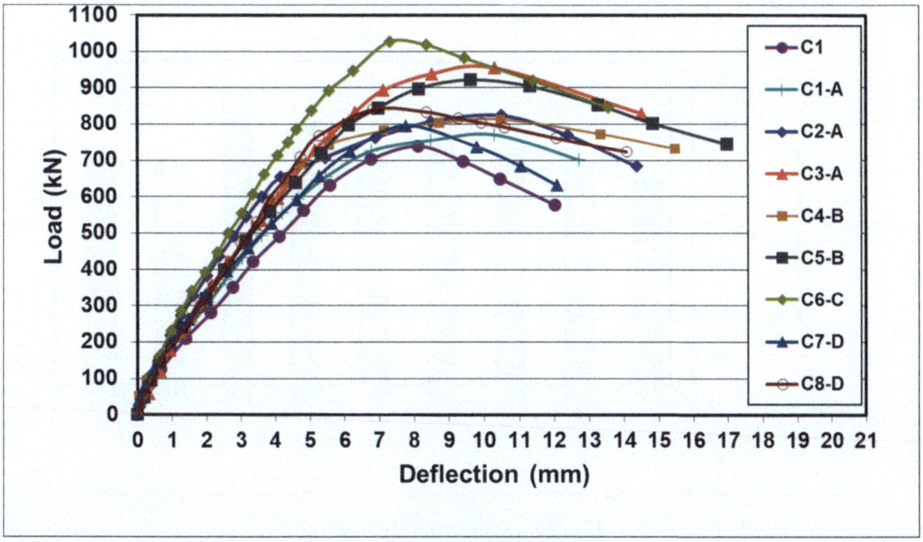

Fig. 18. Comparison between load displacement of all columns

valuable insights into the performance of the tested columns and can be used to compare the effectiveness of different reinforcement configurations.

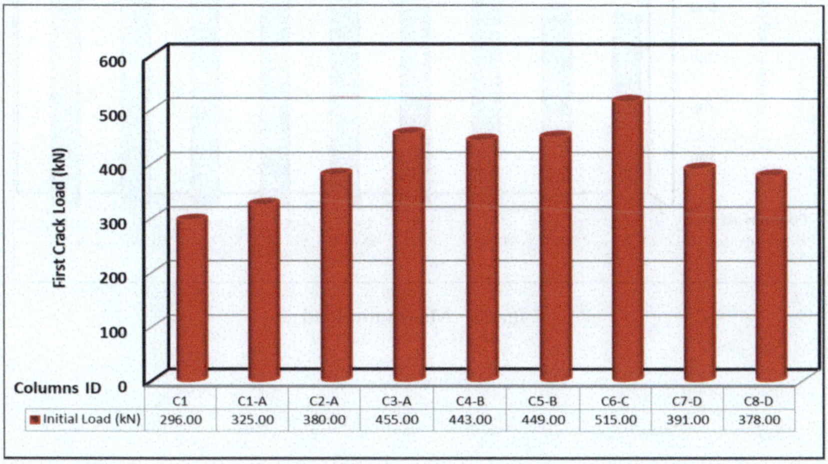

Fig. 19. First crack load

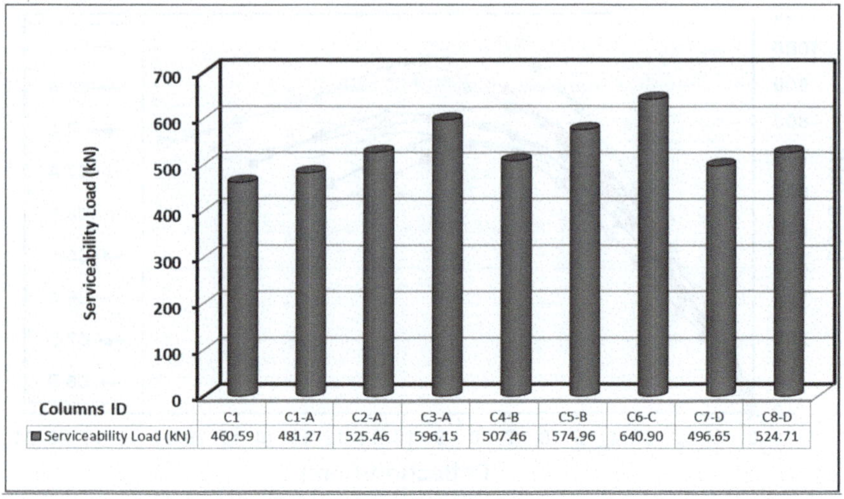

Fig. 20. Serviceability load

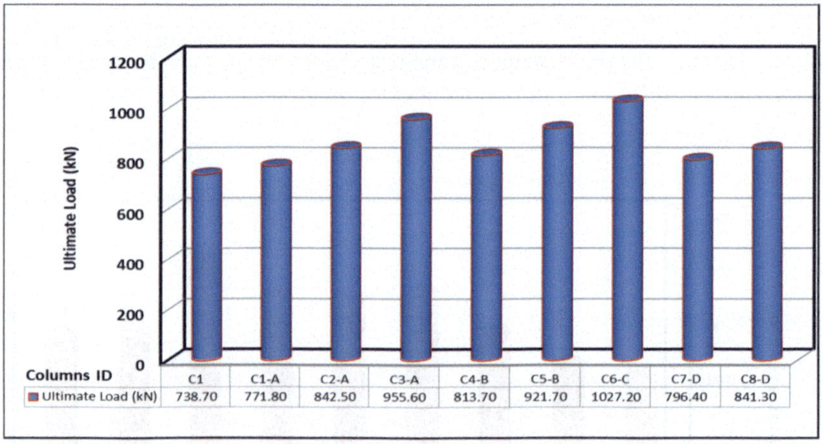

Fig. 21. Maximum load

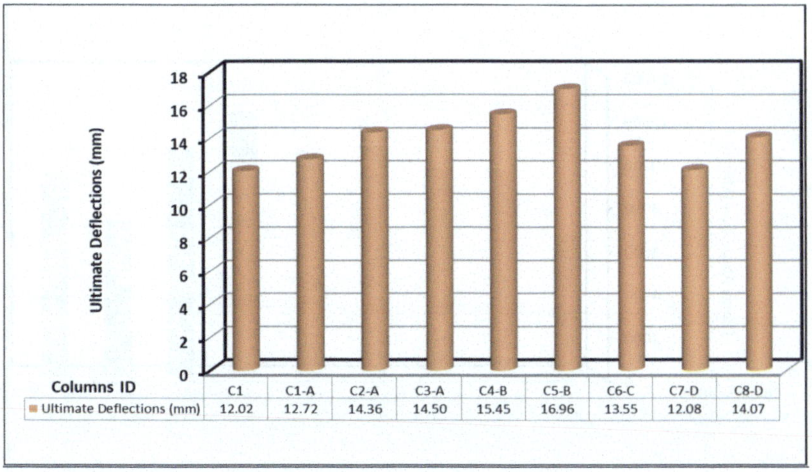

Fig. 22. Maximum displacement

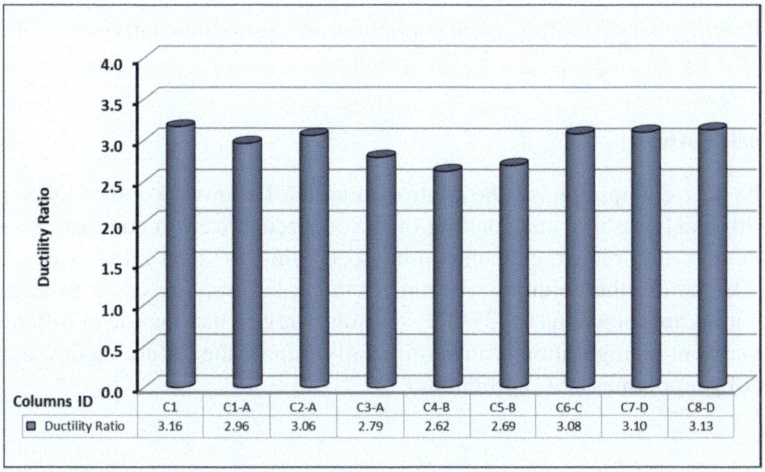

Fig. 23. Ductility ratio

3.4 Energy Absorption

Figures 24 and 25 present the energy absorption values for all of the tested columns. The energy absorption for the control column C1 was 6320.08 kN · mm. For all other columns, the energy absorption was greater than that of the control column C1, with enhancements ranging from 10% to 190%. Group D, which used fiber glass mesh, exhibited the smallest enhancement, while group C, which used tensar-mesh, exhibited the highest enhancement in terms of energy absorption. These results indicate that the

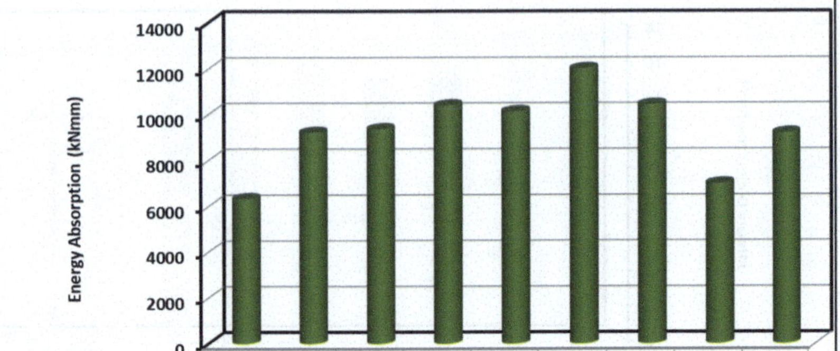

Columns ID	C1	C1-A	C2-A	C3-A	C4-B	C5-B	C6-C	C7-D	C8-D
Energy Absorption (kNmm)	6320.08	9201.48	9353.37	10380.91	10122.72	12008.52	10414.93	6983.16	9179.36

Fig. 24. Energy absorption

use of different types of reinforcement configurations can significantly impact the energy absorption capacity of geopolymer HSC columns.

3.5 Crack Pattern

During the near-collapse phase, the control column C1 showed a compression mode of failure with local crushing and spalling of the concrete cover. In contrast, for all other tested columns, after reaching the maximum load value, the load capacity decreased by 60% to 40% before the columns reached complete failure. The crack patterns for all tested columns are shown in Fig. 25. These results suggest that the use of different types of reinforcement configurations can significantly impact the failure modes and overall behavior of geopolymer HSC columns.

Fig. 25. Cracks patterns

Fig. 25. *(continued)*

C8-D

Fig. 25. *(continued)*

4 Conclusions

The conclusions might be drawn as follow:

1. Based on the experimental results, it can be concluded that Column C6-C, which was reinforced with a non-metallic, one-layer of tensar-mesh, had the highest maximum failure load among all tested columns, without any concrete cover spalling. This suggests that increasing the volume fraction of non-metallic reinforcement had a significant impact on delaying the incidence of crack development in the geopolymer HSC columns, resulting in higher corrosion protection and load-carrying capacity than those columns reinforced with metallic reinforcement. These findings highlight the potential advantages of using non-metallic reinforcement in geopolymer HSC columns, such as improved durability and sustainability, and warrant further investigation in this area.
2. The experimental results also showed that expanded or welded ferrocement geopolymer columns exhibited greater maximum failure loads than the control column C1. Furthermore, the use of expanded or welded wire mesh had a considerable effect on the maximum failure loads of the columns, with welded wire mesh exhibiting an almost 29% increase in load-carrying capacity compared to expanded wire mesh. These findings suggest that the use of ferrocement reinforcement can significantly enhance the strength and load-carrying capacity of geopolymer HSC columns, and

that the choice of wire mesh configuration can impact the performance of the column. Therefore, it is important to carefully consider the type and configuration of ferrocement reinforcement when designing geopolymer HSC columns.
3. Based on the experimental results, it can be concluded that column C7-D, which was reinforced with one layer of fiber glass mesh, had the smallest maximum failure load compared to the control column C1. This suggests that the use of fiber glass mesh as reinforcement may not be as effective in enhancing the load-carrying capacity of geopolymer HSC columns compared to other types of reinforcement, such as ferrocement or non-metallic reinforcement. However, it is important to note that the use of fiber glass mesh may still provide other benefits, such as improved corrosion resistance and durability, and further investigation is needed to fully evaluate the effectiveness of fiber glass mesh reinforcement in geopolymer HSC columns.
4. Overall, the experimental results suggest that ferrocement geopolymer columns can be successfully used as an alternative to traditional reinforced concrete (RC) columns, with potential economic and environmental benefits. These findings are particularly relevant in both developed and developing countries, where there is a growing need for sustainable and cost-effective construction materials and techniques. The use of geopolymer materials can help reduce the carbon footprint of the construction industry, and the use of ferrocement reinforcement can enhance the durability and load-carrying capacity of geopolymer HSC columns. Therefore, the established ferrocement geopolymer columns have the potential to be a viable alternative to traditional RC columns, with potential benefits in terms of sustainability, durability, and cost-effectiveness.

References

1. El-Sayed, T.A., Shaheen, Y.B., Mohamed, F.H., Abdelnaby, R.M.: Performance of ferrocement composites circular tanks as a new approach for RC tanks. Case Stud. Construct. Mater. **19**, e02228 (2023)
2. El-Sayed, T.A., Shaheen, Y.B., AbouBakr, M.M., Abdelnaby, R.M.: Behavior of ferrocement water pipes as an alternative solution for steel water pipes. Case Stud. Construct. Mater. **18**, e01806 (2023)
3. El-Sayed, T.A., Deifalla, A.F., Shaheen, Y.B., Ahmed, H.H., Youssef, A.K.: Experimental and numerical studies on flexural behavior of GGBS-based geopolymer ferrocement beams. Civil Eng. J. **9**(3), 629–653 (2023)
4. Ghareeb, K.S., Ahmed, H.E., El-Affandy, T.H., Deifalla, A.F., El-Sayed, T.A.: The novelty of using glass powder and lime powder for producing UHPSCC. Buildings **12**(5), 684 (2022)
5. El-Sayed, T.A.: Improving the performance of UHPC columns exposed to axial load and elevated temperature. Case Stud. Construct. Mater. **15**, e00748 (2021)
6. El-Sayed, T.A.: Axial compression behavior of ferrocement geopolymer Hsc columns. Polymers **13**(21), 3789 (2021)
7. El-Sayed, T.A.: Performance of porous slabs using recycled ash. Polymers **13**(19), 3319 (2021)
8. Palomo, A., Grutzeck, M., Blanco, M.: Alkali-activated fly ashes: a cement for the future. Cem. Concr. Res. **29**(8), 1323–1329 (1999)

9. Van Jaarsveld, J., Van Deventer, J., Lukey, G.: The effect of composition and temperature on the properties of fly ash-and kaolinite-based geopolymers. Chem. Eng. J. **89**(1–3), 63–73 (2002)
10. El-Sayed, T.A.: Performance of heavy weight concrete incorporating recycled rice straw ash as radiation shielding material. Prog. Nucl. Energy **135**, 103693 (2021)
11. Erfan, A.M., Abd Elnaby, R.M., Elhawary, A., El-Sayed, T.A.: Improving the compressive behavior of RC walls reinforced with ferrocement composites under centric and eccentric loading. Case Stud. Construct. Mater. **14**, e00541 (2021)
12. Bakharev, T.: Durability of geopolymer materials in sodium and magnesium sulfate solutions. Cem. Concr. Res. **35**(6), 1233–1246 (2005). [13] Bakharev, T.: Resistance of geopolymer materials to acid attack. Cem. Concr. Res. **35**(4), 658–670 (2005)
13. El-Sayed, T.A., Erfan, A.M., Abd El-Naby, R.M.: Influence of rice, wheat straw ash & rice husk ash on the properties of concrete mixes. Jokull **67**(5), 103–119 (2017)
14. El-Sayed, T.A., Erfan, A.M., Abd El-Naby, R.M.: Recycled rice & wheat straw ash as cement replacement materials. J. Eng. Res. Rep. **5**(2), 1–9 (2019)
15. El-said, A., Awad, A., El-Sayed, T.A., Özkılıç, Y.O., Deifalla, A., Tawfik, M.: Crack sliding model for non-shear FRP-reinforced slender concrete elements under shear (2023)
16. Kaushik, S.K., Prakash, A., Singh, K.K.: Inelastic buckling of ferro-cement encased columns. In: Nedwell and Swamy (eds.) Proceedings of the Fifth International Symposium on Ferrocement, pp. 327–341. E& FN Spon, London (1994)
17. Nedwell, P.J., Ramesht, M.H., Rafei-Taghanaki, S.: Investigation into the repair of short square columns using ferro-cement. In: Nedwell and Swamy (eds.) Proceedings of the Fifth International Symposium on Ferrcement, pp. 277–285. E& FN Spon, London (1994)
18. Aslam, F., AbdelMongy, M., Alzara, M., Ibrahim, T., Deifalla, A.F., Yosri, A.M.: Two-way shear resistance of FRP reinforced-concrete slabs: data and a comparative study. Polymers **14**(18), 3799 (2022)
19. Erfan, A.M., Hassan, H.E., Hatab, K.M., El-Sayed, T.A.: The flexural behavior of nano concrete and high strength concrete using GFRP. Constr. Build. Mater. **247**, 118664 (2020)
20. Erfan, A.M., Ahmed, H.H., Mina, B.A., El-Sayed, T.A.: Structural performance of eccentric ferrocement reinforced concrete columns. Nanosci. Nanotechnol. Lett. **11**(9), 1213–1225 (2019)
21. Erfan, A.M., Algashb, Y.A., El-Sayed, T.A.: Experimental and analytical behavior of HSC columns reinforced with basalt FRP bars. Int. J. Sci. Eng. Res. **10**(9), 240–260 (2019)
22. El-sayed, T.A., El-sakhawy, N.R., El-dosoky, A.I., Elkelany, M.E.A.: Methods for achievement uniform stresses distribution under the foundation. Int. J. Civil Eng. Technol. **7**, 45–66 (2016)
23. El-Sayed, T.A., Erfan, A.M.: Improving shear strength of beams using ferrocement composite. Constr. Build. Mater. **172**, 608–617 (2018)
24. E.C.P. 203/2018. Egyptian Code of Practice: Design and Construction for Reinforced Concrete Structures, Cairo (2018)

Unlocking the Path to Zero Energy Buildings: Exploring Key Determinants for Success

Elsayed Salem[1](✉), Emad Elwakil[1], and Mohamed Hegab[2]

[1] School of Construction Management, Purdue University, Indiana, USA
{esalem,eelwakil}@purdue.edu

[2] Department of Civil Engineering and Construction Management, California State University, Northridge, USA
Mhegab@csun.edu

Abstract. Numerous nations grapple with energy challenges stemming from the depletion of fossil fuel reservoirs. Concurrently, the harmful ramifications of fossil fuel utilization manifest as carbon emissions, precipitating climate change by elevating global temperature thresholds. Some countries have plans to reduce fossil energy consumption and increase renewable and clean energy production. The Building Sector consumes a lot of energy compared to other sectors. Therefore, many countries have a target to achieve Net Zero Energy Building (ZEB), which means the total energy production is equal to the total energy consumption for a given building or a group of buildings by the end of each year. This study aims to investigate different variables that could be used to reach the goal of a zero-energy building. By considering energy factors when developing the building, this goal helps to lower the overall cost of energy use. In addition, it tries to reduce carbon emissions by balancing renewable energy sources with the primary energy source. The most potential factors were collected, then filtered and classified into five groups: Building Location, Building Design, Building Operation, Energy Sources, and Cost based on the impact of each parameter. For instance, Building Design includes building spaces, occupation, and HVAC Systems. This study concluded that net ZEB design depends more on building location, area, occupation, and residents' behavior. Also, the study found that ZEB is in the developing stage and still needs a lot of work; therefore, many studies are required to cover this idea and standardize the ZEB concept.

Keywords: ZEB · ZEB target · nZEB · zero energy building · nearly zero energy building · net zero energy building · performance of nZEB · building energy · energy parameters · energy factors · low energy building · efficient energy building · sustainable building · renewable energy · green building

1 Introduction and Background

A critical problem facing countries is shortage of fossil energy. Building sector consumes more energy compared to other sectors. For instance, Commercial and residential buildings in the US, Europe, and China consume 40% of primary energy, with renewable

A. S. Mosallam et al. (Eds.): ICATH 2023, SCI, pp. 298–306, 2024.
https://doi.org/10.1007/978-3-031-76557-5_23

energy being seen as a solution to global warming, air pollution, and energy security issues. Therefore, many countries have plans to achieve ZEB target that means the total energy production equal to total energy consumption for a building or group of buildings each year. The net ZEB concept combines energy-efficient and renewable technologies to address energy and environmental issues in the construction industry, aiming to achieve net zero energy (Deng et al. 2014; Hajare and Elwakil 2020). There are several potential factors that can impact the ability to achieve zero energy building performance (Ahn et al. 2017; Alyami et al. 2021; Deng et al. 2014; Lin and Chen 2022). Some of these factors include: Building location, Building design, Renewable energy sources, Energy consumption patterns, and Budget. These are just a few of the many potential factors, therefore, it is important to carefully consider these factors when designing and constructing a zero-energy building.

1.1 Problem Statement

Due to the shortage and destructive effect of fossil energy, buildings will change significantly from conventional to energy-efficient and nearly zero-energy building (ZEB). Several studies have been done in the literature, as those by [1–4], showed that the construction industry is not ready to adopt the nZEB, especially given the high expense of retrofitting older facilities. Designing a strategy is essential for creating nZEB objectives, definitions, and ideas [1, 4–12]. Factors found in this research stands as a recommendation for possible actions that may be made to in the future construction of net-zero energy buildings.

1.2 Study Objectives

This research intends to review the status of research on performance of net ZEB and identify which factors could achieve net ZEB concept.

1.3 ZEB Definitions

The global consensus on the definition of a Zero Energy Building (ZEB) is largely based on the balance between weighted energy supply and demand. Four definitions, including net-zero site energy, net-zero source energy, net-zero energy costs, and net-zero energy emissions, are widely used worldwide [1]. Numerous attempts have been made internationally to establish a single definition of Net-Zero Energy Buildings (NZEBs), but there is no single definition, highlighting the need for a standard approach in energy performance calculation [13].

2 Research Methodology and Data Collection

The methodology applied in this study is based on an extensive review of the studies related to potential factors for achieving ZEB, as illustrated in Fig. 1 and Fig. 2.

Purdue Library, Science Direct, Scopus, and Google Scholar were the key resources used to find relevant publications for this study.

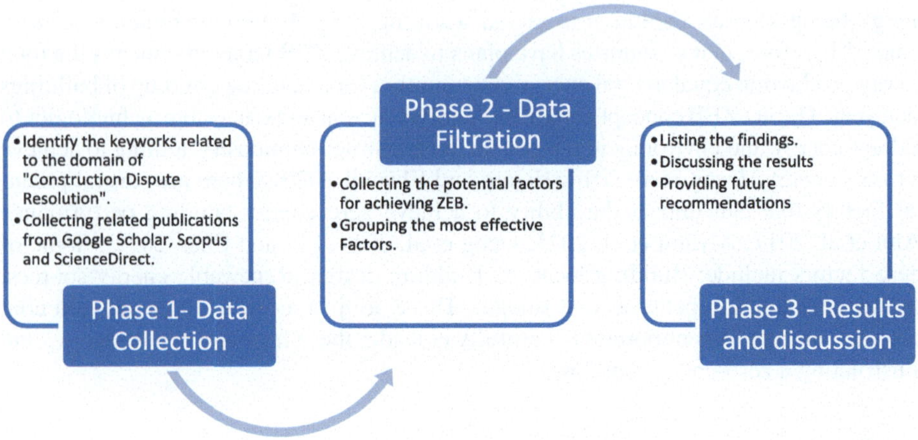

Fig. 1. Study Methodology Flowchart

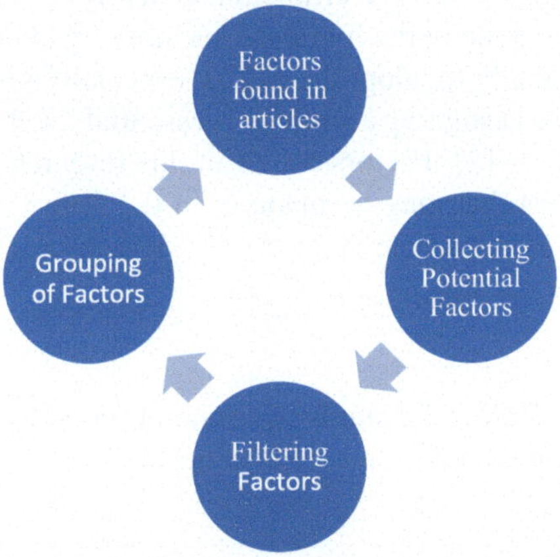

Fig. 2. The process of collecting potential factors.

3 Data Collection: Identifying the Keyworks Related to the Domain

A review of newly released research on the effectiveness of nZEB was conducted in order to understand the breadth of the issue. The Scopus database was used to collect the data. The time period was 10 years, from 2014 to 2023. The data findings were then filtered and plotted in the following figures after further processing. Each keyword's importance was determined by its frequency of use and relationship to the study's central question (Figs. 3 and 4).

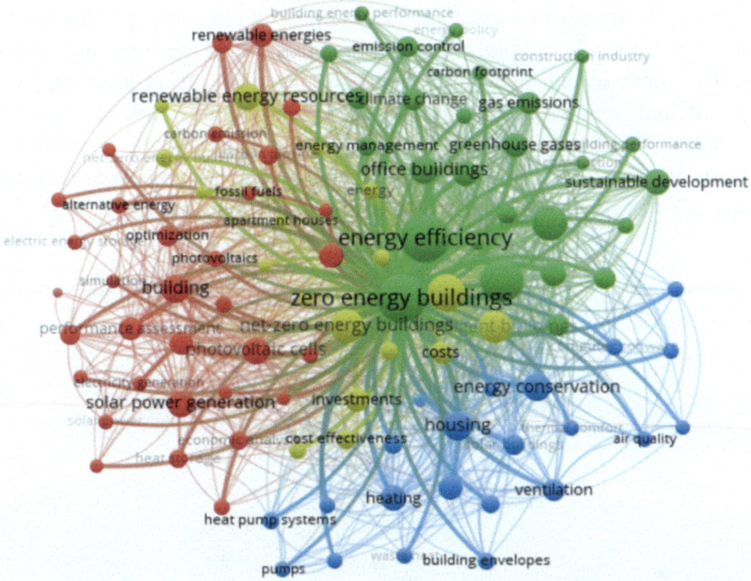

Fig. 3. Network of most used keywords in papers related to performance of nZEB in 10 Years (2014–2023). (Scopus Database)

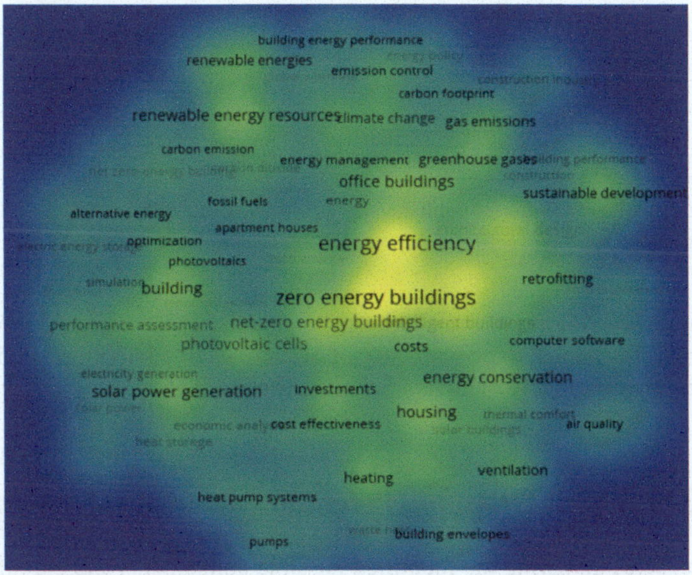

Fig. 4. Most used keywords in papers related to performance of nZEB in 10 Years (2014–2023). (Scopus Database)

At first look, the most often used keyword appeared to be energy efficiency. Significant mentions were made of apartment homes, office buildings, housing, and building envelopes. The topic of energy sources, such as solar energy production, heat pump systems, fossil fuel, renewable energy sources, photovoltaics, and alternative energy, attracted interest. Environmental aspects including greenhouse gas emissions, carbon footprint, emission management, and climate change were the subject of several research. Additionally, a lot of paperwork covered expenses and investments (Fig. 5).

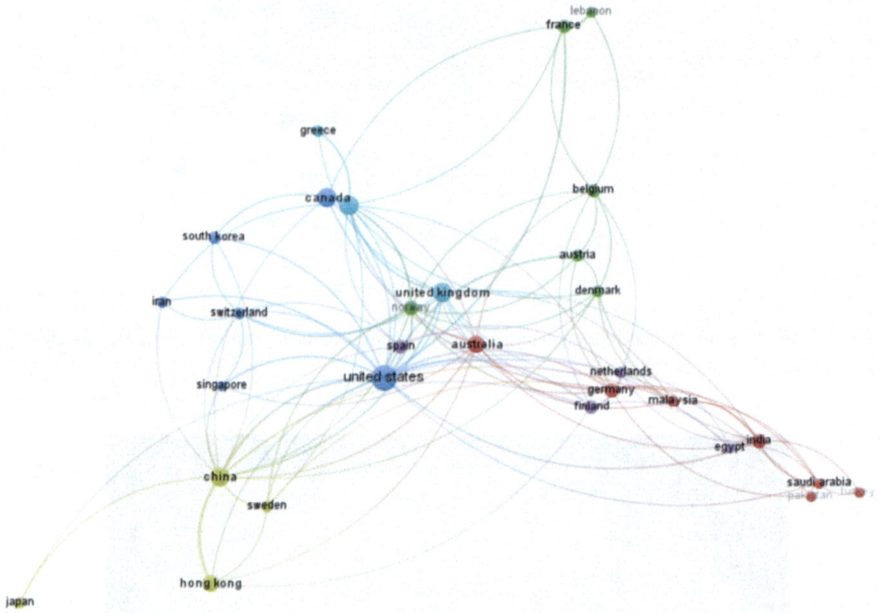

Fig. 5. Network of Top 30 countries published on performance of nZEB in 10 Years (2014–2023). (Scopus Database)

To show the evolution the publications, Scopus data base was filtered by papers including keywords related to performance of net zero energy buildings and in scope of Engineering, Energy, or Environmental Science. The resulted data are shown in Table 1, and Fig. 6.

4 Data Filteration: Grouping Potential Factors for Achieving ZEB

This stage involves gathering the identified possible components and classifying them into categories according to their point of origin, underlying causes, and outcomes as shown in Table 2.

Table 1. Number of published papers that discussed performance of net ZEB since 2003. (Scopus Database)

Year	2003	2004	2005	2006	2007	2008	2009	2010	2011	2012	2013	2014	2015	2016	2017	2018	2019	2020	2021	2022	2023
Number of Papers	1	1	3	1	5	8	9	17	53	45	51	58	76	60	84	70	94	70	83	97	42

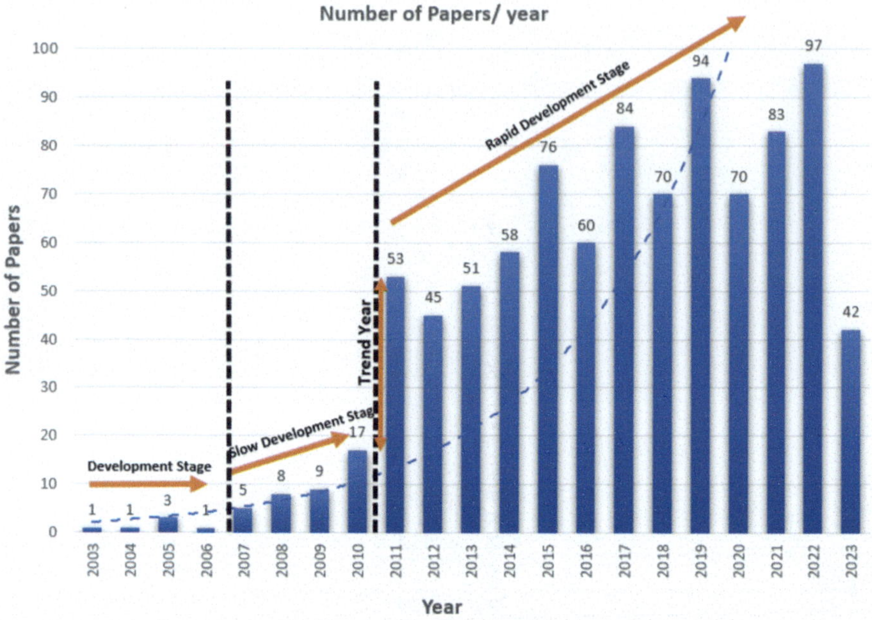

Fig. 6. Number of published papers that discussed performance of net ZEB since 2003. (Scopus Database)

Table 2. Potential Factors for achieving nZEB by category.

Group Number	Group Name	Potential Factors for achieving nZEB
1	Location	· Climate region · Wind Speed
2	Building Design	(1) Forecasting Simulation for Energy demand (2) Building Spaces (3) Building Occupation (4) Heating ventilation and air conditioning (HVAC) System
3	Building Operation	(1) Occupant behavior (2) Energy consumption patterns
4	Energy sources	(1) Wind (2) Solar Energy (3) Grid Energy (4) Earth Heat Pump (5) Energy Storage Devices
5	Cost	(1) Budget and funding

5 A Roadmap for Future nZEB Design and Operation Improvement

Many studies in literature introduced a road map for achieving nZEB target for instance, [14] proposed a road map for achieving significant energy efficiency improvements in non-residential buildings. In their study [15] introduced a road map for improving roof and façade windows in nZEB houses in Europe. For renewable energy generation, wind and PV systems are commonly used. The roof should be designed to maximize PV site area [16].

6 Conclusion

This study provided an overview of the nZEB target's status and suggested possible variables for accomplishing it. The study started out with defining the concept of nZEB and then moving on to nZEB target Boundaries. To illustrate the significance of the study, a bibliographic analysis of the literature on the performance of nZEB was then completed. At the conclusion of the investigation, the variables that emerged were categorized into 5 groups, as shown in Table 2.

The following research gaps were identified throughout this study:

- There is a lack of published research on nZEB performance in various climatic areas.
- The design of a nZEB lacks uniformity.
- Finding precise instructions for designers to follow in order to reach the nZEB objective is challenging.
- The number of studies measuring the amount of energy saved by using nZEB technology is quite small.

The concept of nZEB is currently being worked on and will need a lot of work. While implementing the nZEB idea in the construction sector, stakeholders, designers, and owners require clear rules and standard standards to follow. Since nZEB has several definitions, the limits should be established before the idea is implemented. At the design stage, the building's use should be taken into account.

This study was confined to this objective since it was primarily concerned with the performance of nZEB and was carried out to identify potential variables for reaching nZEB goals. Since there are gaps in the information currently accessible, more study is required. The necessary investigations needed to fill the research gaps may be summed up as follows:

- More research is required to standardize the nZEB.
- The influence of climatic region on nZEB achievement.
- nZEB design depending on building area and occupancy.
- Residents' actions and how they affect nZEB.
- Energy demand forecasting simulation.

References

1. Zhang, J., Zhou, N., Hinge, A., Feng, W., Zhang, S.: Governance strategies to achieve zero-energy buildings in China. Build. Res. Inf. **44**, 604–618 (2016). https://doi.org/10.1080/09613218.2016.1157345
2. Nematchoua, M.K., Orosa, J.A., Ricciardi, P., Obonyo, E., Sambatra, E.J.R., Reiter, S.: Transition to zero energy and low carbon emission in residential buildings located in tropical and temperate climates. Energies (Basel) **14** (2021). https://doi.org/10.3390/en14144253
3. Godin, K., Sapinski, J.P., Dupuis, S.: The transition to net zero energy (NZE) housing: an integrated approach to market, state, and other barriers. Clean. Respons. Consump. **3** (2021). https://doi.org/10.1016/j.clrc.2021.100043
4. Attia, S., et al.: Overview and future challenges of nearly zero energy buildings (nZEB) design in Southern Europe. Energy Build. **155**, 439–458 (2017). https://doi.org/10.1016/J.ENBUILD.2017.09.043
5. Salem, E., Elwakil, E.: The challenges of transition from traditional to net zero energy buildings. In: 2023 IEEE Conference on Technologies for Sustainability, SusTech 2023, pp. 104–110 (2023). https://doi.org/10.1109/SUSTECH57309.2023.10129611
6. Salem, E., Elwakil, E.: Minimizing the cost gap between net zero energy and conventional buildings. In: 2023 IEEE Conference on Technologies for Sustainability, SusTech 2023, pp. 49–54 (2023). https://doi.org/10.1109/SUSTECH57309.2023.10129570
7. Karlsson, A., et al.: ZenN nearly zero energy neighborhoods common barriers and challenges in current nZEB practice in Europe D.1.1. Report IVL Rapport C 38
8. Chan, A.P.C., Darko, A., Olanipekun, A.O., Ameyaw, E.E.: Critical barriers to green building technologies adoption in developing countries: the case of Ghana. J. Clean. Prod. **172**, 1067–1079 (2018). https://doi.org/10.1016/J.JCLEPRO.2017.10.235
9. Du, P., Zheng, L.Q., Xie, B.C., Mahalingam, A.: Barriers to the adoption of energy-saving technologies in the building sector: a survey study of Jing-jin-tang. China. Energy Policy. **75**, 206–216 (2014). https://doi.org/10.1016/J.ENPOL.2014.09.025
10. Scognamiglio, A., Garde, F., Røstvik, H.N.: How net zero energy buildings and cities might look like? New challenges for passive design and renewables design. Energy Procedia. **61**, 1163–1166 (2014). https://doi.org/10.1016/J.EGYPRO.2014.11.1044
11. Dadzie, J., Runeson, G., Ding, G., Bondinuba, F.K.: Barriers to adoption of sustainable technologies for energy-efficient building upgrade—semi-structured interviews. Buildings **8**, 57 (2018). https://doi.org/10.3390/BUILDINGS8040057
12. Makvandia, G., Safiuddin, M., Reda, F., Berardi, U.: Obstacles to developing net-zero energy (NZE) homes in greater toronto area (2021). https://doi.org/10.3390/buildings
13. Feng, W., et al.: A review of net zero energy buildings in hot and humid climates: experience learned from 34 case study buildings. Renew. Sustain. Energy Rev. **114** (2019). https://doi.org/10.1016/j.rser.2019.109303
14. Franco, A., Miserocchi, L., Testi, D.: Energy efficiency in shared buildings: quantification of the potential at multiple scales. Energy Rep. **9**, 84–95 (2023). https://doi.org/10.1016/J.EGYR.2022.11.142
15. Skarning, G.C.J., Hviid, C.A., Svendsen, S.: Roadmap for improving roof and façade windows in nearly zero-energy houses in Europe. Energy Build. **116**, 602–613 (2016). https://doi.org/10.1016/j.enbuild.2016.01.038
16. Lin, B., Chen, Z.: Net zero energy building evaluation, validation and reflection – a successful project application. Energy Build. **261** (2022). https://doi.org/10.1016/j.enbuild.2022.111946

Experimental Evaluation of FRP Composite Systems for Confinement of Axially Loaded Unreinforced Concrete Cylinders

Marwan Youssef[1](✉), Ashraf Abdel-Khalek Agawa[2], and Faisal Abdulaziz Almegren[3]

[1] Department of Civil Engineering and Construction Engineering Management, California State University, Long Beach, CA 90840, USA
marwan.youssef@csulb.edu

[2] Department of Civil Engineering, Valley Higher Institute for Engineering and Technology, Cairo, Egypt

[3] Department of Civil and Environmental Engineering, University of California, Irvine, CA 92697, USA

Abstract. The strength and ductility of confined concrete has been proven to far exceed that of unconfined concrete. Confined concrete is achieved by implementing means that would provide lateral pressure exerted on the concrete core in order to increase load carrying capacity and ductility. Over the years, several techniques have been developed to provide confinement for reinforced concrete (RC) axial members in the form of RC, steel and polymeric composite external jackets. This paper presents a summary of results of a multi-phase study that aims at assessing structural behavior of RC short columns confined with either E-glass/epoxy and carbon/epoxy fiber reinforced polymer (FRP) composite jackets with different thicknesses. All tests were based upon procedures of the International Code Council Evaluation Services (ICC-ES) Acceptance Criteria for Concrete and Reinforced and Unreinforced Masonry Strengthening Using Fiber Reinforced Composite Systems (AC125). The FRP composite materials used in this evaluation conform to the ICC-ES AC-85 acceptance criteria. Displacement, strains and loads were continuously monitored and recorded using a calibrated data acquisition system throughout the tests. A summary of different semi-empirical models to predict mechanical behavior of confined columns is presented with a focus on Marwan et al. universal model. A discussion on the performance of each group is presented.

Keywords: Fiber-reinforced-polymer (FRP) composites · axial members · reinforced concrete · confinement

1 Introduction

In the early 1990's, FRP composites were introduced to the construction industry as valuable alternative materials and systems. Some of the applications include repair and rehabilitation of existing structures (e.g., Mosallam et al., Ramadan et al). One of the

A. S. Mosallam et al. (Eds.): ICATH 2023, SCI, pp. 307–317, 2024.
https://doi.org/10.1007/978-3-031-76557-5_24

successful pioneering application is the use of FRP composite jackets for enhancing both strength and ductility of deficient reinforced concrete columns in both buildings and bridges. A number of studies have been published validating effectiveness of this technique (e.g., Youseff et al., Haroun et al., and others). In this paper, results of an experimental investigation assessing the effectiveness of using two types of polymeric composite jackets to improve both strength and ductility of unreinforced short axial concrete members is presented.

2 Descriptions of Experimental Program

This section describes the experimental program conducted in this study. All compression tests were performed using a calibrated ACCU-TEK 225-ton compression tester, Model CT7500 (refer to Fig. 1).

2.1 Specimen Preparation

A total of twenty-four standard 152.4 mm × 304.80 mm ASTM C39 concrete cylinders were poured and tested to failure in order to evaluate the effect of applying both carbon/epoxy (CFRP), and E-glass/epoxy (GFRP) composite jackets, in the hoop direction, in enhancing the confinement and increasing the axial loading capacity. All cylinders were allowed to cure for 28 days prior applications of FRP composite jackets. For confined specimens, a thin coat of low-viscosity epoxy primer was applied after careful surface preparation. The appropriate numbers of saturated carbon/epoxy (CFRP) laminates or E-glass/epoxy (GFRP) were then applied, with a an overlap splice of 101.60 mm. The cylinders confined by carbon/epoxy laminates were wrapped using 1, 2 and 3 plies, while specimens confined with the E-glass/epoxy laminates were wrapped with 2, 3 and 4 plies. For each type of FRP composite jacket, three control (as-built) cylinders were tested. For all specimens both the top and bottom surfaces were capped by sulfur-capping compound to ensure a leveled surface to minimize eccentricity.

2.2 Test Matrix

Table 1 shows the test matrix of the tested specimen. As mentioned earlier, a total of eighteen standard concrete cylinders were confined with carbon/epoxy and E-glass/epoxy laminates. In addition, a total of six control specimens were tested up to failure; three for the carbon/epoxy and three for the E-glass/epoxy jacketed specimens for comparison.

2.3 Instrumentation

Each specimen was instrumented using rosette strain gauges at mid height of the specimen to measure the axial strain as well as the horizontal (hoop) strain in the jacket. Two LVDTs, 180-degree apart were attached to each specimen to measure the axial deformation. The load was measured using a 690-kPa calibrated pressure transducer. A computerized data acquisition system was used to collect the test both the displacement and the strain data.

Table 1. Test Program of Specimens

Specimen Cross Section	No. of Control Specimens	No. of Jacketed Specimens					
		No. of Plies of CFRP Composites			No. of Plies of GFRP Composites		
		1	2	3	2	3	4
Cylinder (152 mm × 305 mm)	6	3	3	3	3	3	3

2.4 Test Description

The test was conducted by first setting the specimen in the test machine centered under the compression cylinder. The wires for the strain gages and the LVDTs were attached to the data acquisition system. The LVDT's were then zeroed out in preparation to start the test, then the load was applied at a loading rate equal to 2,000 lb per second (8.9 kN/s). Figure 1 shows the test setup for a 4-ply GFRP jacketed specimen. The test followed ASTM C39 standard.

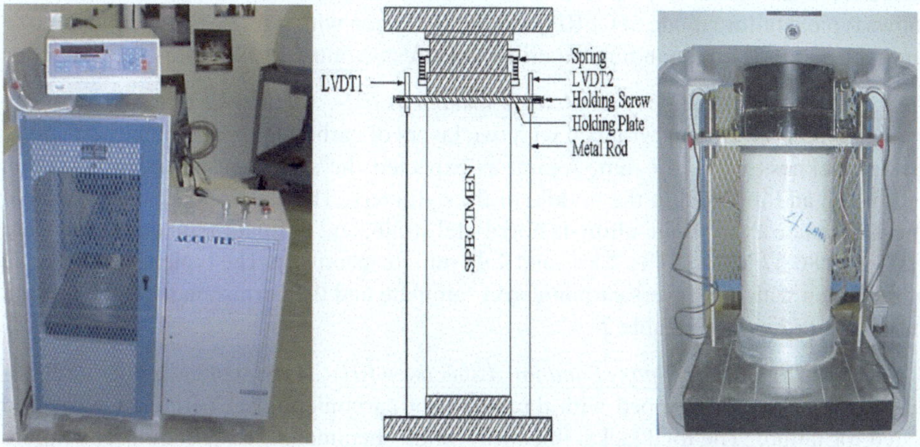

Fig. 1. Typical Test Setup

3 Experimental Results

3.1 Control Specimens

As stated earlier, six control specimens were used to determine the baseline value for comparison purposes. Three of which were poured at the same time as those wrapped by the carbon/Epoxy laminates. The remaining three were poured with those wrapped by E-glass/ epoxy laminates. The average 28-day compressive strength of the concrete used for all short cylinder specimens, f'_c, was 38.28 MPa. The results from the three specimens were averaged out to get representative value. The failure of the control specimens was noticed to be a shear failure. The average ultimate axial strain of control specimens was 0.0056 mm/mm. The stress-strain curve and the associated failure modes of representing samples of control (as-built) specimens are presented in Table 2.

3.2 Experimental Results of Carbon/Epoxy (CFRP) Retrofit System

As stated earlier, specimens confined with different numbers of CFRP plies applied at the hoop directions were tested (refer to Table 1). The goal is to increase the confinement ratio of the specimens and to study the behavior improvement represented by increased axial load, axial strain, and horizontal strain, because of applying an external composite jacket system.

3.2.1 Single Ply Carbon/Epoxy Laminated Specimens $[0]_1$: A single layer of carbon/epoxy laminate was used to wrap three specimens. The thickness of the materials was 0.584 mm. The specimens failed in a sudden and brittle rupture mode that was initiated at the mid-height of the specimen where the jacket ruptured. The failure load, ultimate axial stress, ultimate axial strain, ultimate horizontal strain, and ultimate axial deflection were 1,103 kN, 60.48 kN, 2.5%, 2.3%, and 4.80 mm, respectively. Table 2 shows typical failure mode of CFRP jacketed specimen with a single layer carbon/epoxy laminate and the corresponding experimental stress-strain curves for this specimen.

3.2.2 Two-Ply Carbon/Epoxy Laminated Specimens $[0]_2$: Another set of tested cylinders included three specimens wrapped with two layers of carbon/epoxy laminates. The total jacket thickness was 1.17 mm. Again, as expected the specimen failure was sudden and brittle and initiated in the middle of the cylinders. The failure load, ultimate axial stress, ultimate axial strain, ultimate horizontal strain, and ultimate axial deflection were 1,579 kN, 86.57 MPa, 2.7%, 2.1%, and 4.93 mm, respectively. The typical failure mode of specimen with two layers carbon/epoxy laminate and the corresponding stress-strain curves are presented in Table 2.

3.2.3 Three-Ply Carbon/Epoxy Laminated Specimen $[0]_3$: Three standard concrete cylinder specimens were wrapped with three plies of carbon/epoxy (CFRP) laminates and tested for failure. The total jacket thickness for all specimens was 1.75 mm. The failure mode was sudden and brittle and was initiated at the mid-height of the specimens. Due to the increase in the number of layers of the composite materials, a cracking noise was heard prior to the rupture. The failure load, ultimate axial stress, ultimate axial strain, ultimate horizontal strain, and ultimate axial deflection were 2,028.3 kN, 136.64 MPa, 3.9%, 2.36%, and 8.737 mm, respectively. Table 2 shows the failure mode of jacketed specimen with three layers carbon/epoxy laminates along with the stress-strain relations for the axial and the hoop directions.

Table 2. Experimental Results Summary for Control and CFRP Jacketed Specimens (***1 ksi = 6.89 MPa***)

Specimen's Description	Failure Mode	Stress-Strain Curve
Control Specimens		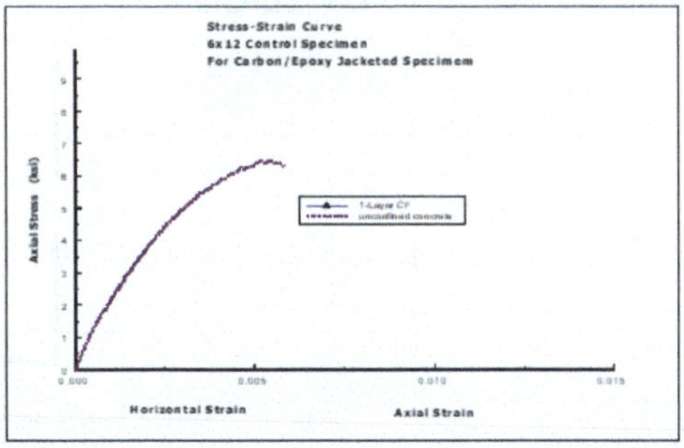
A Single-Ply CFRP Jacket Specimen $[0^{\circ}]_1$		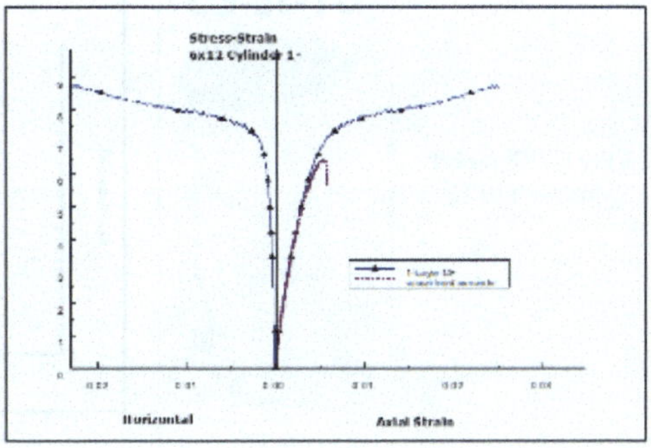
2-Ply CFRP Jacket Specimen $[0^{\circ}]_2$		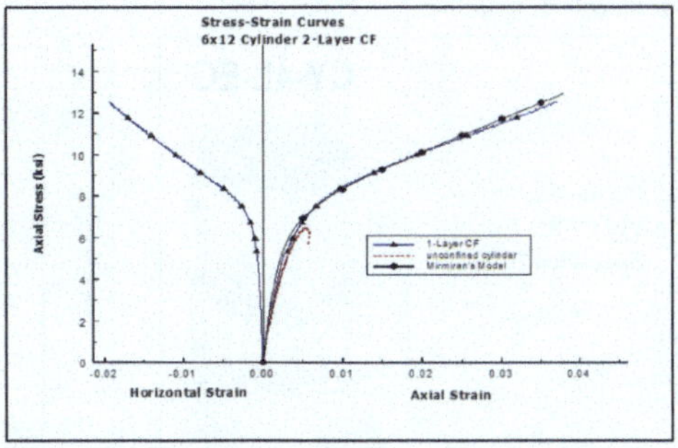
3-Ply CFRP Jacket Specimen $[0^{\circ}]_3$	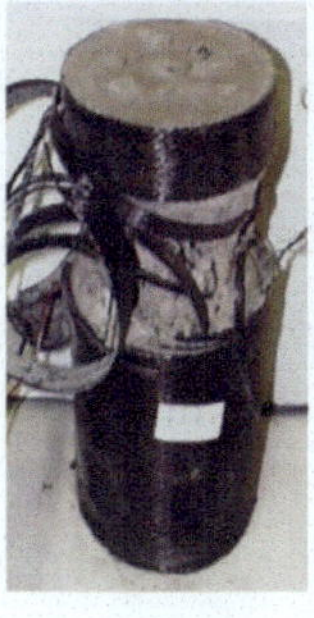	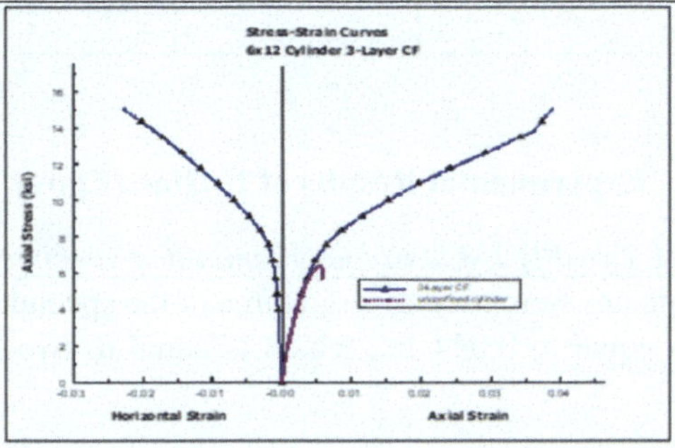

Table 3. Experimental Results Summary for GFRP Jacketed Specimens (*1 ksi = 6.89 MPa*)

Specimen's Description	Failure Mode	Stress-Strain Curve
2-Ply GFRP Jacket Specimen $[0^{o}]_2$		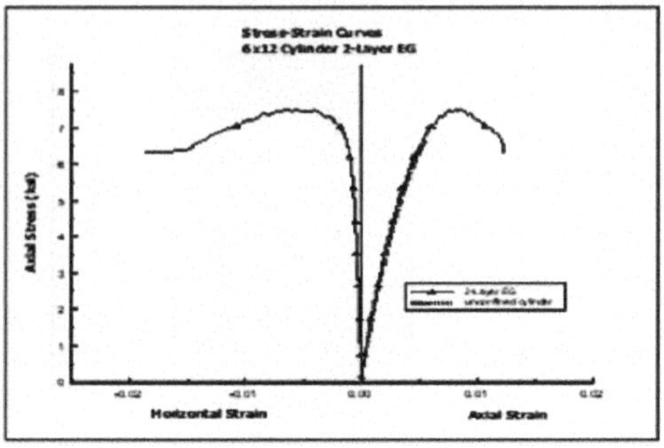
3-Ply GFRP Jacket Specimen $[0^{o}]_3$	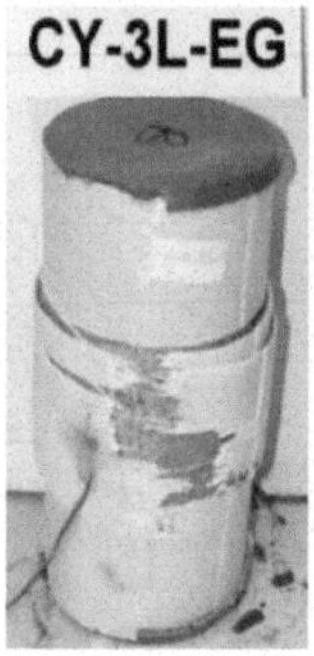	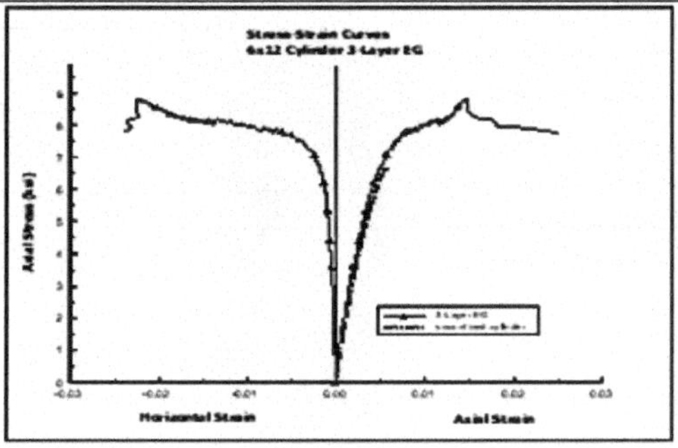
Specimen's Description	**Failure Mode**	**Stress-Strain Curve**
4-Ply CFRP Jacket Specimen $[0^{o}]_4$		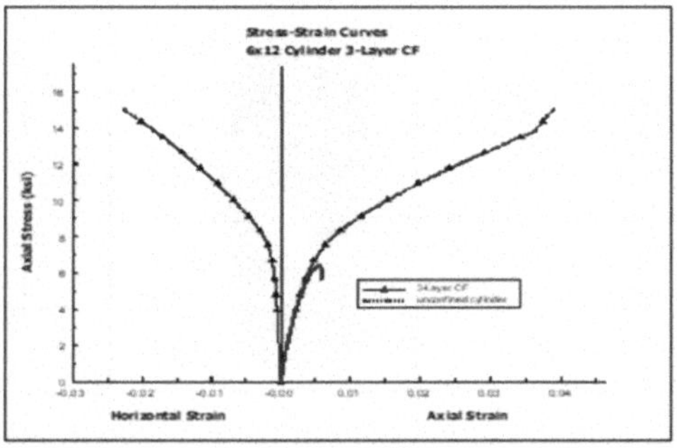

3.3 Experimental Results of E-Glass/Epoxy Retrofitted Specimens

3.3.1 Two-Ply E-glass/epoxy laminated specimens $[0^{o}]_2$: Two layers of E-glass/Epoxy laminates were used to wrap three of the specimens. The thickness of the materials used was equal to 0.044 in., which is equal to two layers of E-glass/epoxy laminate. The specimen failure was sudden and brittle rupture at a load of 942.98 kN. Table 3 shows experimental results of these specimens.

3.3.2 Three-Ply E-Glass/Epoxy Laminated Specimens $[0^o]_3$: Three layers of E-glass/epoxy jackets were used to wrap three of the specimens. The jacket thickness of this specimen group is 1.676 mm which is equivalent to three layers of E-glass/epoxy laminate. The specimen failure mode was sudden and brittle in nature. The failure was initiated due to the tensile rupture of the composite laminates at the mid-height of the specimen. The failure load, ultimate axial stress, ultimate axial strain, ultimate horizontal strain, and ultimate axial deflection for this specimen were 1,112 kN, 60.96 MPa, 2.5%, and 2.5%. Table 3 shows the failure mode of jacketed specimen with three layers E-glass/epoxy laminates and the corresponding stress-strain relations in both axial and hoop directions.

3.3.3. Four-Ply E-glass/Epoxy Laminated Specimen $[0^o]_4$: Four layers of E-glass/epoxy laminates were also used to wrap three of the specimens. The composite jacket thickness of this specimen is 2.24 mm, which is equivalent to the thickness of four layers of E-glass/epoxy laminates. Like the other specimens, the failure was sudden and brittle in nature. The failure was initiated due to the composite laminates tensile rupture at about specimen's mid-height of the. The ultimate failure load was 1,445.6 kN that corresponds to 79.25 MPa of axial stresses. The axial and hoop strains at failure were 3.12%, and 3.27%, respectively. Table 3 presents the typical failure mode of jacketed specimen with four layers E-glass/Epoxy laminates and the stress-strain relations in both axial and hoop directions.

4 Analytical Modeling & Design Models for Frp Confined Axial Members

Several confinement models describing the stress-strain behavior of RC columns strengthened with FRP composite jackets were developed over the years. The following are some of the models that are commonly used for modeling FRP confined RC columns:

1- *Mander Model (1988), developed for concrete confined with steel jacket,*
2- *Hoppel et a. (1994) Model,*
3- *Samaan et al. Model (1998),*
4- *Moran-Pantelides Model (2002^{a,b}),*
5- *Youssef et al. Model (2003), and recently*
6- *Zeng et al. Model (2023), and*
7- *Shayanfar et al. (2023).*

As compared to all published model, Youssef et al. Model (2003) provide a universal treatment and prediction to confined axial members, both reinforced and unreinforced, with different geometries such as circular, elliptical, octagonal, hexagonal, square, and rectangular cross sections. The following paragraphs summarize Youssef et al. model (2003).

4.1 Youssef et al. Universal Model (2003)

The following section describes the general equations of Youssef et al. Model (2003) that is based on stress-strain idealization presented in Fig. 2.

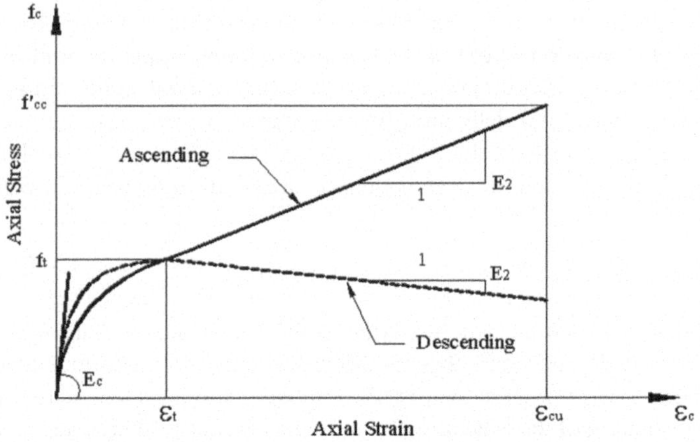

Fig. 2. Youssef et al. Model for FRP-Confined Concrete

- *First Region,* where $0 \le \boldsymbol{\varepsilon_c} \le \boldsymbol{\varepsilon_t}$ and $\boldsymbol{E}_2 > 0$,

$$f_c = E_c \varepsilon_c \left[1 - \frac{1}{n} \left(1 - \frac{E_2}{E_c} \right) \left(\frac{\varepsilon_c}{\varepsilon_t} \right)^{n-1} \right] \tag{1}$$

where:

$$n = \frac{(E_c - E_2)\varepsilon_t}{E_c \varepsilon_t - f_t} \tag{2}$$

- *Second Region,* where $\boldsymbol{\varepsilon_t} \le \boldsymbol{\varepsilon_c} \le \boldsymbol{\varepsilon_{cu}}$ for both $\boldsymbol{E}_2 < 0$ and $\boldsymbol{E}_2 > 0$,

$$\boldsymbol{f_c} = \boldsymbol{f_t} + \boldsymbol{E}_2(\boldsymbol{\varepsilon_c} - \boldsymbol{\varepsilon_t})$$

The general terms of the confinement model are as follows:

$$\frac{f_{cc\prime}}{f_{c\prime}} = 0.85 \left[\alpha + \beta \left(\frac{f_l}{f_{c\prime}} \right)^{\gamma} \right] \tag{3}$$

$$\varepsilon_{cu} = 0.85 \left[\lambda + 0.260 \left(\frac{f_l}{f_{c\prime}} \right) \left(\frac{f_{fu}}{E_f} \right)^{\frac{1}{2}} \right] \tag{4}$$

$$\frac{f_t}{f_{c\prime}} = 0.85 \left[1.0 + \eta \left(\frac{\rho_f E_f \varepsilon_{jt}}{f_{c\prime}} \right)^{\frac{5}{4}} \right] \tag{5}$$

$$\varepsilon_t = 0.85\left[\mu + \psi\left(\frac{\rho_f E_f \varepsilon_{jt}}{f_c\prime}\right)^{\frac{6}{7}}\left(\frac{f_{fu}}{E_f}\right)^{\frac{1}{2}}\right] \quad (6)$$

$$E_2 = \frac{f_{cc}\prime - f_t}{\varepsilon_{cu} - \varepsilon_t} \quad (7)$$

where: α, β, γ, η, μ, *and* ψ are shape factors that take into effect the actual shape of the column and are given in Table 4. Other parameters in the above expressions are clarified in the subsequent design examples.

Table 4. Youssef et al. Model's Shape Factors

Column's Cross Section	α	β	γ	λ	η	μ	ψ
Circular	1	2.25	1.25	0.00337	3	0.00274	0.117
Rectangular	0.5	1.225	0.6	0.00433	1.135	0.0020	0.078

Using Youssef et al. model, the stress-strain curves for different specimens were calculated and compared to experimental results. As shown in Fig. 3, an excellent correlation between model's and experimental results is achieved.

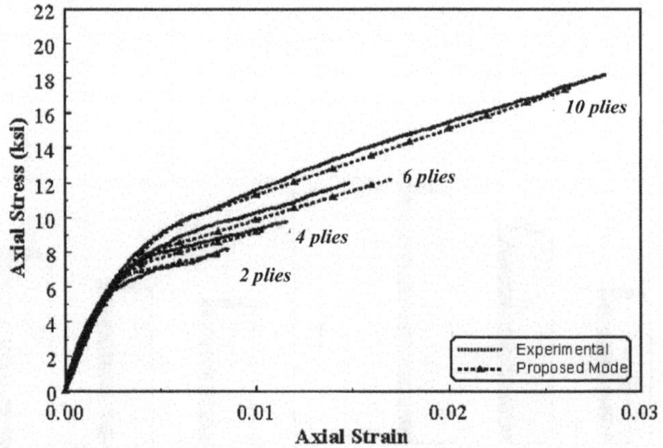

Fig. 3. Model vs. Experimental Stress-Strain Curves for Circular Columns with Different Confinement Reinforcement Ratios (*1 ksi = 6.89 MPa)*

5 Conclusions

Table 5 and Fig. 4 present a summary of the experimental results for all specimens. As shown, results indicated that the use of composite jackets resulted in an appreciable improvement of the behavior of the retrofitted specimen. This included a large increase in the axial load, axial stress, axial strain, as well as the horizontal hoop strain. Also, study results indicated that Youssef et al. model (2003) can be used to predict the behavior of FRP confined concrete members with high degree of accuracy.

6 Impact of Research on Humanities

The results of this research will have a positive impact on humanity through offering a sustainable alternative to retrofit existing buildings and bridges suffered from damage due to earthquakes and corrosion due to harsh environment. Thus, it will promote financial savings, sustainable and high-resilience structures that will contribute to saving lives and properties.

Table 5. Summary of Experimental Results

Specimen Description	Axial Load (kN)	Axial Stress (MPa)	Axial Strain (mm/mm)	Horizontal Strain (mm/mm)
Carbon/Epoxy	814.00	44.62	0.0058	----
Control	1,103.16	60.50	0.025	0.023
1 Ply CFRP	1,579.12	86.60	0.037	0.021
2 Plies CFRP	1890.50	103.64	0.039	0.0236
3 Plies CFRP				
E-glass/Epoxy	836.26	45.84	0.0056	---
Control	943.03	51.70	0.0140	0.0251
2 Plies GFRP	1,112.06	60.70	0.025	0.025
3 Plies GFRP	1,445.70	79.25	0.0312	0.0327
4 Plies GFRP				

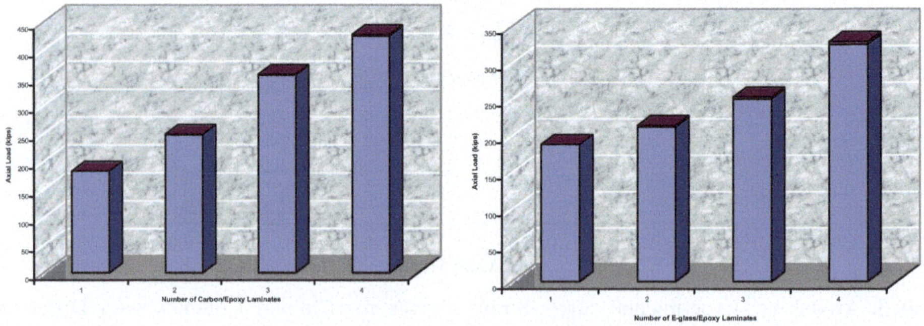

Fig. 4. Effect of Number of Carbon/Epoxy laminates on the Axial Strength of Unreinforced Concrete Cylinders (***1 ksi = 6.89 MPa***)

Acknowledgement. The authors would like to acknowledge Professor Ayman Mosallam, University of California, Irvine for his kind mentorship and valuable contributions to all aspects of this study.

References

1. Acceptance Criteria (AC125), 2010 - Concrete and Reinforced and Unreinforced Masonry Strengthening Using Externally Bonded Fiber-reinforced Polymer (FRP) Composite Systems, International Code Council Evaluation Service (ICC-ES), Brea, CA, USA
2. ACI Committee 440: Guide for the design and construction of externally bonded FRP systems for strengthening concrete structures (2017). ISBN 9781945487590
3. ASTM C39/C39M: Standard Test Method for Compressive Strength of Cylindrical Concrete Specimens. ASTM International, West Conshohocken, USA (2016)
4. Haroun, M.A., Mosallam, A.S., Feng, M.Q., Elsanadedy, H.M.: Experimental investigation of seismic repair and retrofit of bridge columns by composite jackets. J. Reinf. Plast. Compos. **22**(14), 1243–1268 (2003)
5. Hoppel, C.R., Bogetti, T.A., Gillespie, J.W., Howie, I., Karbhari, V.M.: Analysis of a concrete cylinder with a composite hoop wrap. In: Proceedings of ASCE Materials Engineering Conference, pp. 191–198. ASCE, New York, N.Y. (1994)
6. Hosotani, M., Kawashima, K., Jun-ichi Hoshikuma: A stress-strain model for concrete cylinders confined by carbon fiber sheets. Report No. TIT/EERG 98-2, 55 p. Tokyo Institute of Technology, Tokyo, Japan (1998). (in Japanese)
7. Mander, J.B., Priestley, M.I., Park, R.: Observed stress-strain behavior of confined concrete. J. Struct. Divis. ASCE **114**(8 August), 1927–1849 (1988)
8. Moran, D.A., Pantelides, C.P.: Variable strain ductility ratio for fiber-reinforced polymer-confined concrete. J. Compos. Constr. ASCE **6**(4), 224–232 (2002a)
9. Moran, D.A., Pantelides, C.P.: Stress-strain model for fiber-reinforced polymer-confined concrete. J. Compos. Constr. ASCE **6**(4), 233–240 (2002b)
10. Mosallam, A.S., Bayraktar, A, Elmikawi, M, Pul, S, Adanur, S.: Polymer composites in construction: an overview. SOJ Mater. Sci. Eng. (2014). https://symbiosisonlinepublishing.com/materialsscience-engineering/materialsscience-engineering07.php
11. Ramadan, A.S., Elgendi, E.O.: A review of optimization techniques and algorithms used for FRP applications in civil engineering. J. Eng. Appl. Sci. **70**(1), 1–49 (2023)
12. Samaan, M., Mirmiran, A., Shahawy, M.: Model of concrete confined by fiber composites. J. Struct. Eng. ASCE **124**(9 September), 1025–1031 (1998)
13. Shayanfar, J., Barros, J.A., Rezazadeh, M.: Analysis-oriented model for partially FRP-and-steel-confined circular RC columns under compression. Eng. Struct. **276**, 115330 (2023)
14. Youssef, M.N., Feng, M.Q., Mosallam, A.S.: Stress–strain model for concrete confined by FRP composites. Compos. B Eng. **38**(5–6), 614–628 (2007)
15. Zeng, J.J., Liao, J., Zhu, D.H., Li, P.D.: Axial compressive behavior and design-oriented model for large-rupture-strain (LRS) FRP-confined concrete in rectangular columns. J. Build. Eng. 106925 (2023)

Author Index

A. S. Mosallam et al. (Eds.): ICATH 2023, SCI, pp. 319–320, 2024.
https://doi.org/10.1007/978-3-031-76557-5

The manufacturer's authorised representative in the EU is Springer Nature Customer Service Centre GmbH, Europaplatz 3, 69115 Heidelberg, Germany. If you have any concerns regarding our products, please contact ProductSafety@springernature.com

Printed and bound by CPI Group (UK) Ltd, Croydon, CR0 4YY
01/10/2026
02225940-0003